# Perspective Grid Sourcebook

# Perspective Grid Sourcebook

## Computer Generated Tracing Guides for Architectural and Interior Design Drawings

Ernest Burden

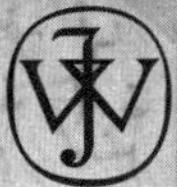

JOHN WILEY & SONS, INC.

New York • Chichester • Weinheim • Brisbane • Singapore • Toronto

# ACKNOWLEDGMENTS

This book would not have been possible without the contributions of a number of talented people, and I am thankful to all who participated in its development and execution.

My son Ernest III developed the grids for sketching in Part One of the book. As a working perspective artist who uses the computer to create architectural perspectives, his understanding of computer systems was invaluable. The Software program from Datacad was appreciated.

My friend Hans Christian Lischewski, an associate in the firm of Perkins & Will and Manager of Professional Applications and Information Systems in their New York office, guided the early development of the large-scale grids in Parts Two, Three and Four. Richard Brennan, architect, worked closely with me in establishing the database and was responsible for fine-tuning the grid plotting.

I thank the employees at Perkins & Will, New York, for the development and final production of all the large-scale grids.

I also thank those involved in the production of the book: Wendy Lochner, my editor, whose early vision and continuing support made it happen; Monika Keano, for assistance in the preparation of the graphics; the VNR production staff, for assistance in the assembly and printing of this unusual project.

And finally, I thank my daughter Chelsea for her assistance, and Joy, for her early encouragement and continuing commitment during the many years it took to complete this project.

---

This book is printed on acid-free paper. ♾



**Library of Congress Cataloging-in-Publication Data:**
Burden, Ernest F. 1934–
Perspective grid sourcebook: computer generated tracing grids for architectural and interior design drawings/Ernest Burden.
p. cm.
Includes index.
ISBN 0-471-28866-7
1. Architectural drawing—Data processing. 2. Interior decoration—Data processing. 3. Perspective—Data processing.
I. Title.
NA2728.B87 1991 92-26860
720'.28'40285—dc20
Printed in the United States of America

10 9 8 7 6 5 4 3 2

# CONTENTS

# PREFACE

The electronic revolution has changed every aspect of our daily lives, especially the way we work and present our ideas. It has radically changed the way architects, engineers, and designers approach every aspect of their practice. This includes office practice, production, and the presentation of design concepts. The traditional tools of communication have all been replaced by electronic ones. The replacements for the "cut and paste" approach have greatly impacted both the printed word and the projected image. For the printed word we now use word processing; for graphic layout we use desktop publishing; for drafting we use CADD systems; and for presentations we use three-dimensional solid modeling and animated visual simulations, which can represent ideas and spaces with near photographic realism. We have even created simulated realities, or cyberspaces, as they are called, wherein the viewer can experience an environment created by the computer and interact within that space.

With the potential of electronic media expanding every day, with no end in sight, are artists becoming extinct? Is traditional drawing by hand an anacronism? Not at all. There will always be a need for the freehand sketch, the artist's vision, the personal touch. However, if we are going to emulate the computer in its applications, the tools we use should be produced within that system. These computer-generated perspective grids are an answer to this need.

When appropriate software becomes available, you will be able to sketch out any design in three dimensions, using any one of the sketching grids, input that sketch with its perspective coordinates into your computer, and have it translated into an orthographic image of plan and elevation. It will eventually be possible for the computer to convert any sketch back into an orthographic design database, using your sketch as the input. Similar technology that can translate sketch plans into hard line drawings already exists.

These grids are another design and visualization tool created entirely by electronic means. They were never touched by human hands, and now they are yours to use in your own creative way.

# HISTORY OF GRIDS

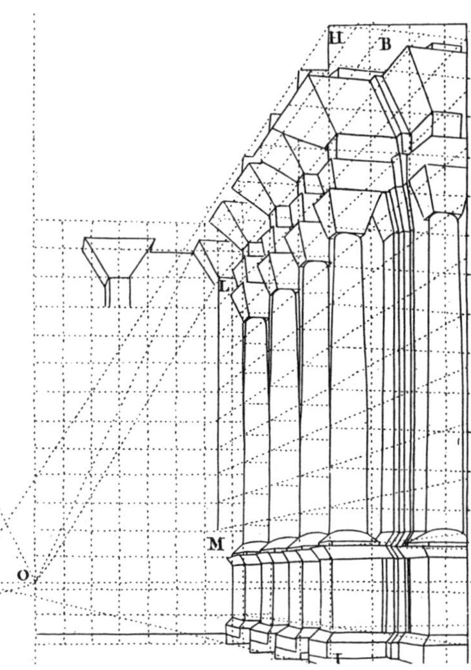


The development of grids as an aid in drawing is widely assumed to have begun during the Renaissance in Italy. Actually, grids were used centuries earlier in ancient Egypt to lay out relief sculpture and to enlarge designs to their full scale on temple walls. They were usually drawn in red ochre directly on the stone.

In the Renaissance, grids were used quite extensively by painters and architects alike. Pespective grids can be traced to such artists as Jan Vredeman de Vries, Andrea dal Pozzo, Bibiena, and others. The invention of the camera obscura, a device with a lens and mirror similar to our modern-day camera, allowed artists to view any scene in its true perspective. This prompted a totally new method for drawing architectural subjects. Artists such as Canaletto would sketch a scene using this device, return to the studio to enlarge the sketch, most likely using a grid, then complete the details by hand.

Most of the perspective methods developed during the Renaissance are still in use in many architectural offices today, and are still taught widely in schools.

Other methods based on photography have since been introduced as an alternative to mechanically plotting a perspective. The camera duplicates objects in the real world in exact detail and in a fraction of the time it takes to mechanically construct a perspective drawing.

Even more recently, the computer was introduced—first as a drafting tool, then as a design tool, and finally as a tool for visualization. It is with this latest development in perspective that this book originated. People who use computers with three-dimensional capabilities have many options for viewing their designs from many angles, distances, and viewpoints. A similarly wide variety of images were developed for this book.

The computer image is the projection of three-dimensional data onto a plane that records images similar to those that a camera would depict, as opposed to one that is mechanically constructed. The computer can generate views we may not be able to conceive of. It can even create views that we could not realistically see within our limited cone of vision. Perspective grids in the past have been produced primarily by mechanically plotting them by hand. The accuracy of the images was dependent on the accuracy of the person constructing the grid, and the inherent precision of the drafting tools used to construct them.

The computer sees things in terms of x, y, and z coordinates; each one is strictly controlled within its own axis so that no deviation is possible. This exactness can be verified by noting that all the diagonal shapes created within a perspective view create their own distinct pattern of vanishing points. If the perspective is inaccurate, these diagonals will be the telltale sign. The accuracy of grids developed by the computer is without precedent in the history of perspective grid construction.

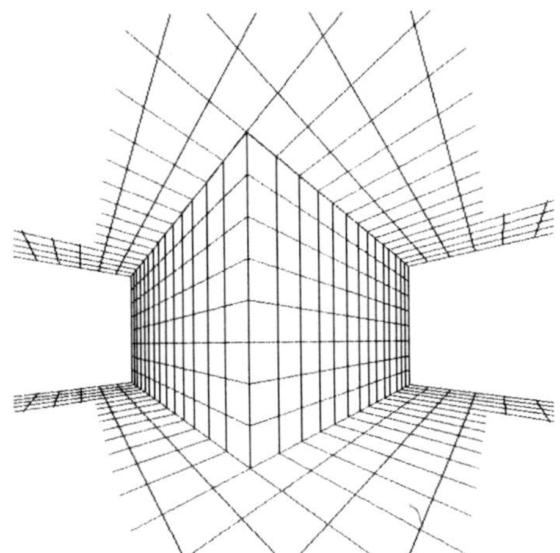

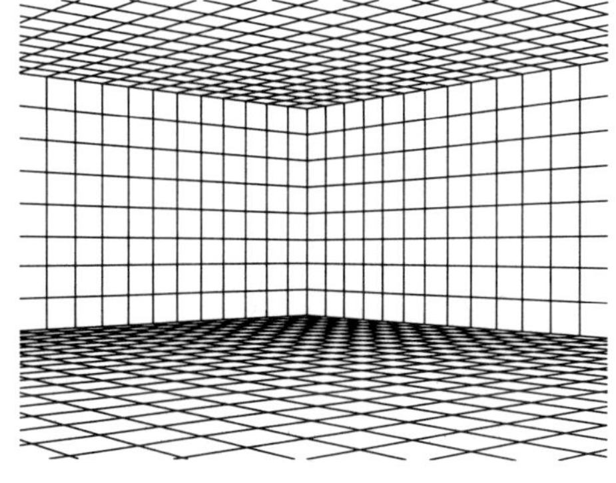

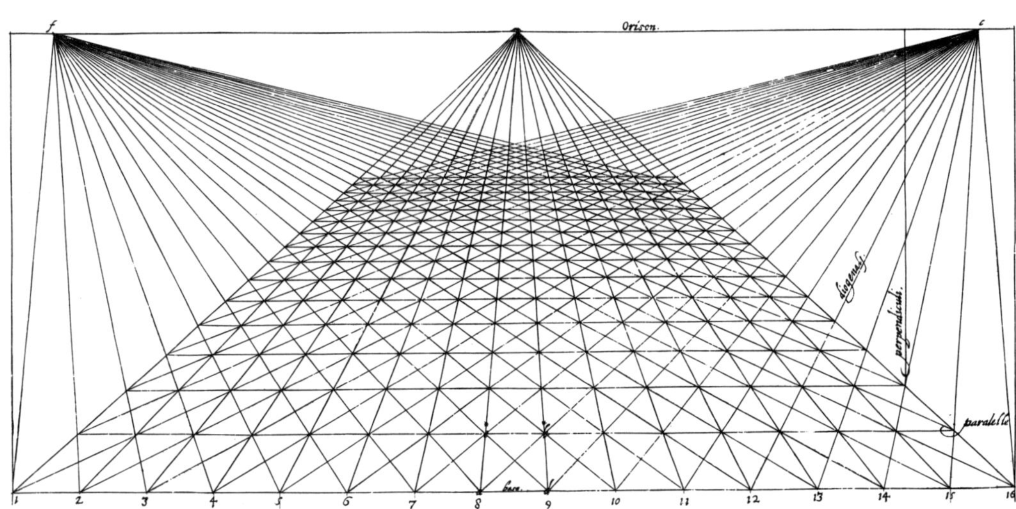

# ELEMENTS OF PERSPECTIVE SYSTEMS

Perspective is a function of one primary element—the station point (SP). Where you stand in relation to the object is the key in determining the convergence of elements within the scene. This is also true in photography. It does not matter whether you shoot a scene with a wide-angle lens, a normal lens, or a telephoto lens from the same spot. If each picture is cropped to the same relative frame and enlarged to the same relative size, the images will be identical in every respect.

Other elements of a perspective system are the picture plane (PP), the area upon which the image is projected; the cone of vision (CV), the extent of the image seen, usually expressed in degrees; and the central visual ray (CVR), the axis of the cone of vision. These elements exist as a fixed system, and a change in one can affect the others. If the central visual ray is not perpendicular to the picture plane, for example, the image will converge to a third vanishing point, and a "three-point" perspective will result.

In photography there are certain types of camera movements that can correct this form of "distortion," such as tilts and swings on large-format view cameras, and perspective control lenses that are made for many small format cameras.

The computer follows the same rules for displaying its internal perspective system, and can display any view that is called for. In fact, one has to be certain that the directions given the computer are correct in terms of where it is looking, and from what eye level (EL), or it may automatically distort the image in this third-dimensional axis.

A changing SP is the only element that changes the degree of perspective convergence. It is essentially the difference between "zooming in" and walking closer to an object. The PP merely determines the size of the image, and the CV how much of the object you can see.

Thus, perspective systems remain as they were when Renaissance artists depicted a view of a building or public plaza and just as architects have constructed their perspective views from that day to the present time, whether mechanically, photographically, or with the computer.

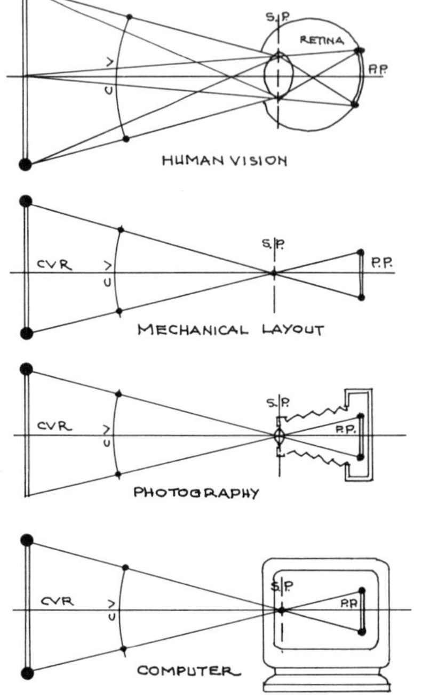


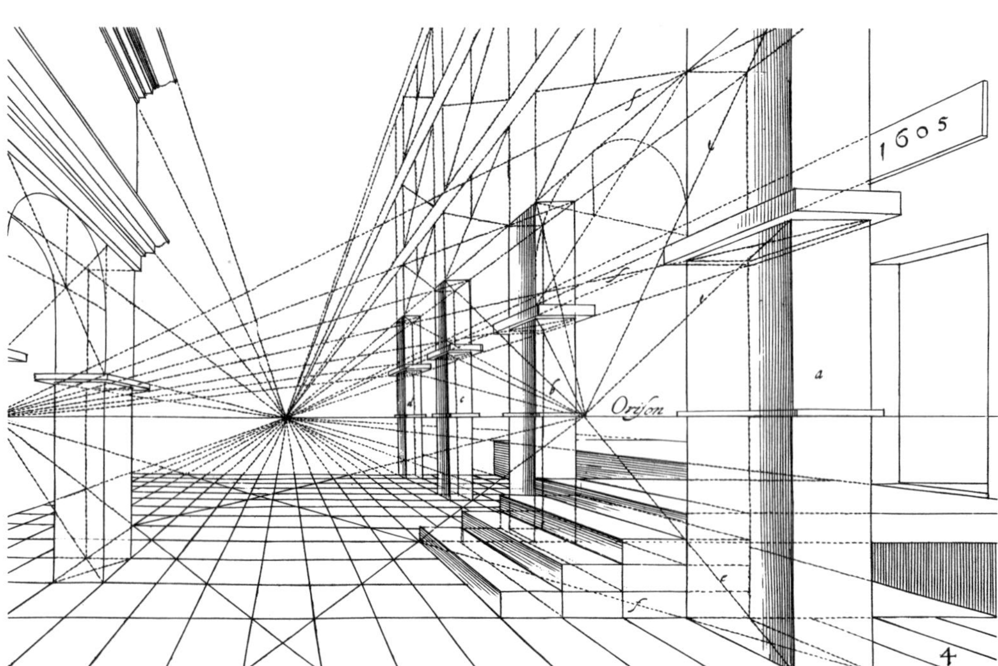

# HOW THE GRIDS WERE DEVELOPED

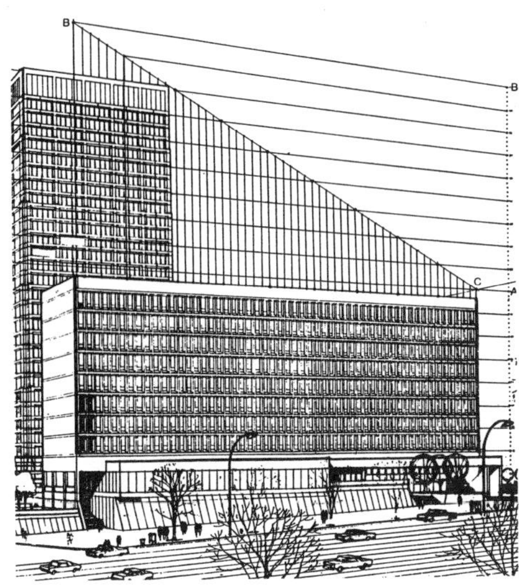

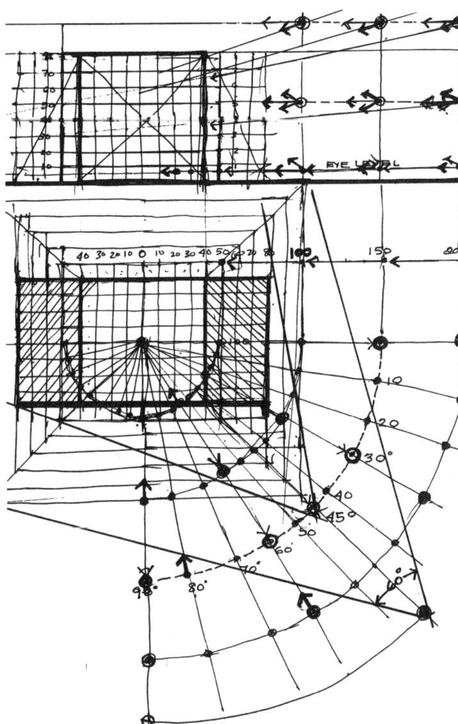


The development of each set of grids evolved from a different combination of determinates and practical considerations. My previous fifteen years of experience as a perspective artist involved the introduction of photographic techniques for constructing perspectives. This technique involved photographing plans and models, which often required the making of a customized grid to match the photographic image. This was also necessary to deal with the distant vanishing points that were inherent in this method. Most traditional methods use close vanishing points, so they will fit on the drawing board. If both points are easily located on the drawing board, there is not as much need for a perspective grid.

The development process began in the conceptual stage with sketches of objects and interior spaces and with photographic investigations of these objects and spaces. The choice of angles for the small-scale grids began with a diagram outlining possible choices. A medium-format camera was used to view these angles in the ground glass. This method was helpful in that it instantly previewed certain situations. For example, it showed that extreme close views of the aerial object were far too distorted, so they were deleted from the selections. It also demonstrated that all 45-degree views from mid-height of the object were not suitable for perspective use.

Since the computer could reproduce these identical viewpoints, angles, and measurable distances, a 2-D database was set up in a PC computer, with a dynamic perspective software program. Each view was then enlarged so that the corner of the object was exactly the same height. Otherwise, the wide-angle view would have filled the page, while the distant view would have appeared tiny. Keep in mind that the perspective was not affected by enlarging or reducing the image.

By using rotation angles of 10, 20, 30, and 40 degrees, plus the mirror images of 50, 60, 70, and 80 degrees for each distance from 50 feet to 400 feet (with the few exceptions noted above) one can study or sketch any design from hundreds of angles or the same design from multiple angles.

The large scale grids began as freehand sketches, depicting typical angles and distances commonly used for constructing perspectives. These sketches were used to establish the computer database as to the angle of view, station point, and eye level. The primary model consisted of a 200-by-200-unit space, with 10-unit major divisions and 2-unit minor divisions. These were set up on a minicomputer system with a large database capacity, since the model would be used for most of the views.

The desired views were first modeled in a 2-D file, and then studied through quick "screen dumps" to check viewpoints and to fine-tune the details of each grid plot. Sample sections of each grid were enlarged on the screen and then plotted to check overlapping lines in the background area. If the lines filled in, the fine-line grid was modified or turned off altogether. Since each module was put on a separate layer, it was easy to control this process.

The objective in the field matrix series of grids was to create a limitless three-dimensional space that was demarcated with multiple vertical measuring devices. The drawing could then start anywhere in space and be controlled easily by the series of vertical measurements. The objective of the multiple grid series was to make larger grids out of adjoining pages. Each multiple grid is one-half the full grid, joined in the center.

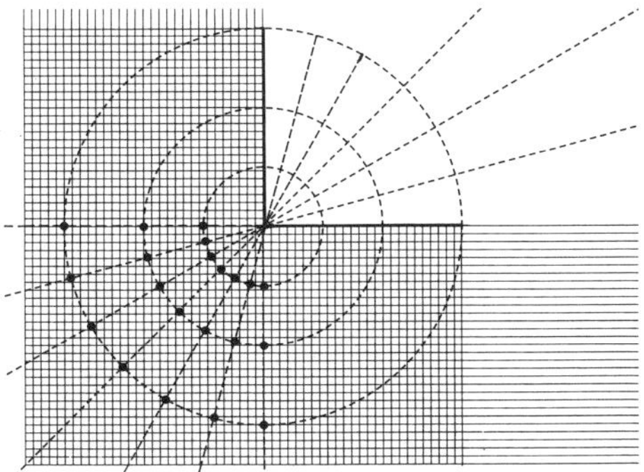

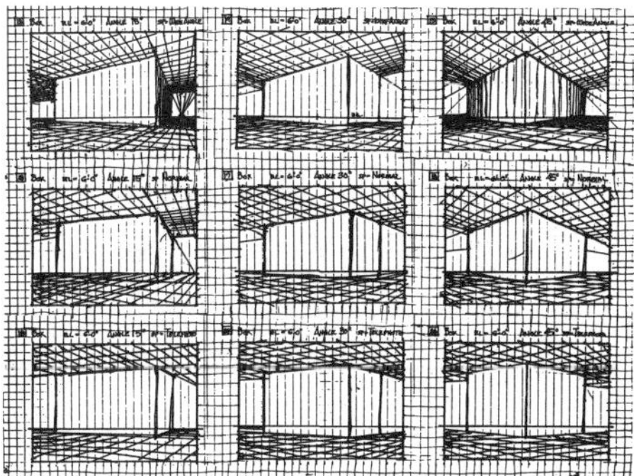

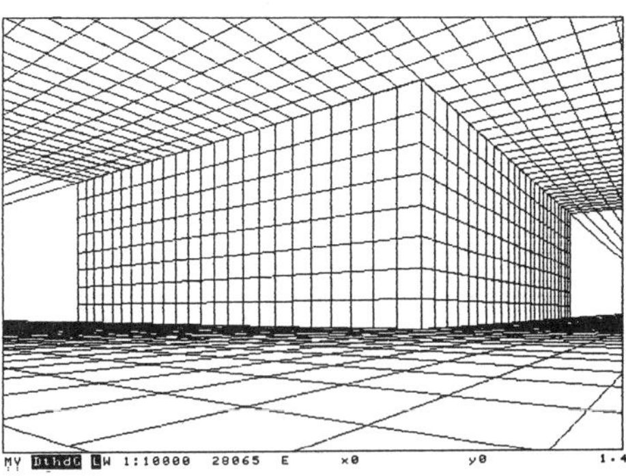

# HOW TO USE THE GRIDS

The grids in this book can be used in preparing quick freehand sketches or fully developed perspectives from closeup to distant views. Some of the floor and ceiling grids in Part Two can be combined with each other, whereas the multiple grids are specifically designed to be joined in the middle.

The pages in the book have been perforated for easy removal for photocopying or for tracing. Any grid can be enlarged by the photocopy process quite inexpensively. On some of the grids you can customize certain areas to fit the needs of your drawing. For example, you can extend lines to vanishing points and create a more transparent plane effect.

Since there are no fixed scale designations you can assign any convenient scale, whether in inches, feet, or metric measurements. A convenient unit of measurement may be 1 foot for small objects, or up to 10 feet for larger objects. Thus, interior rooms can be 8 feet high using a 1-foot unit, or 40 feet high using a 5-foot unit.

On the small grids for sketching in Part One there are no minor subdivisions. However, you can enlarge your favorite view and add the minor subdivisions quite easily. To subdivide a unit into quarters, simply draw in the diagonals within a unit. Where the diagonals cross will be the center of the unit in each direction. Draw lines parallel to each direction, and the resulting unit will now be made up of four units in exact perspective.

To divide each grid module into other divisions, the following methods can be used. At any point along two converging lines, you can slide a scaled ruler parallel to the picture plane. Once you have the new divisions, simply draw them to the vanishing points. If the vanishing points are not on the drawing, you may need to do this procedure at two locations along the same two converging lines, one near and one further back. To find divisions in the direction of the convergence, simply draw a diagonal within the unit being subdivided. Where the diagonal intersects with the converging lines will be the location of the new unit division.

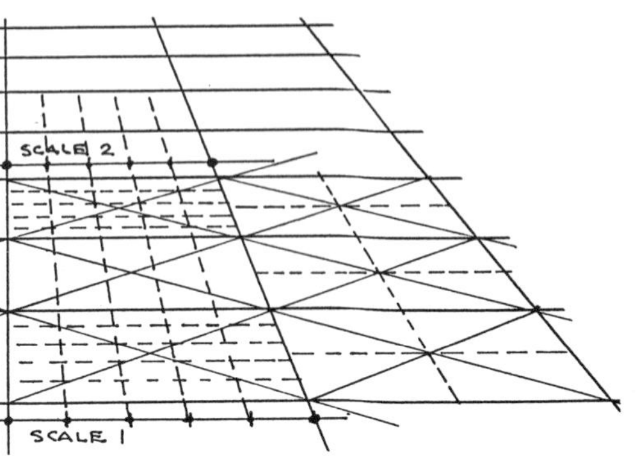


To achieve proper perspective, the grids must be aligned with a T-square and triangle. The vertical lines of the grid must be aligned with the edge of the triangle. The height of each vertical unit is the same as the horizontal dimension. You can determine heights by scaling any horizontal line of known measure. Once this height is established, it can be traced throughout the drawing following the perspective lines.

There are countless other ways to use these grids and to customize them for your own individual situation. The important factor is that the computer-generated base represents the most accurate matrix for your drawing.

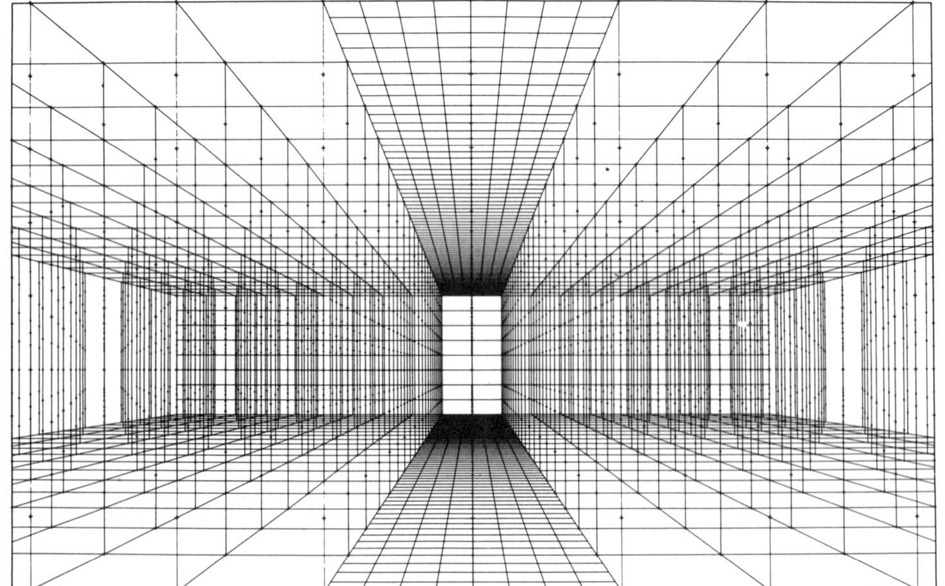

# VISUAL INDEX

This index is a display of the grids in the sequence in which they were conceived as part of the perspective system. It is cross-referenced with each page of the book to assist in finding any specific angle or distance.

The grids for sketching consists of views from 10 degrees to 80 degrees in rotation and distances from 50 feet to 400 feet in 50- and 100-foot increments. Since there are four grids to a page, the visual index will assist in locating any specific grid. The interior grids are one to a page and can be located quite easily. The same is true of the floor and ceiling grids, as well as the entire series of the exterior object. The multiple grids are shown assembled to assist in putting them together.

Care was taken in the design of these grids, and in the "calendar style" format of this book, so that the mirror image of each grid is printed on the reverse side of the page. This not only doubles the usefulness of each viewpoint by presenting its exact opposite image; it also makes it possible to use the grids on a light box without a distracting image showing through from the other side.

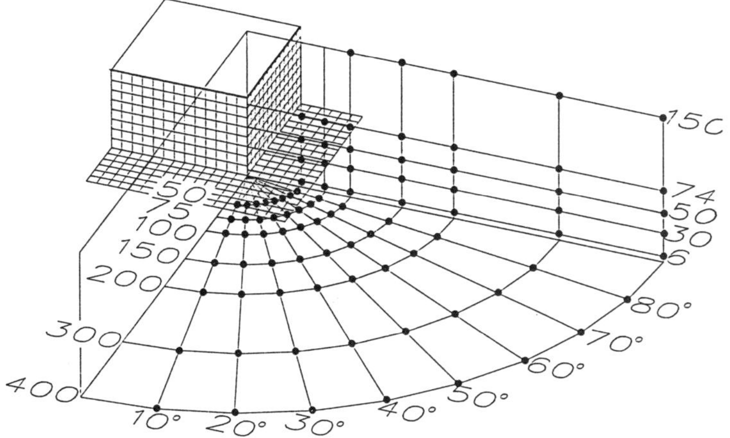

# VISUAL INDEX

**EXTERIOR OBJECT**

Eye Levels/30 - 50
Station Points/50 - 75 - 100 -150 - 200 - 300 - 400
Angles of View/10 - 20 - 30 - 40 - 50 - 60 - 70 - 80

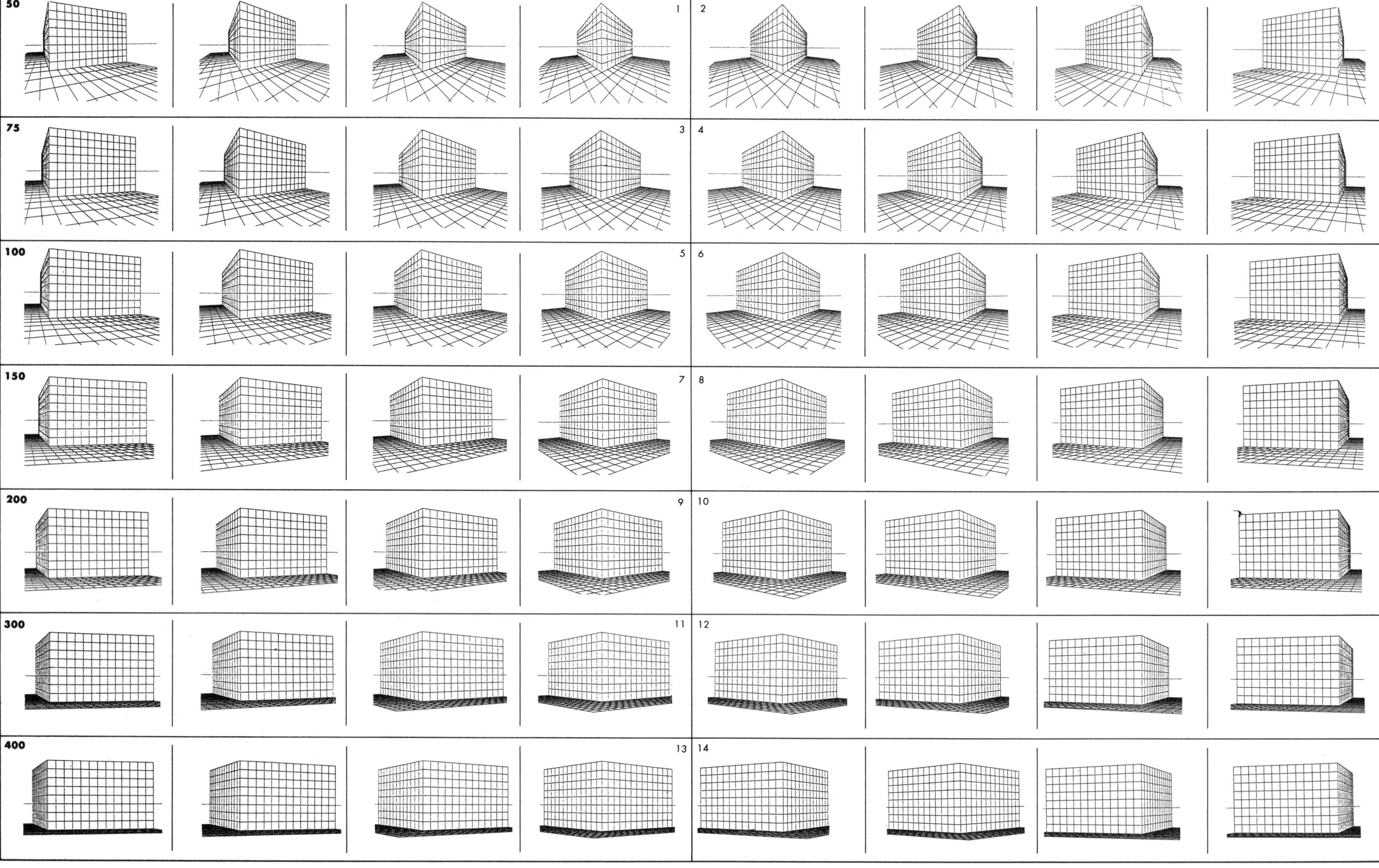

# VISUAL INDEX

**EXTERIOR OBJECT**

Eye Levels/74 - 6
Station Points/50 - 75 - 100 - 150 - 200 - 300 - 400
Angles of View/10 - 20 - 30 - 40 - 50 - 60 - 70 - 80

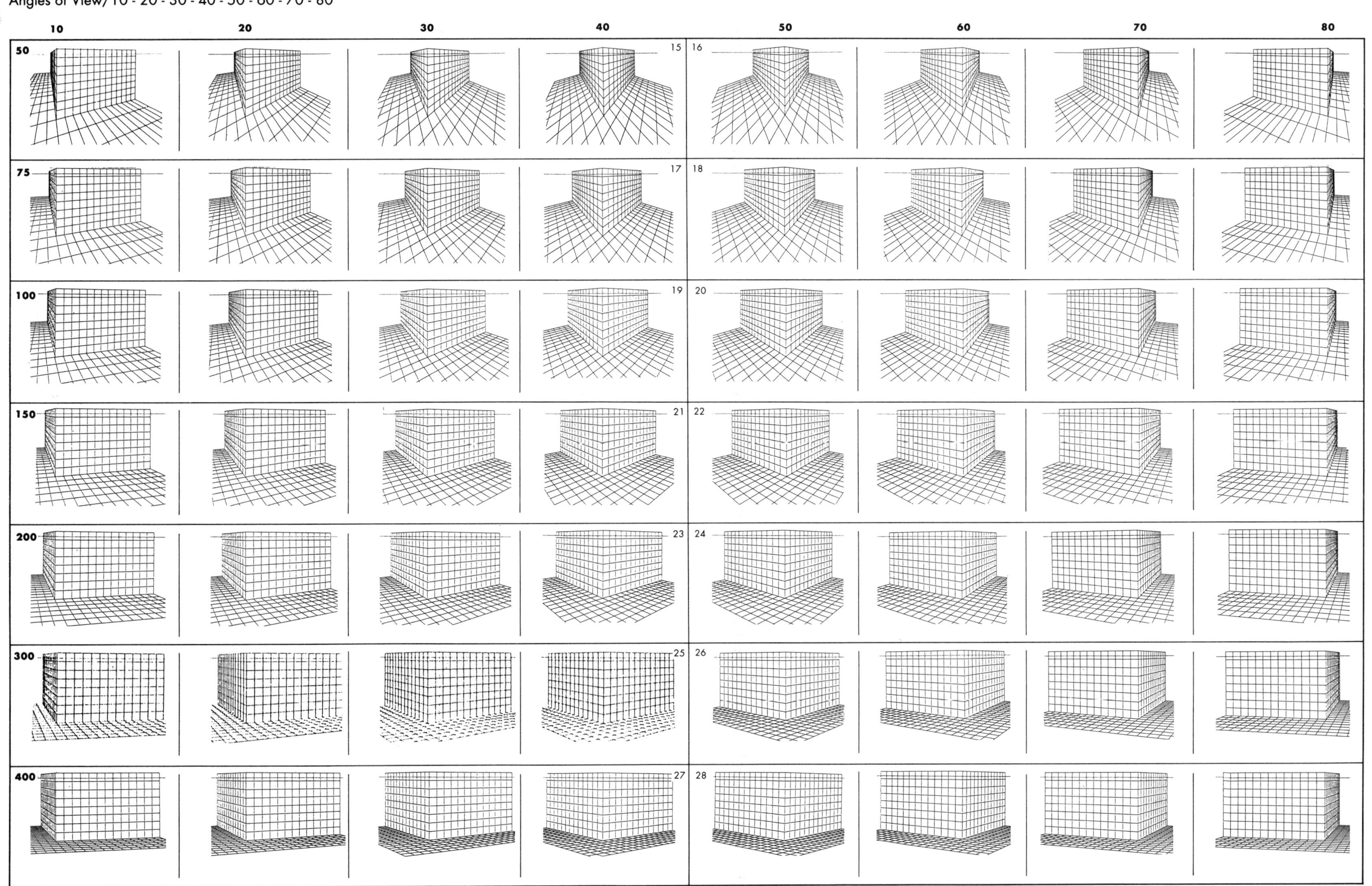

# VISUAL INDEX

**EXTERIOR OBJECT**

Eye Level/150

Station Points/150 - 200 - 300 - 400

Angles of View/10 - 20 - 30 - 40 - 50 - 60 - 70 - 80

**EXTERIOR 3 PT. TOWER**

Eye Levels/6, 30

Station Point/100

Angles of View/10 - 20 - 30 - 40 - 50 - 60 - 70 - 80

| | 10 | 20 | 30 | 40 | 50 | 60 | 70 | 80 |
|---|---|---|---|---|---|---|---|---|
| 100 | | | | 29 | 30 | | | |
| 150 | | | | 31 | 32 | | | |
| 200 | | | | 33 | 34 | | | |
| 300 | | | | 35 | 36 | | | |
| 400 | | | | 37 | 38 | | | |
| 100 | 39 | 41 | 43 | 45 | 46 | 4·4 | 42 | 40 |
| 100 | 47 | 49 | 51 | 53 | 54 | 52 | 50 | 48 |

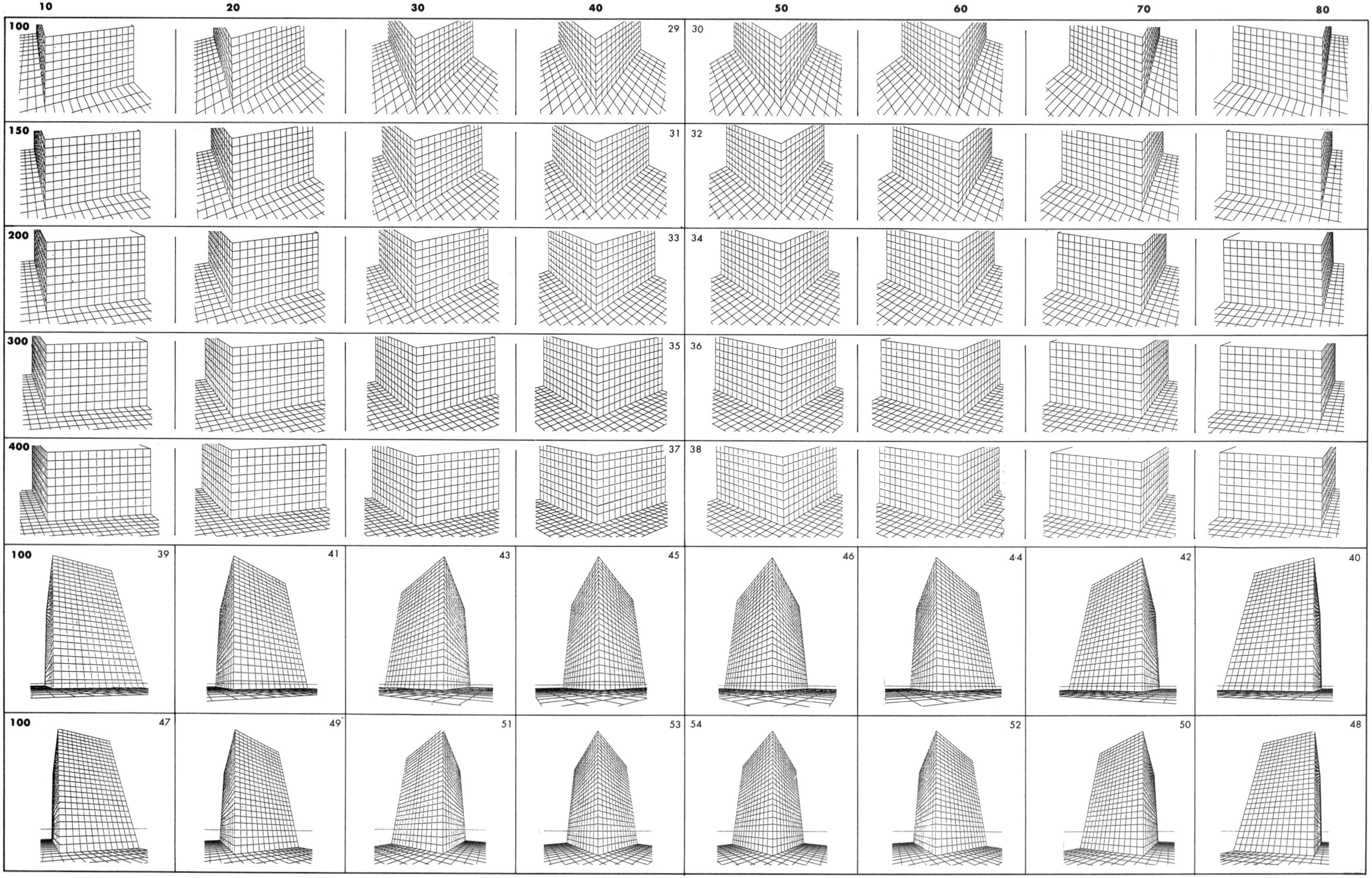

# VISUAL INDEX

**INTERIOR ROOM**

Eye Levels/5.5 - 2.5
Station Points/40, 50
Angles of View/0 - 10 - 20 - 30 - 40 - 50 - 60 - 70 - 80 - 90

**FLOOR/CEILING GRIDS**

Eye Levels/7.5 - 100 - 300
Station Points/100 - 200 - 250
Angles of View/0 - 15 - 30 - 45 - 60 - 75 - 90

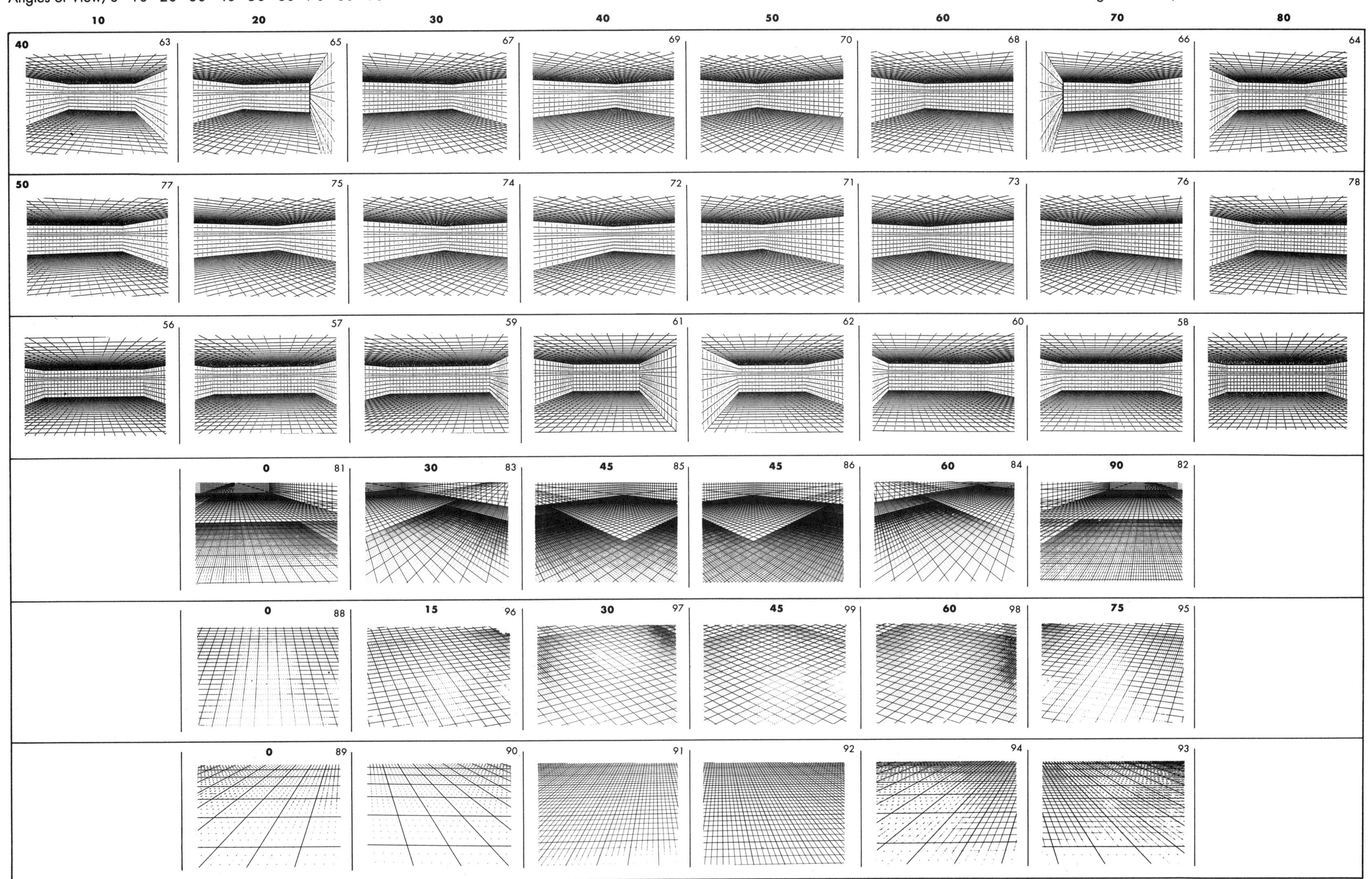

# VISUAL INDEX

**EXTERIOR OBJECT**

Eye Levels/6 - 74,
Station Points/100 - 150 - 200
Angles of View/15 - 30 - 45 - 60 - 75

**EXTERIOR OBJECT**

Eye Levels/ 30 - 50
Station Points/100 - 150 - 200
Angles of View/15 - 30 -

**EXTERIOR 3 PT. TOWER**

Eye Level/24
Station Point/150
Angles of View/15-30-45-60-75

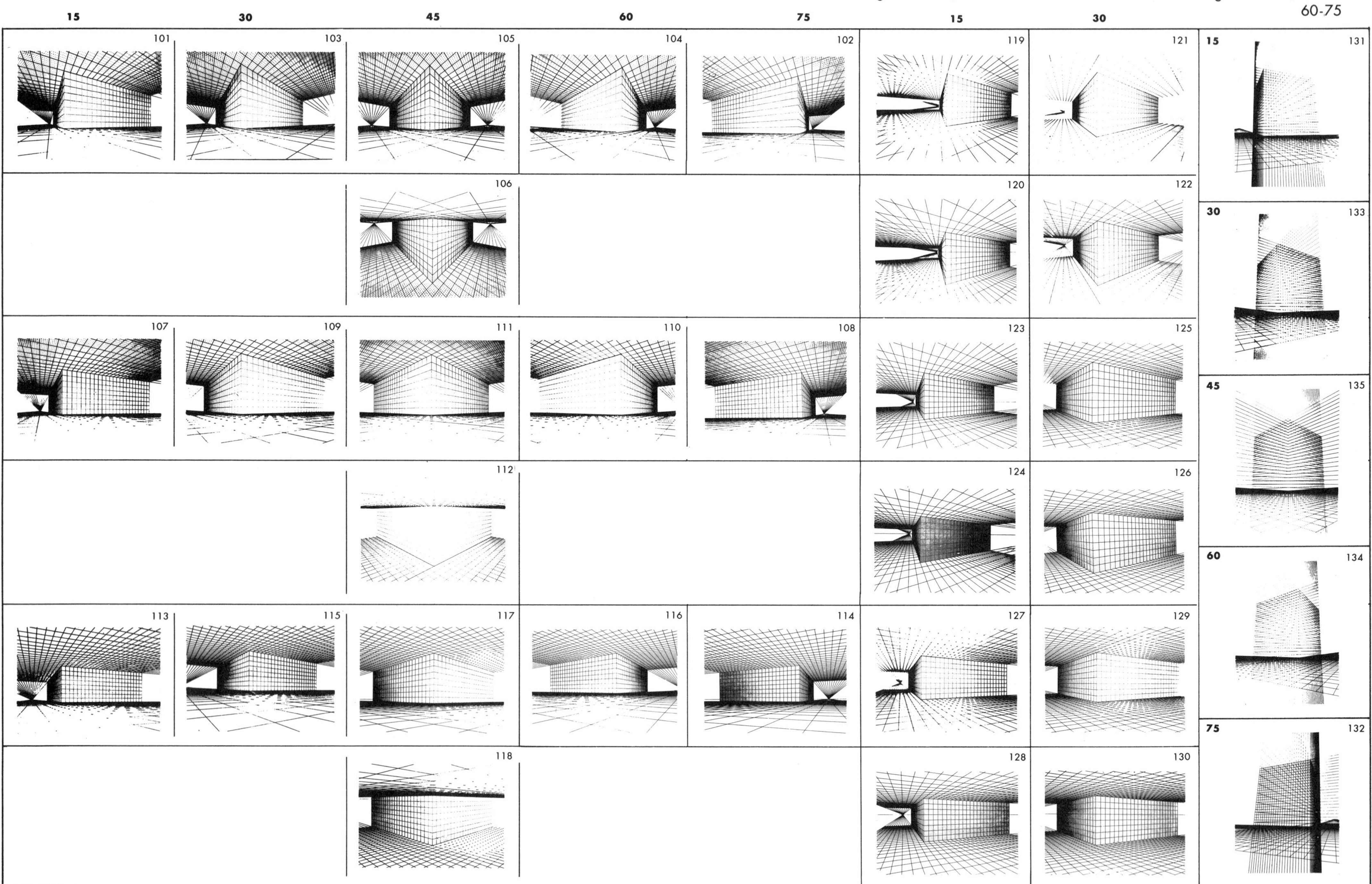

# VISUAL INDEX

| FIELD MATRIX | INTERIOR ROOM | AERIAL PLAN | INTERIOR ROOM | EXTERIOR OBJECT | EXTERIOR TOWER |
|---|---|---|---|---|---|
| Eye Levels/3 - 5 - 6 - 74 | Eye Levels/3, 5 | Eye Level/150 | Eye Levels/3, 5 | Eye Levels/6 - 74 | Eye Levels/24 |
| Station Point/150 | Station Point/100 | Station Point/100 | Station Point/100 | Station Point/100 | Station Point/100 |
| Angles of View/0 - 30 - 45- -60 - 90 | Angles of View/0 - 15 - 30 - 45 | Angle of View/45 | Angles of View/75 - 60 - 45 | Angles of View/30 - 60 | Angles of View/30 - 60 |

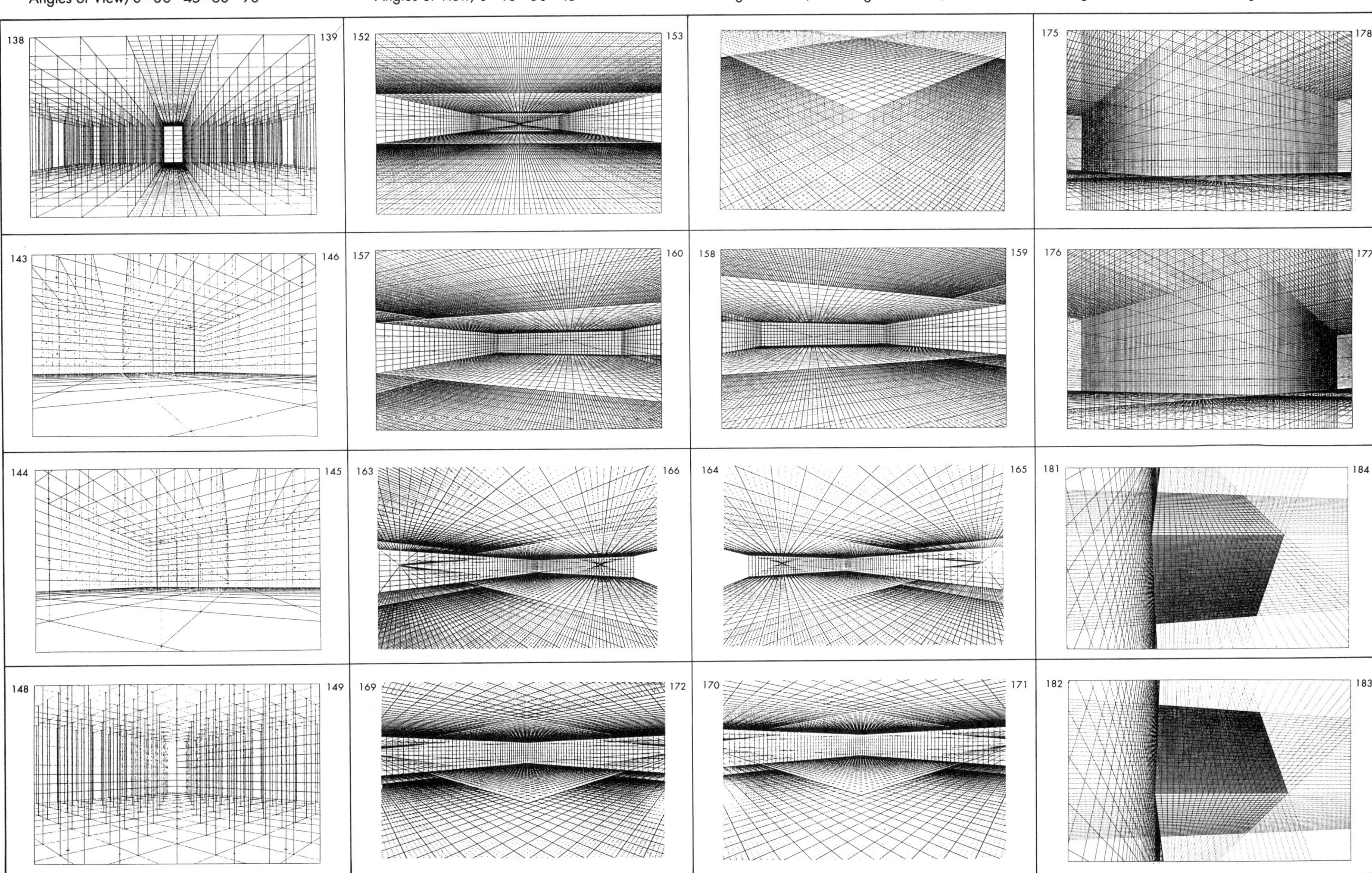

**PART ONE**

# GRIDS FOR SKETCHING

## Exterior

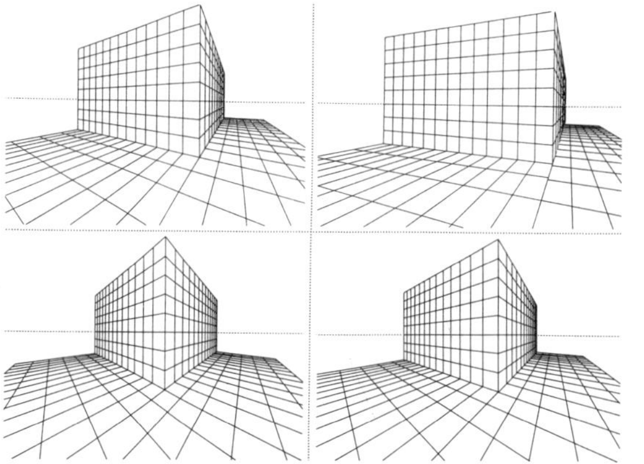

1 Exterior Object SP/50 EL/30 CVR/10–20–30–40

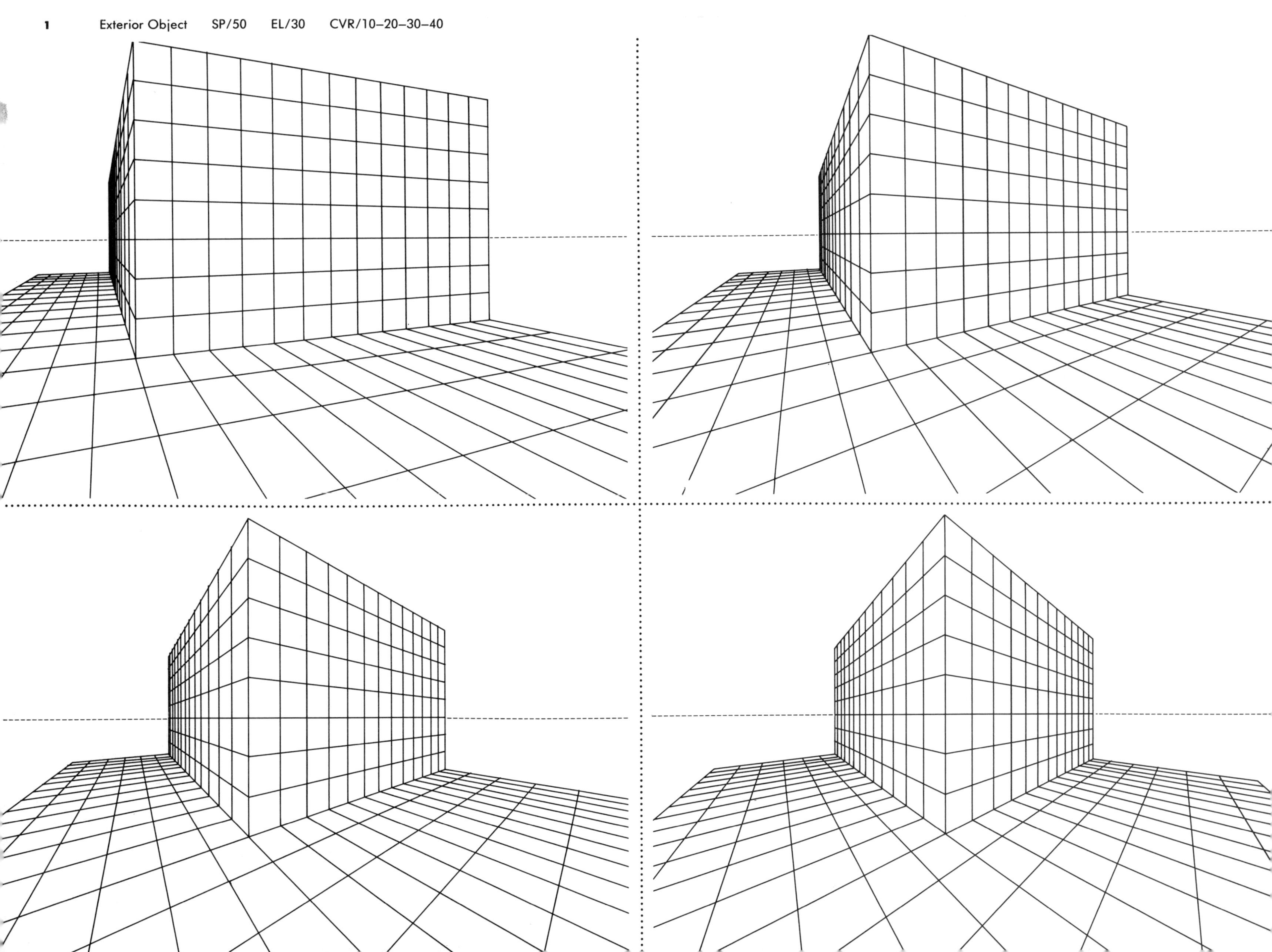

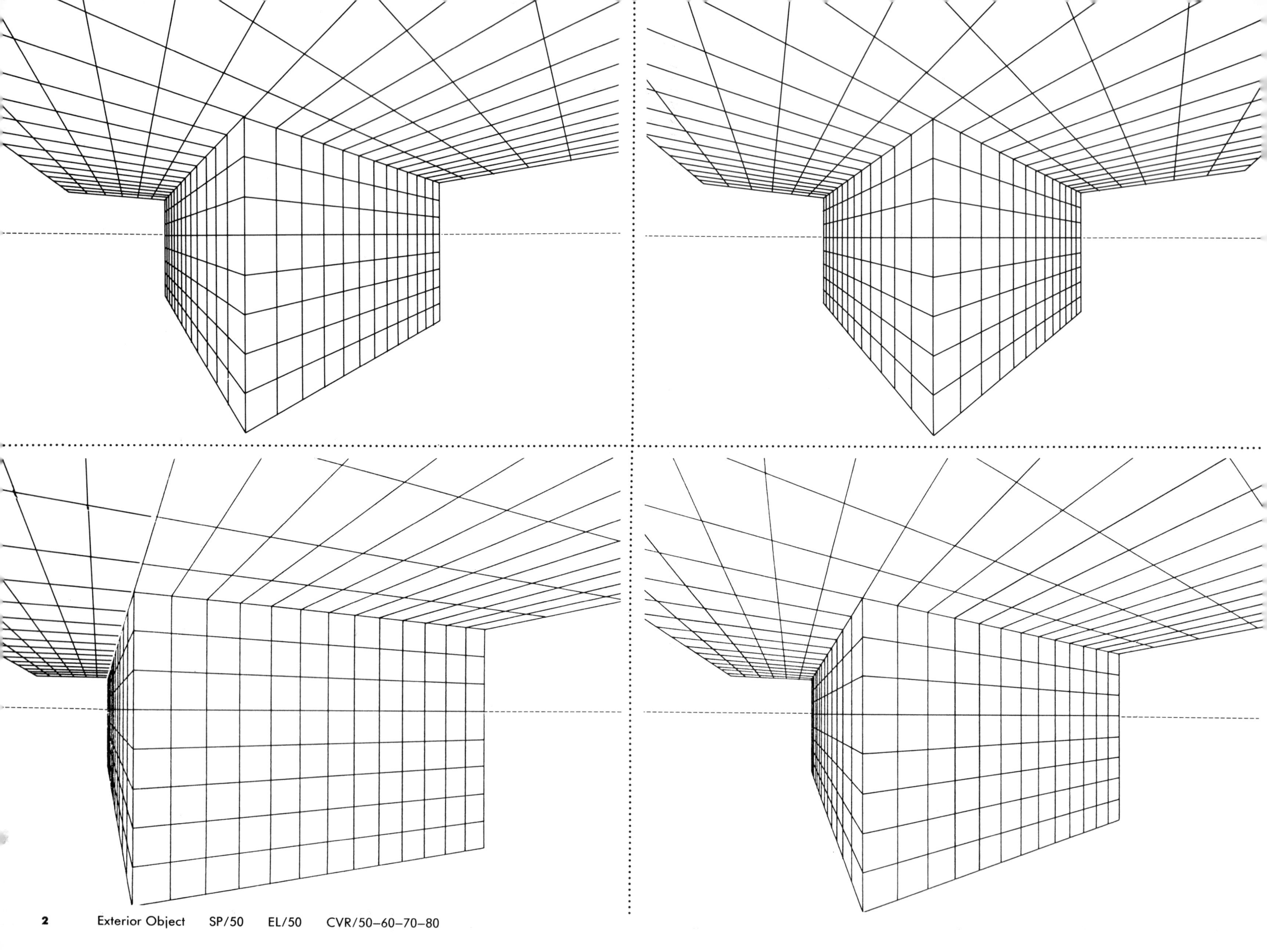

Exterior Object SP/75 EL/30 CVR/10–20–30–40

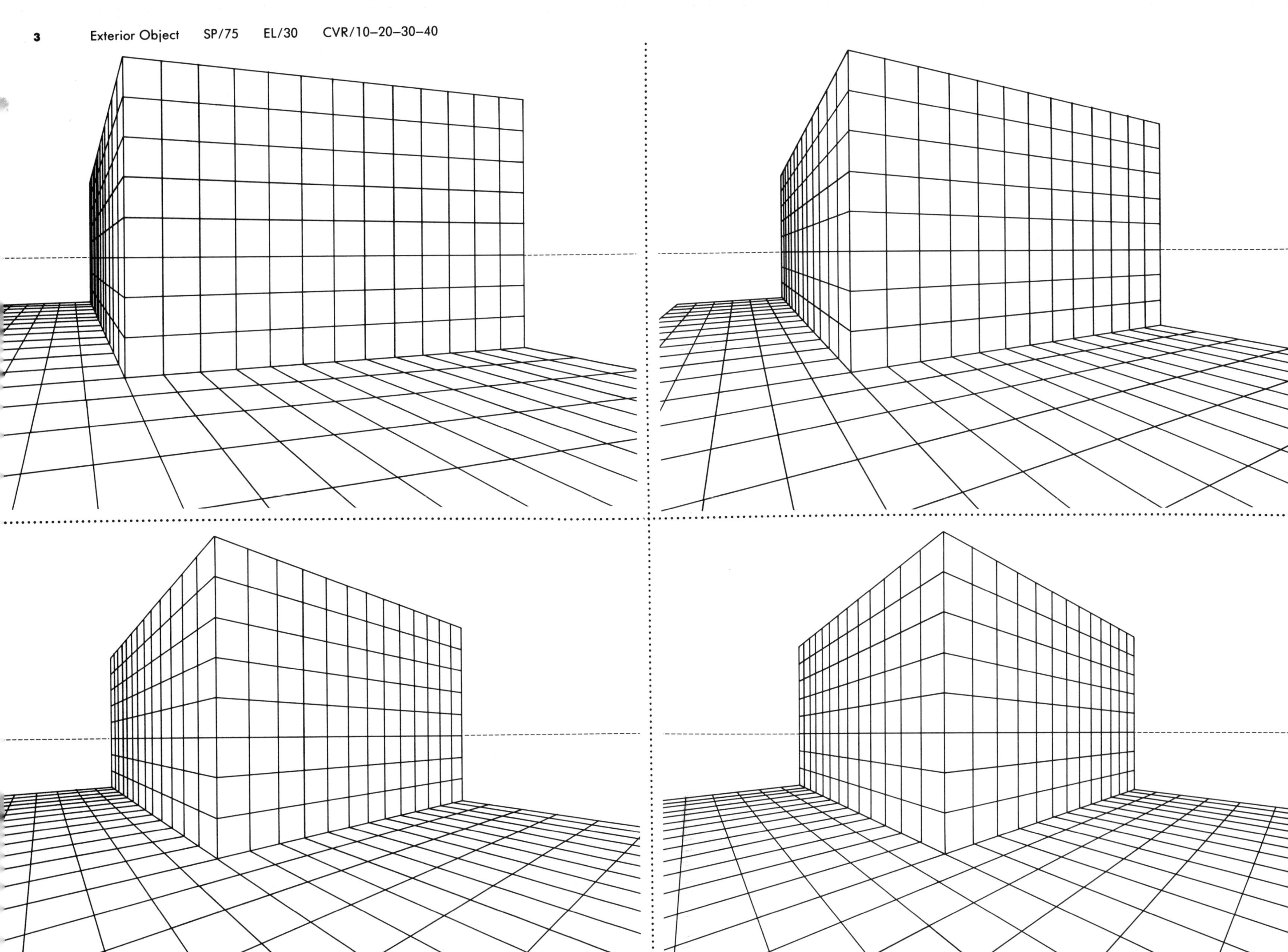

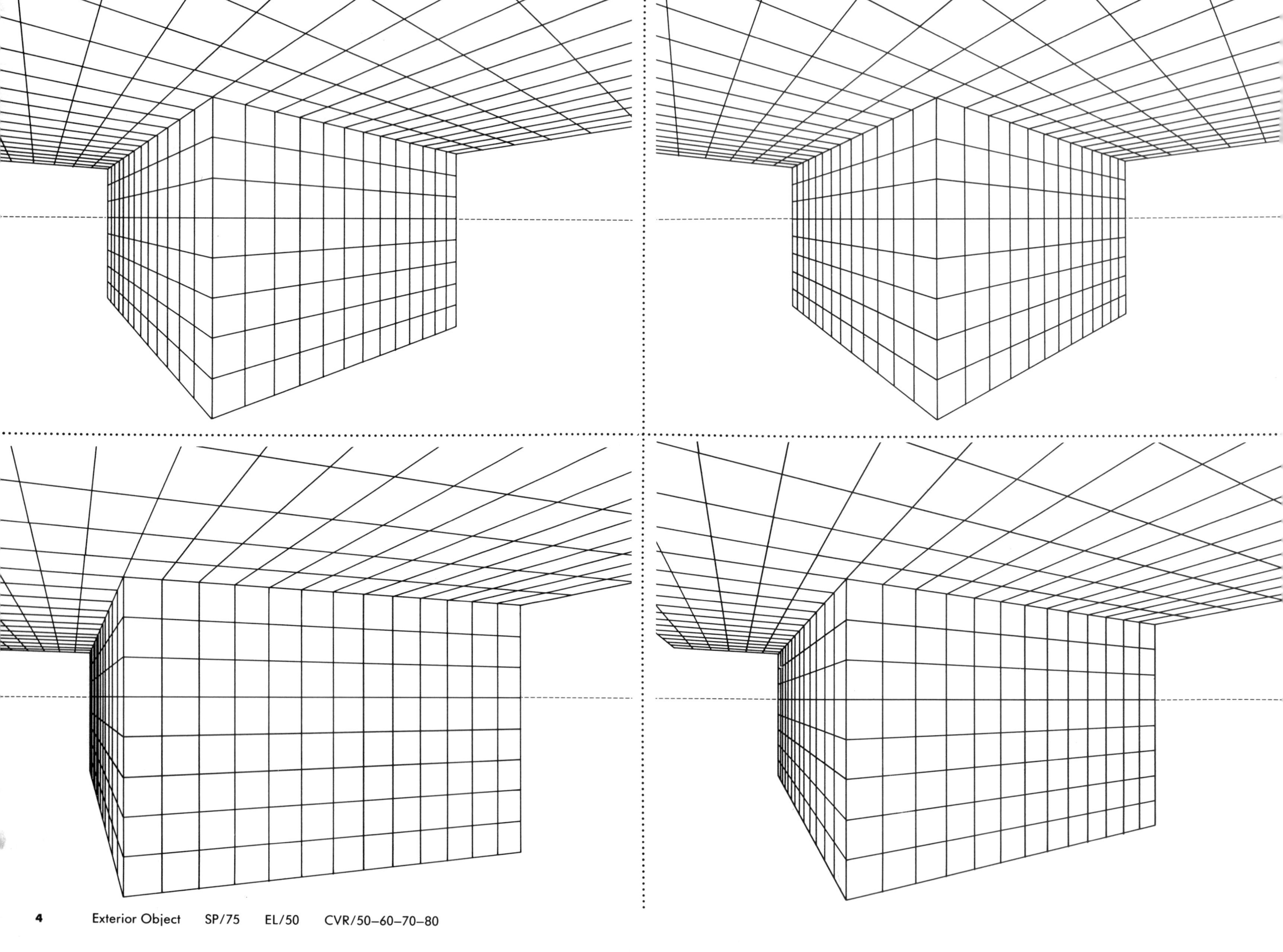

 Exterior Object SP/75 EL/50 CVR/50–60–70–80

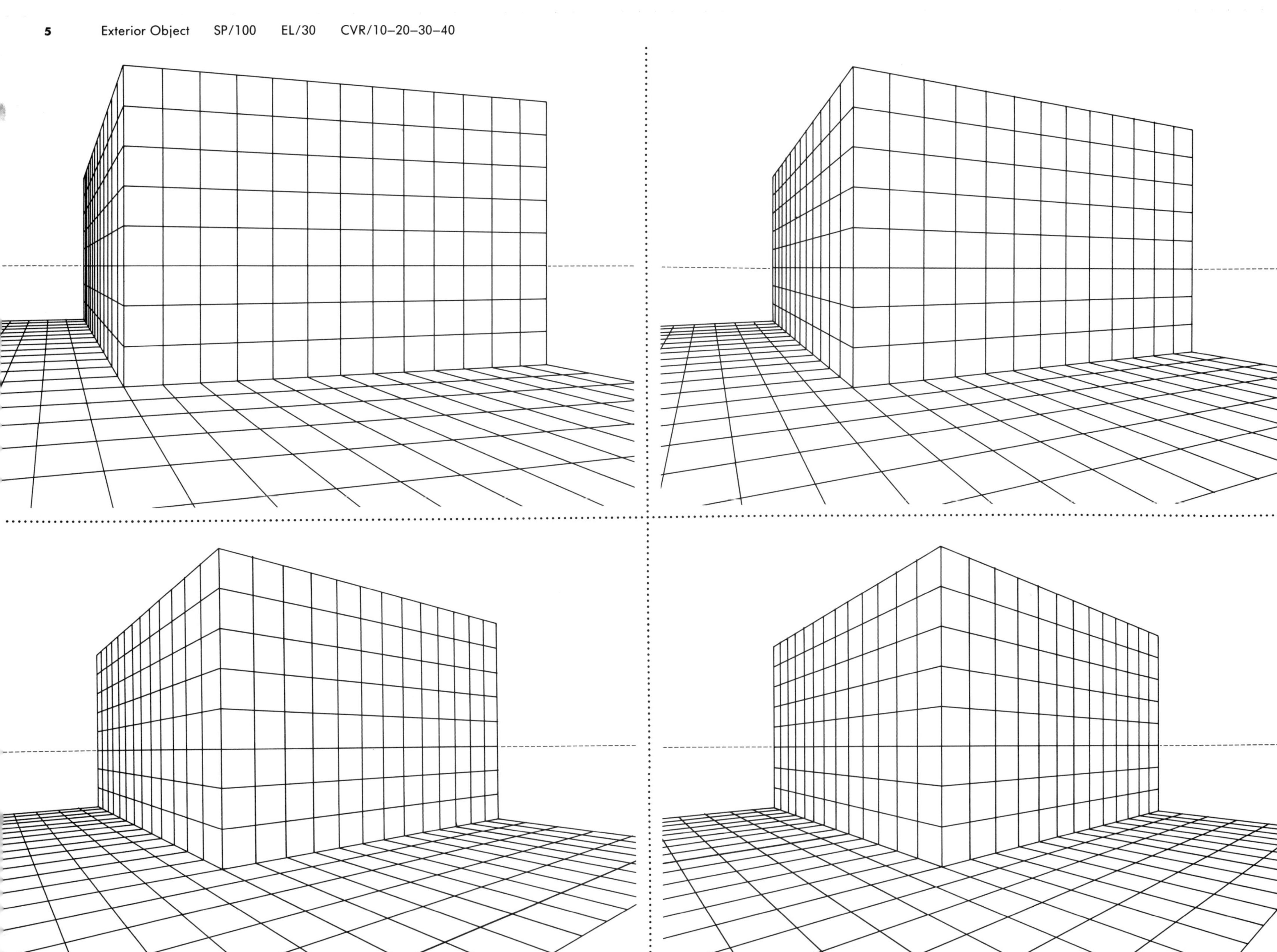

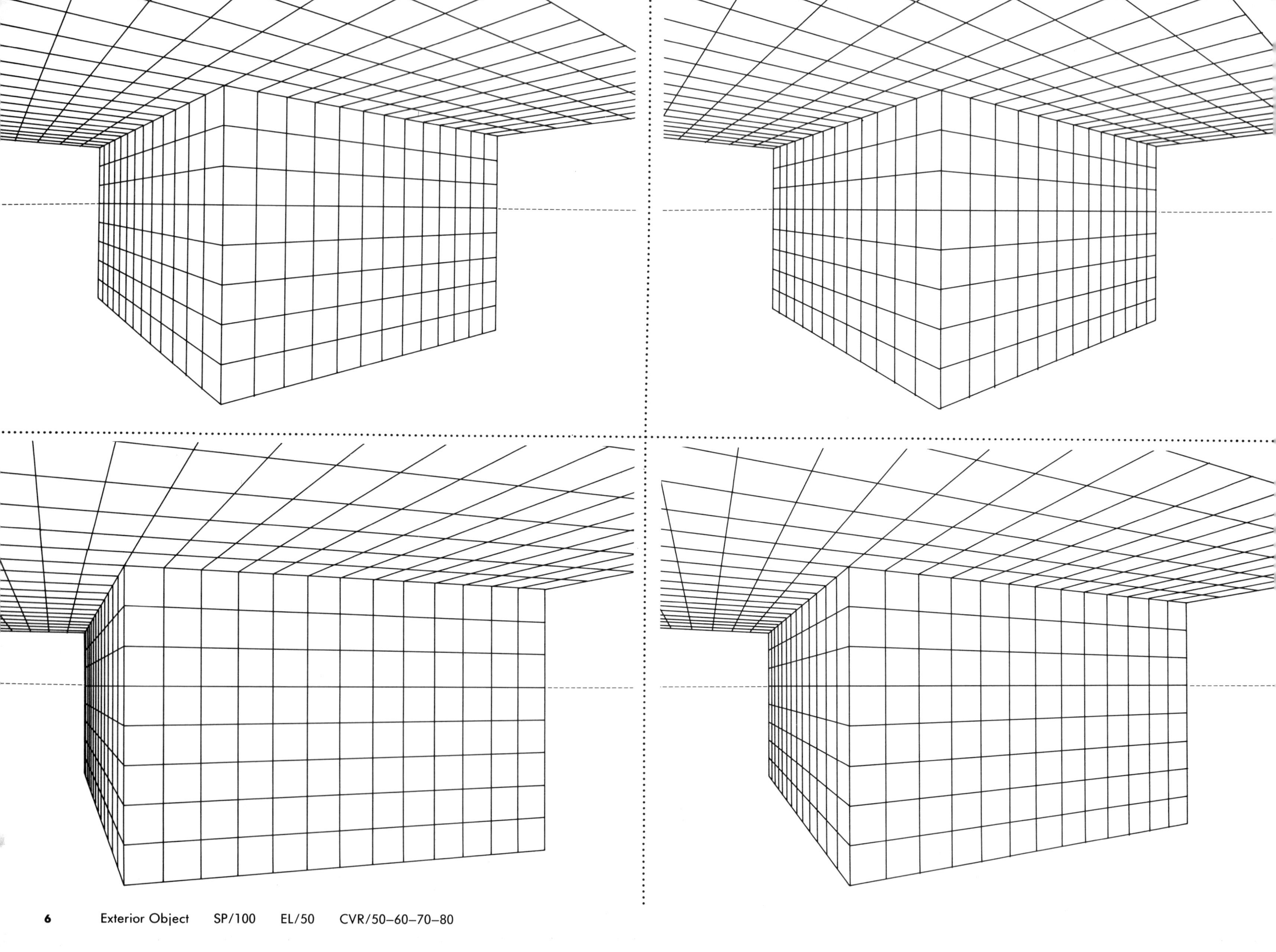

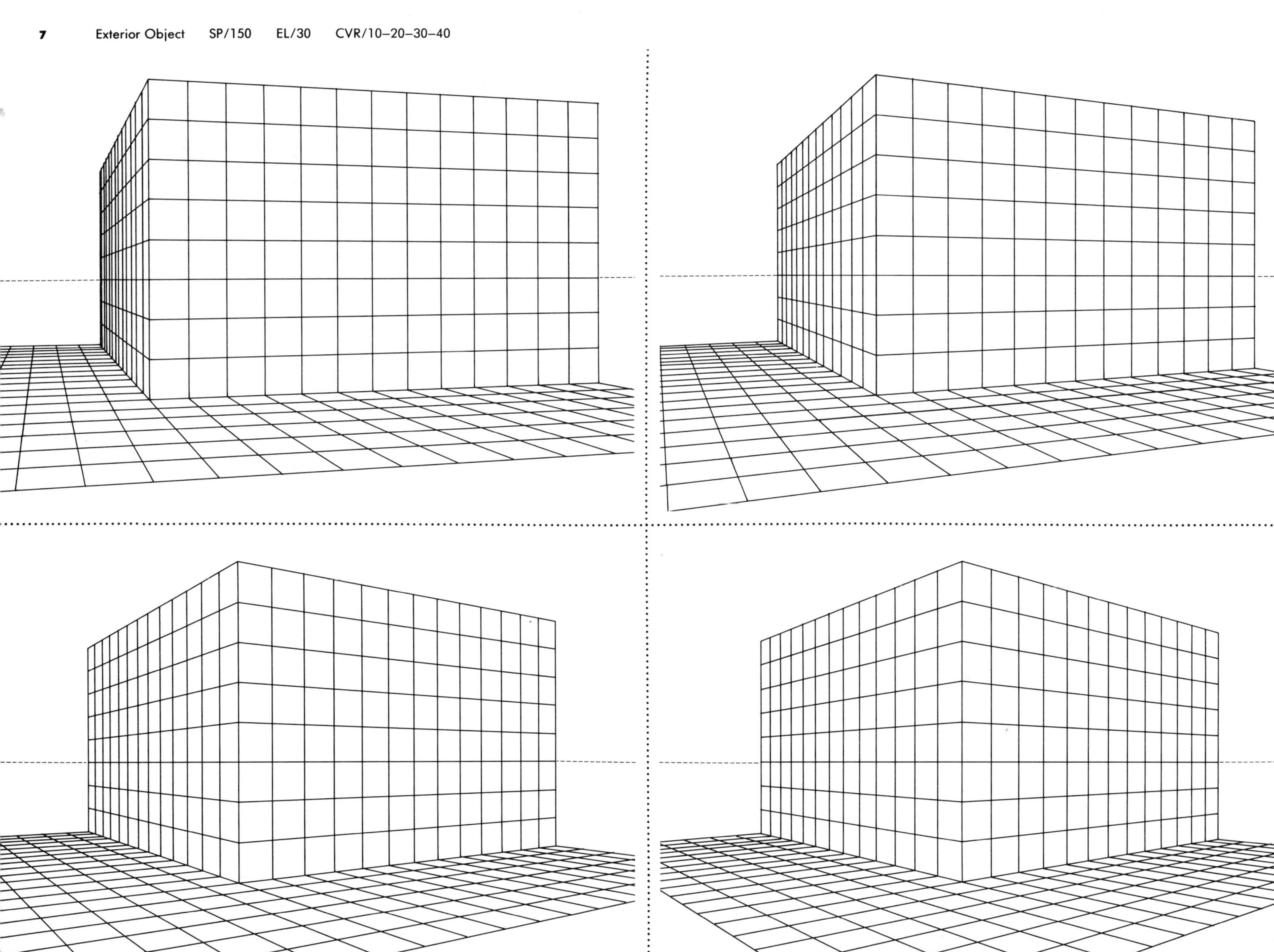

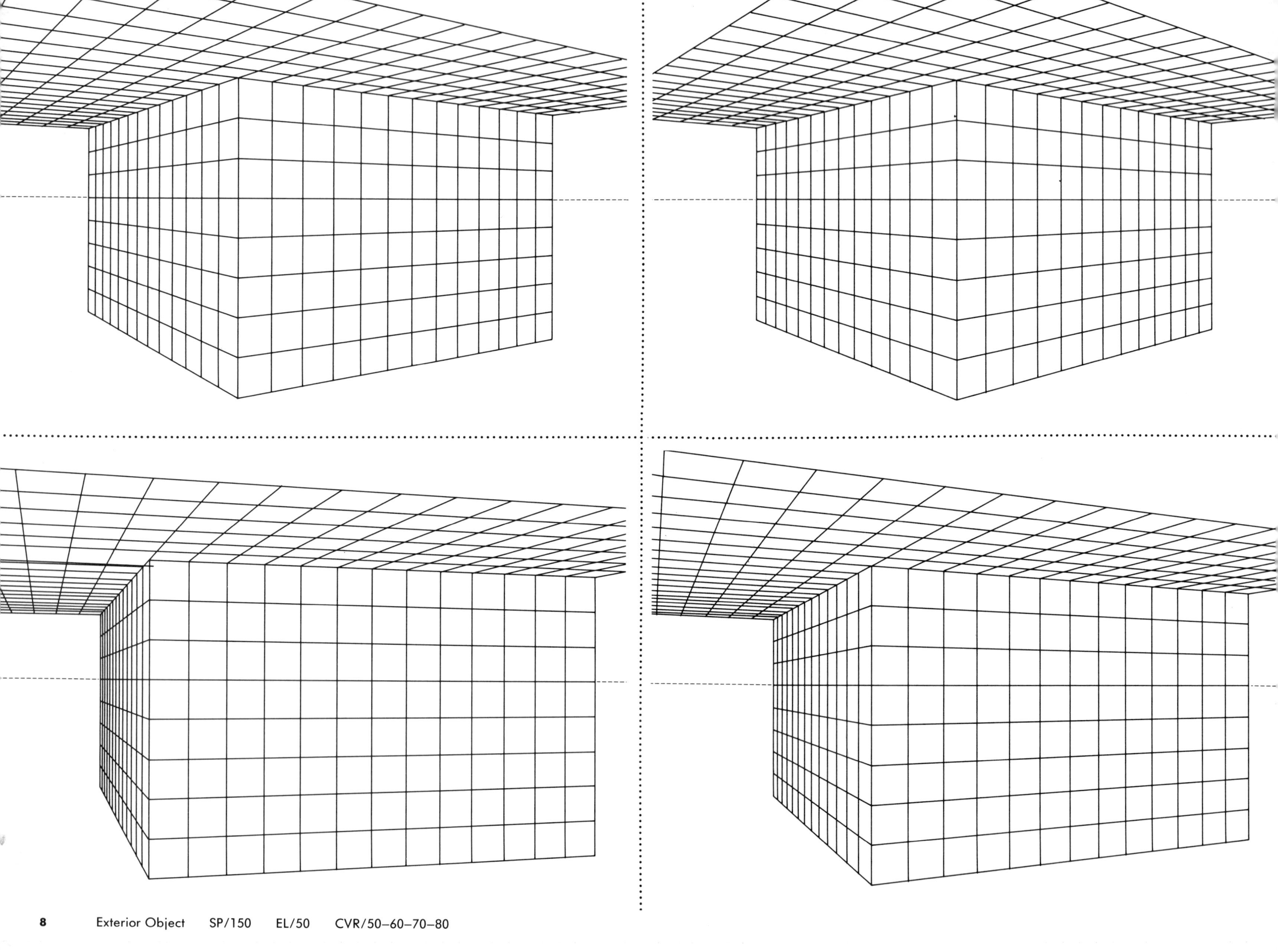

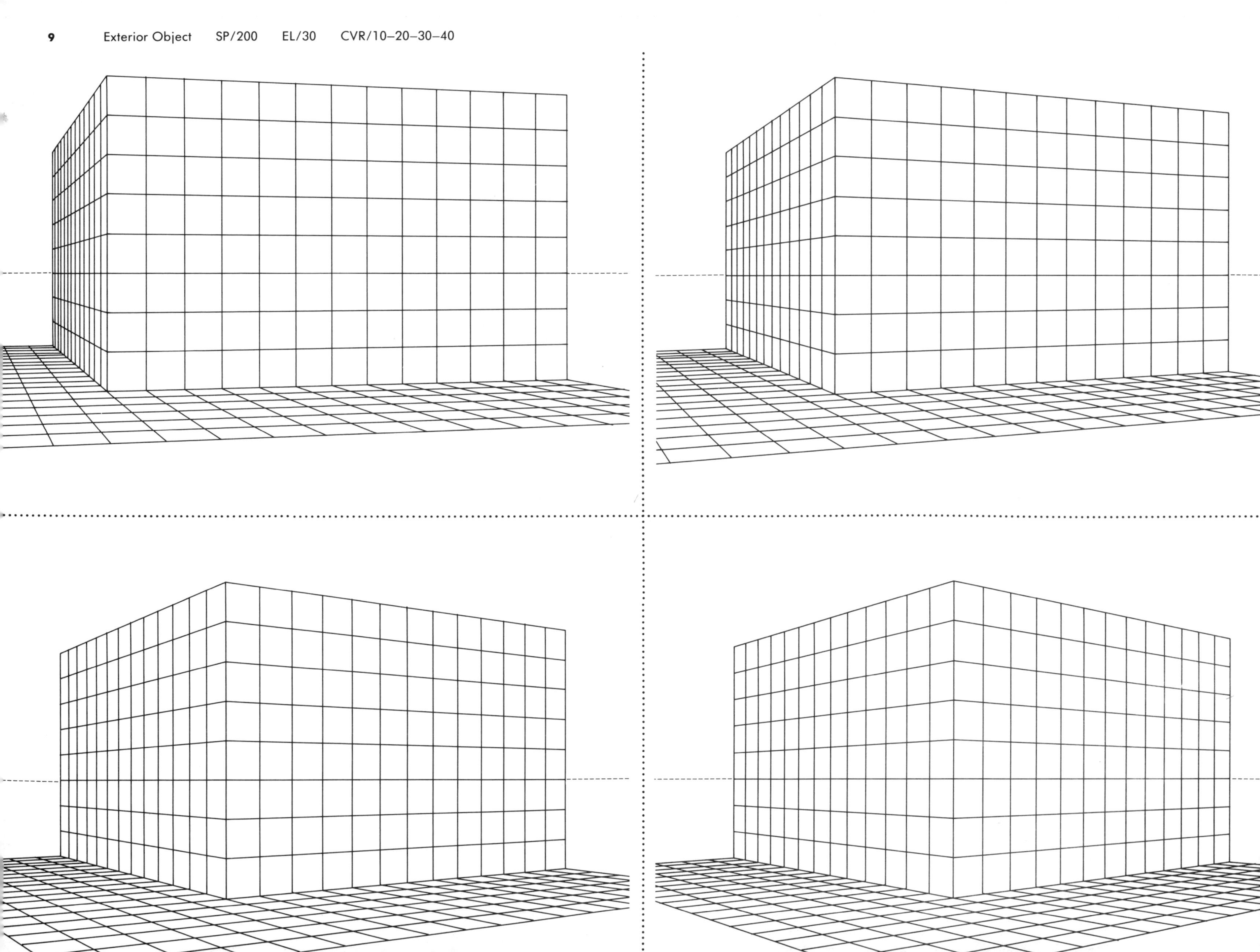

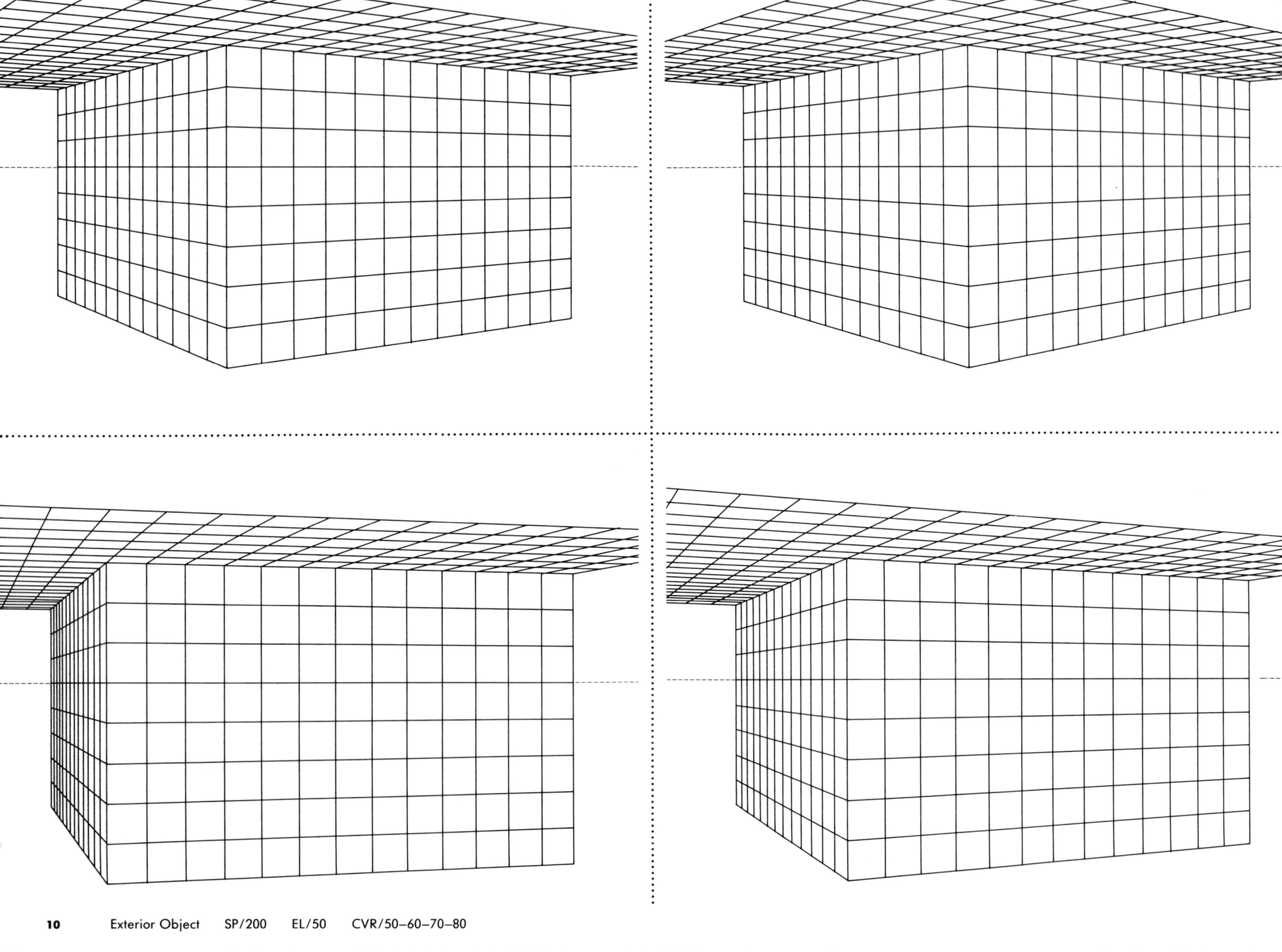

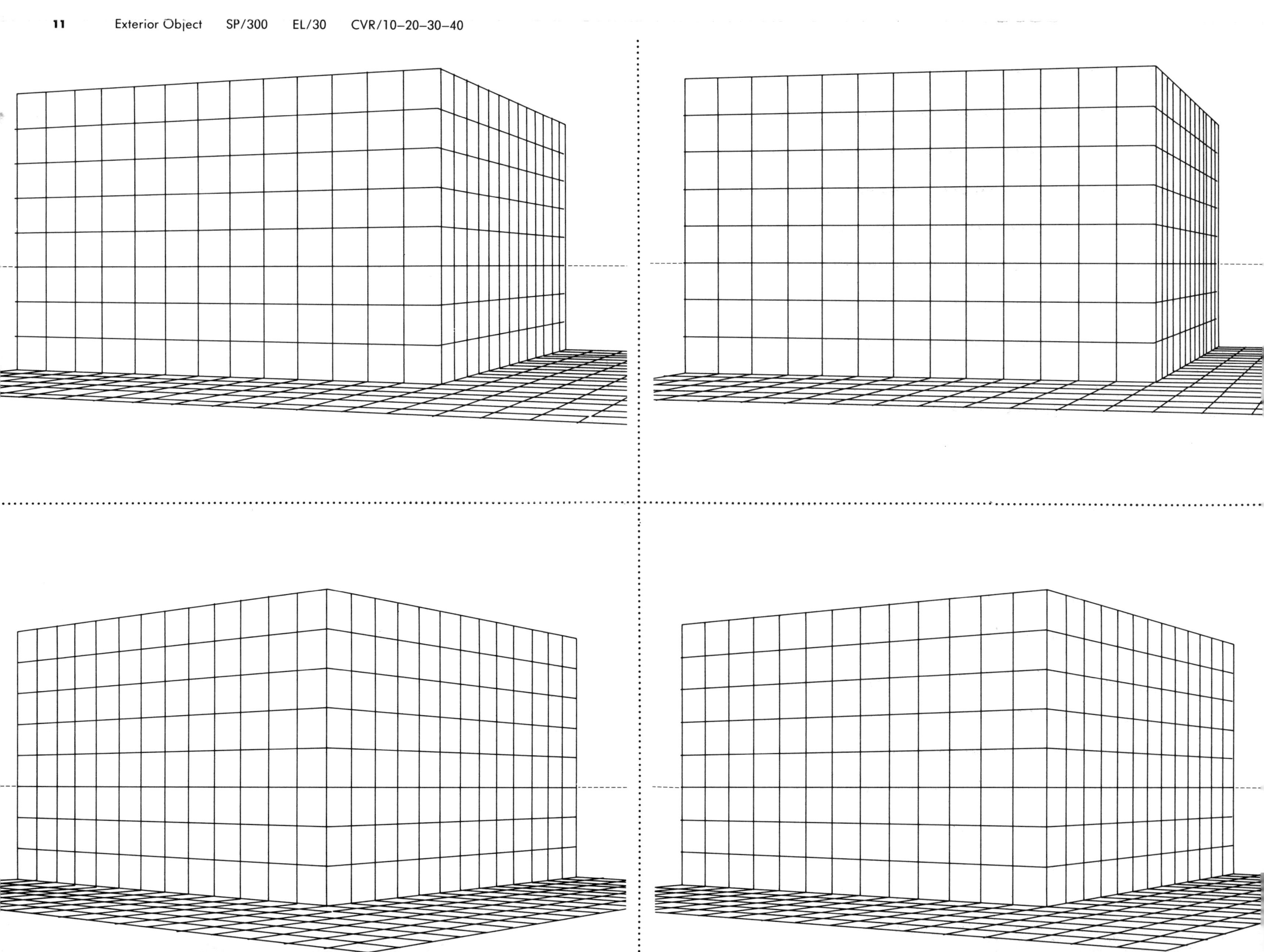

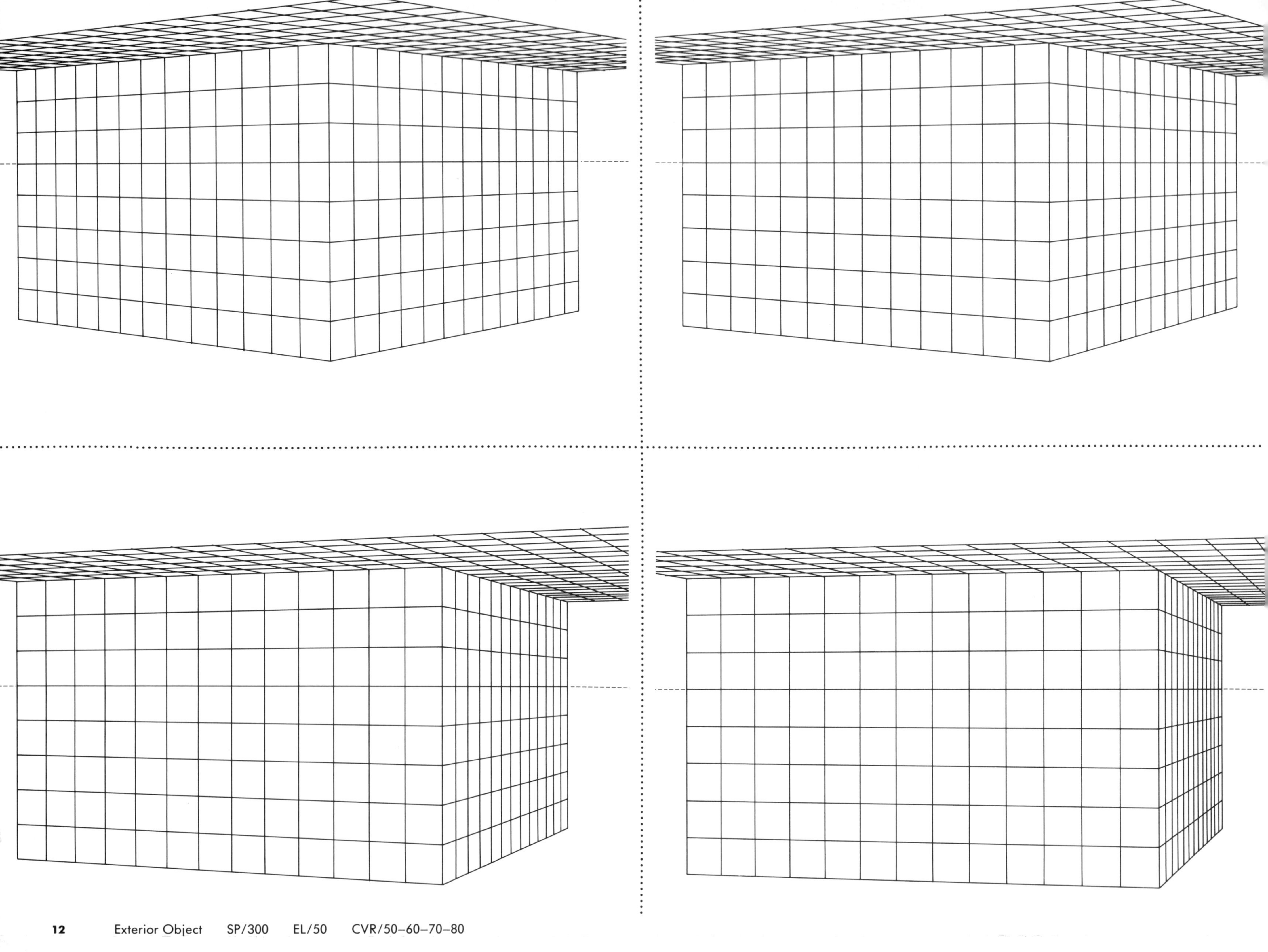

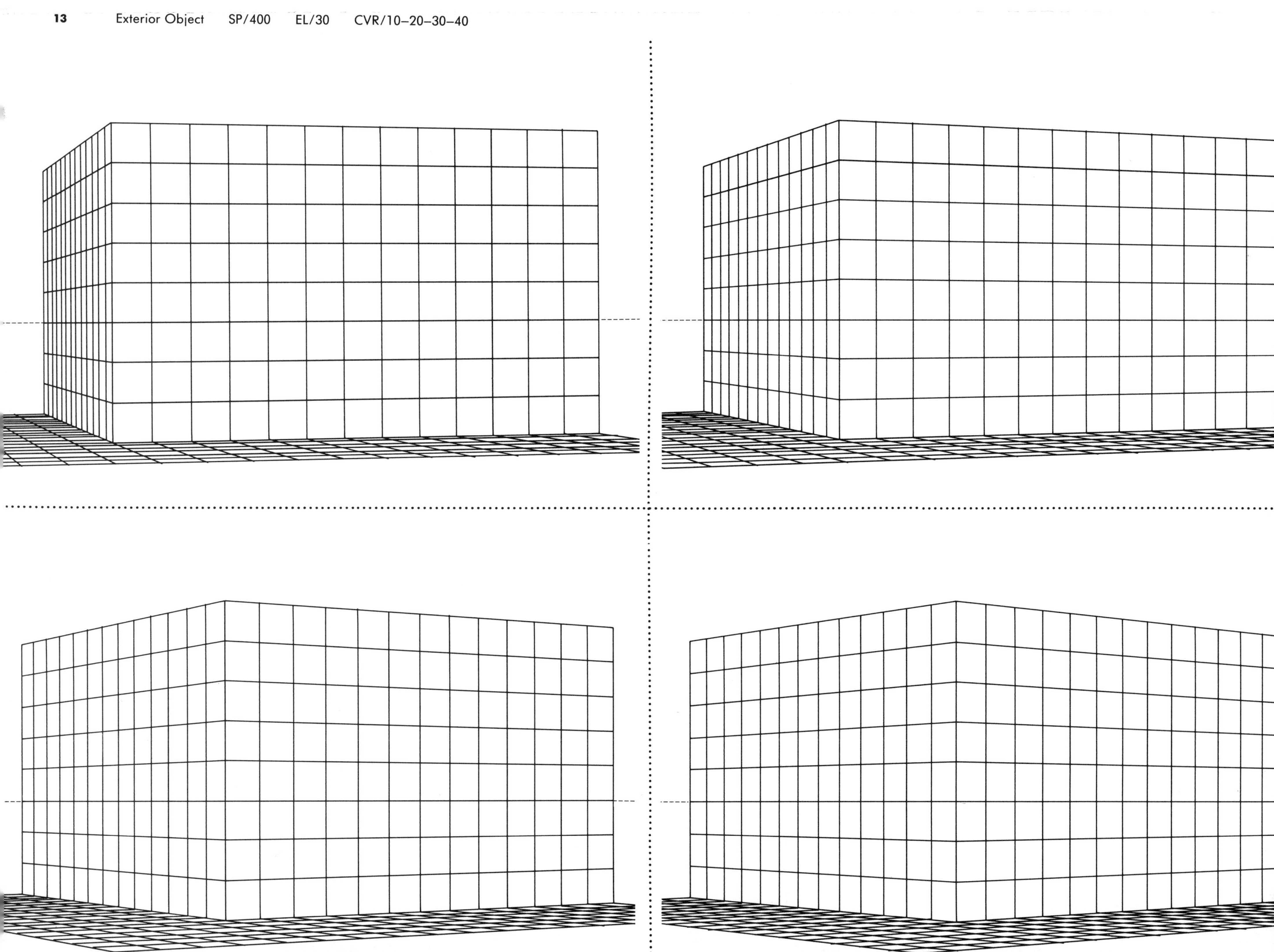

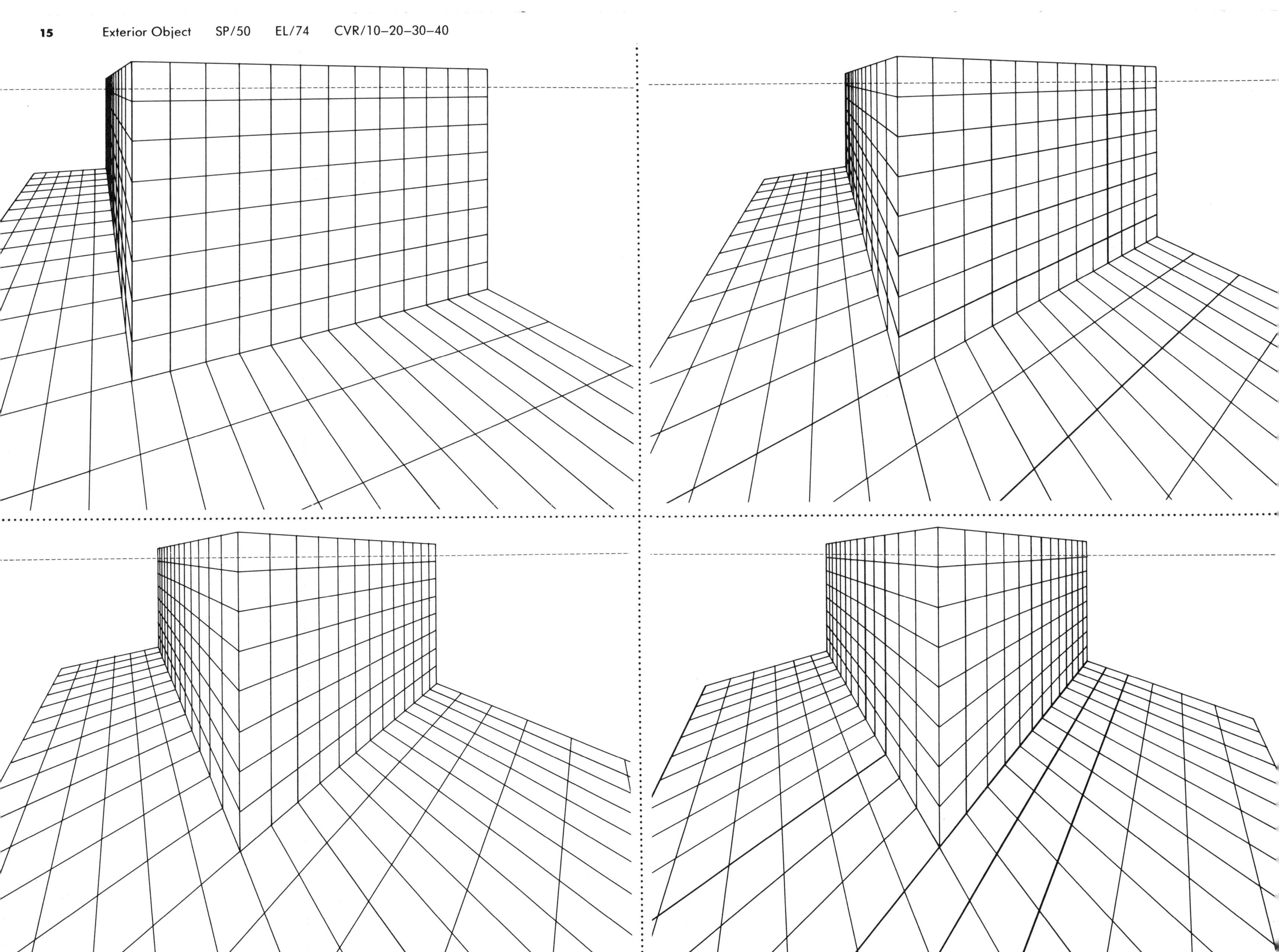

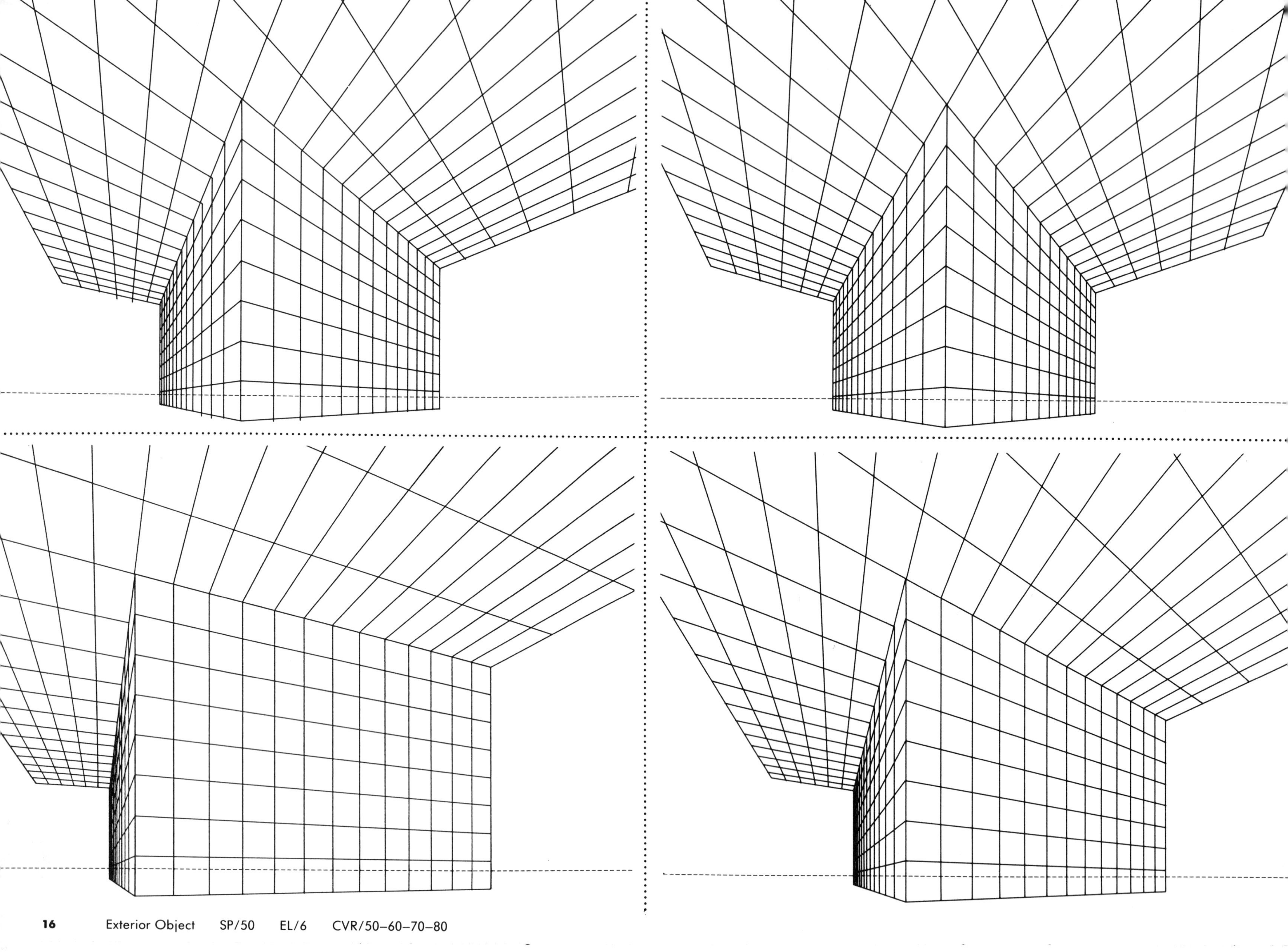

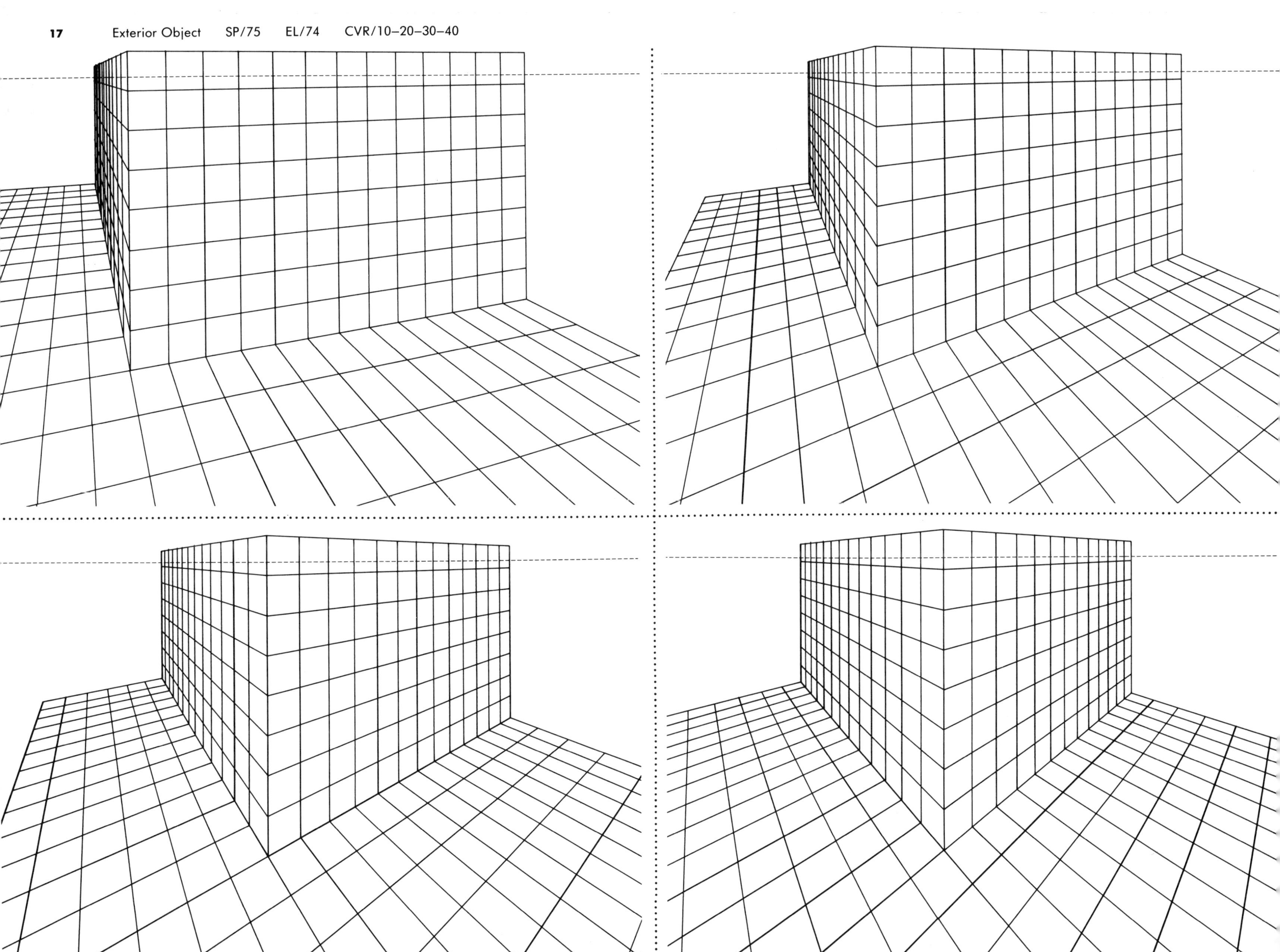

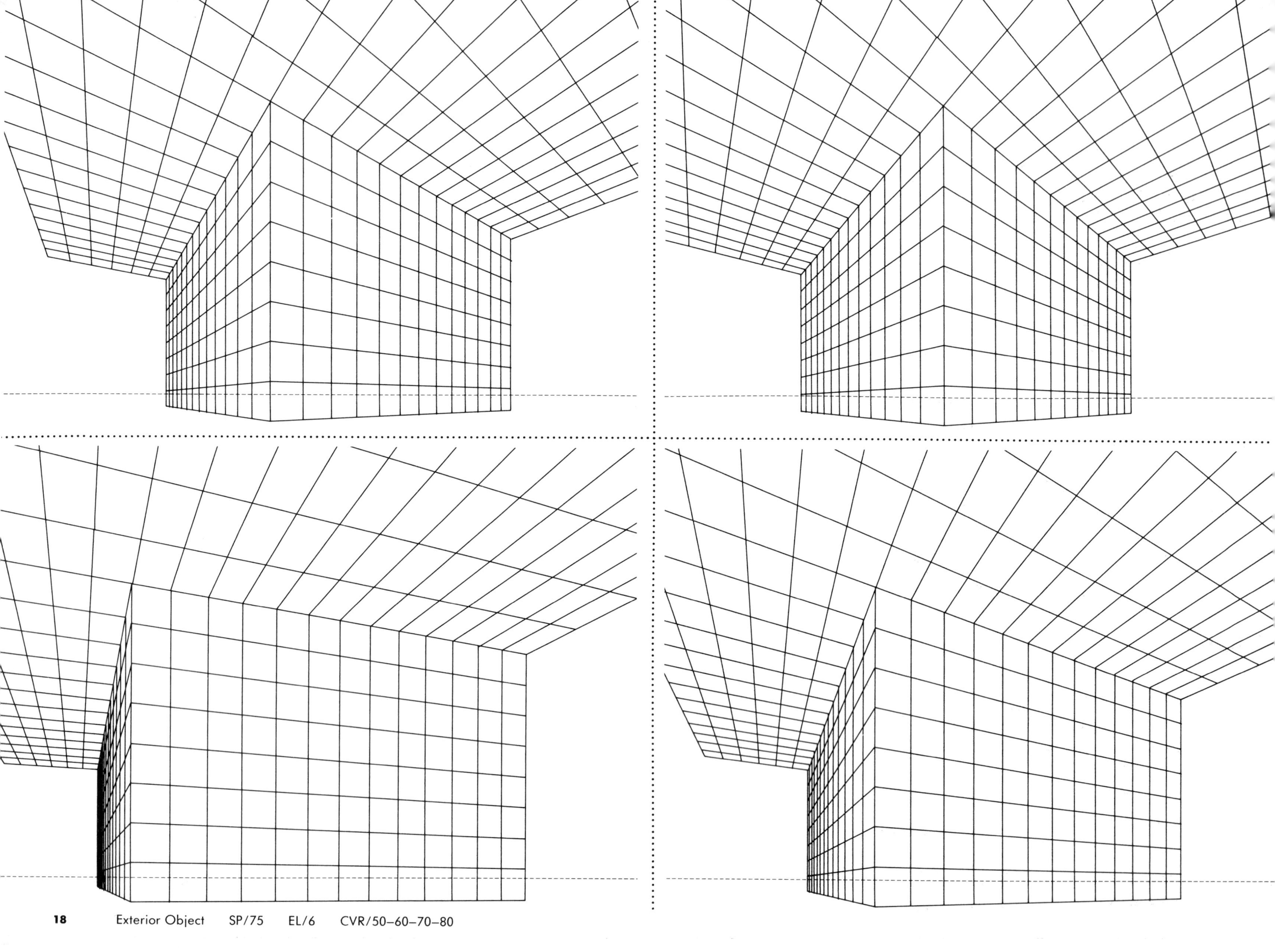

 Exterior Object SP/75 EL/6 CVR/50–60–70–80

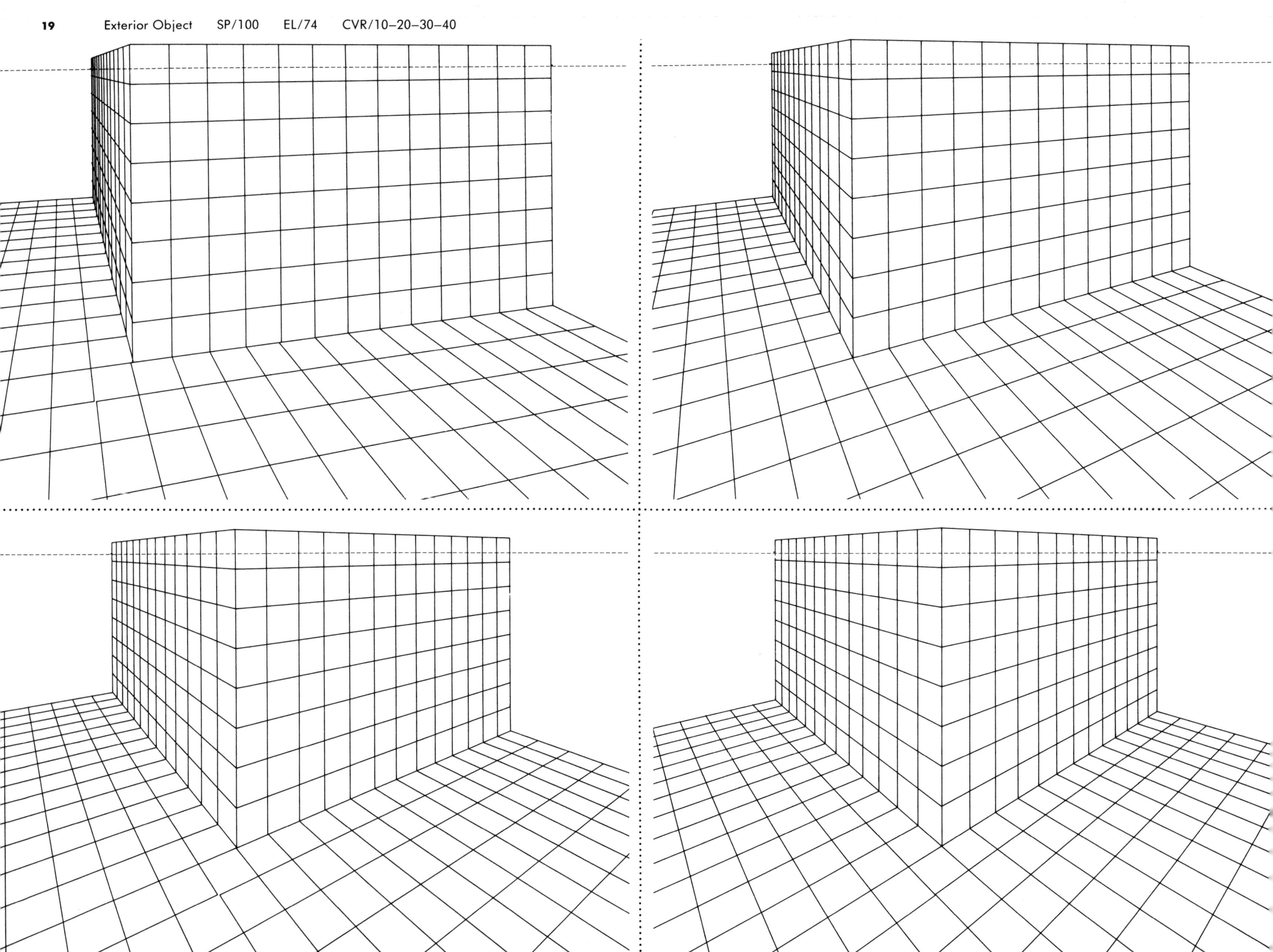

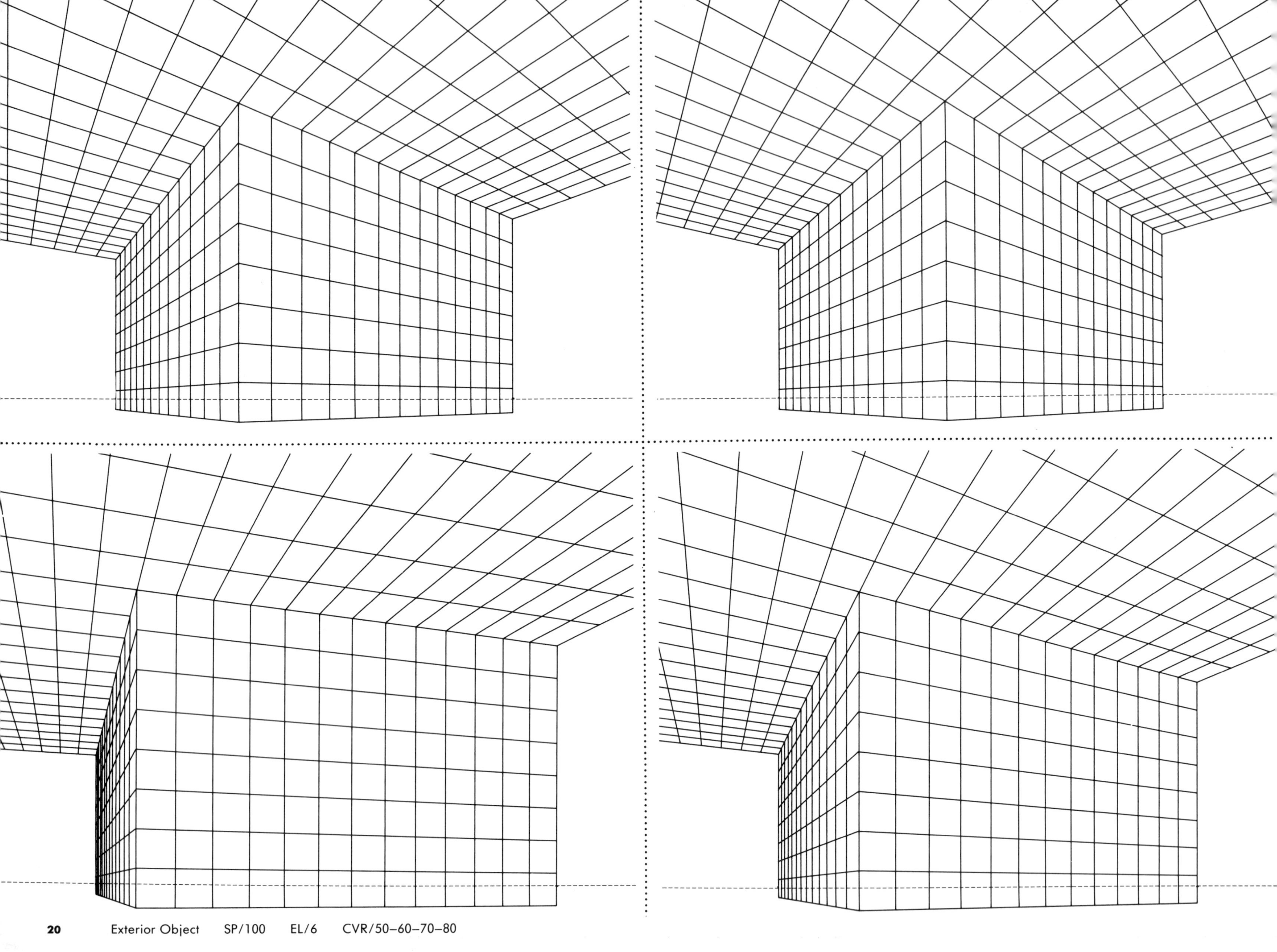

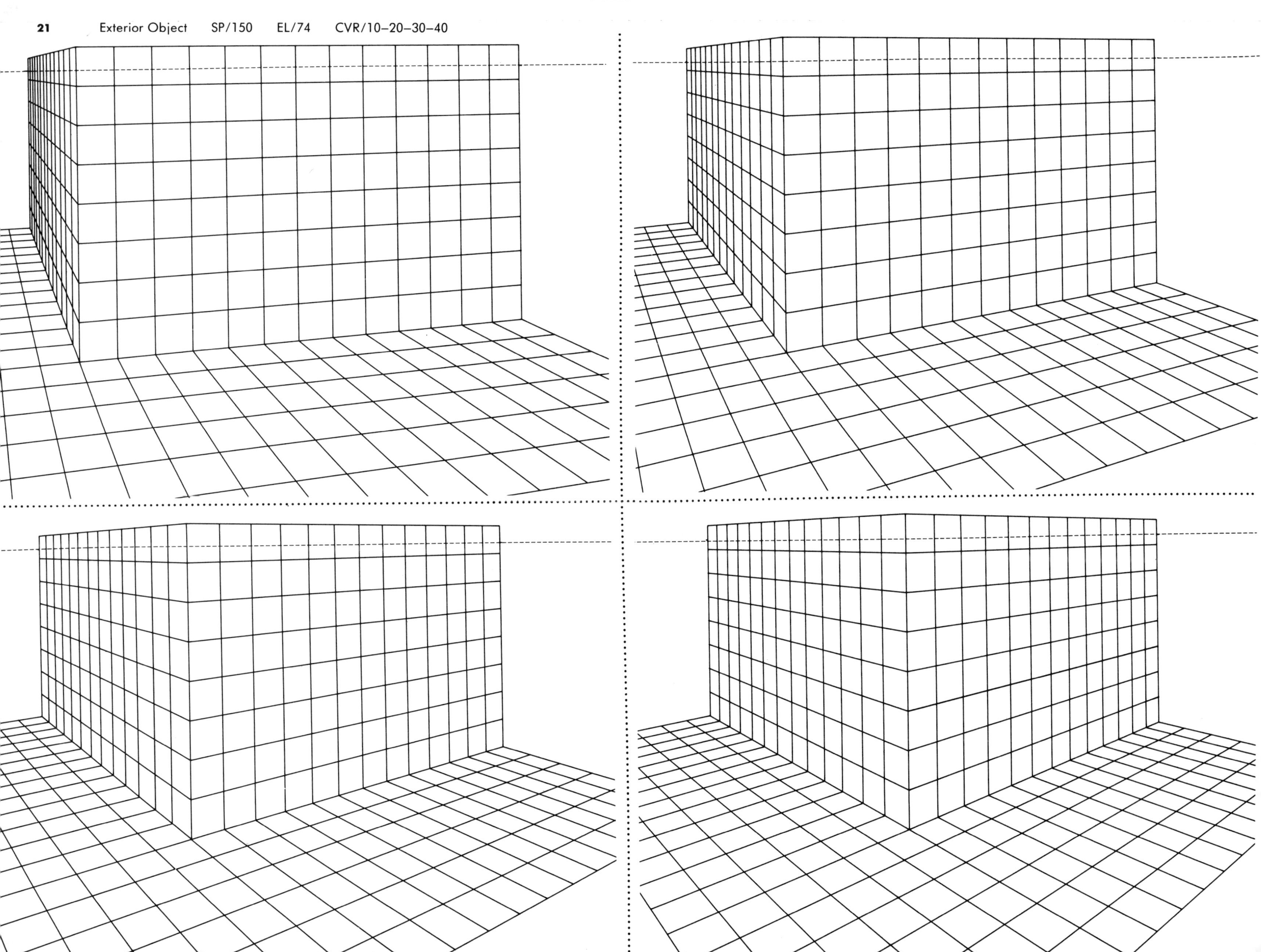

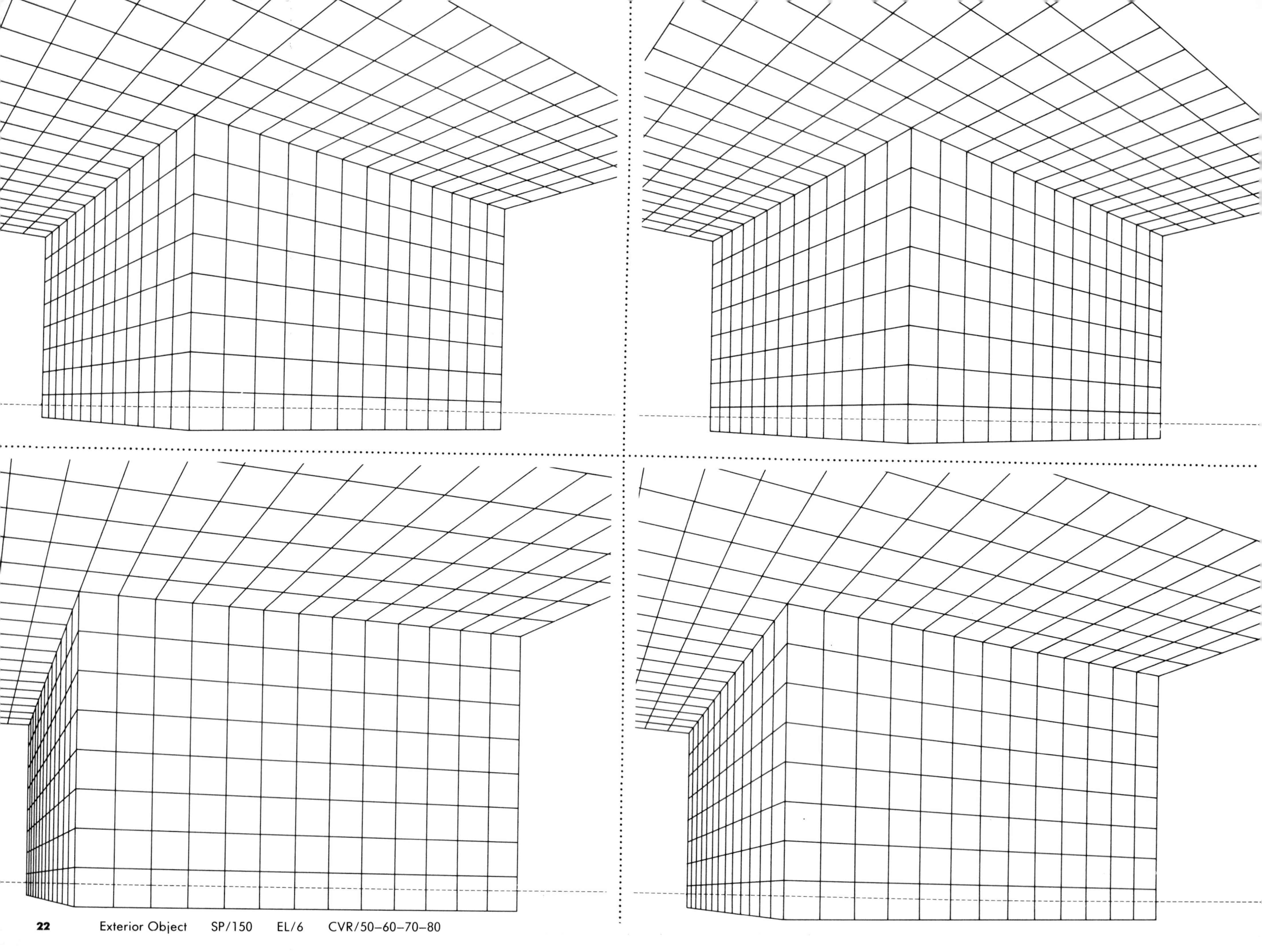

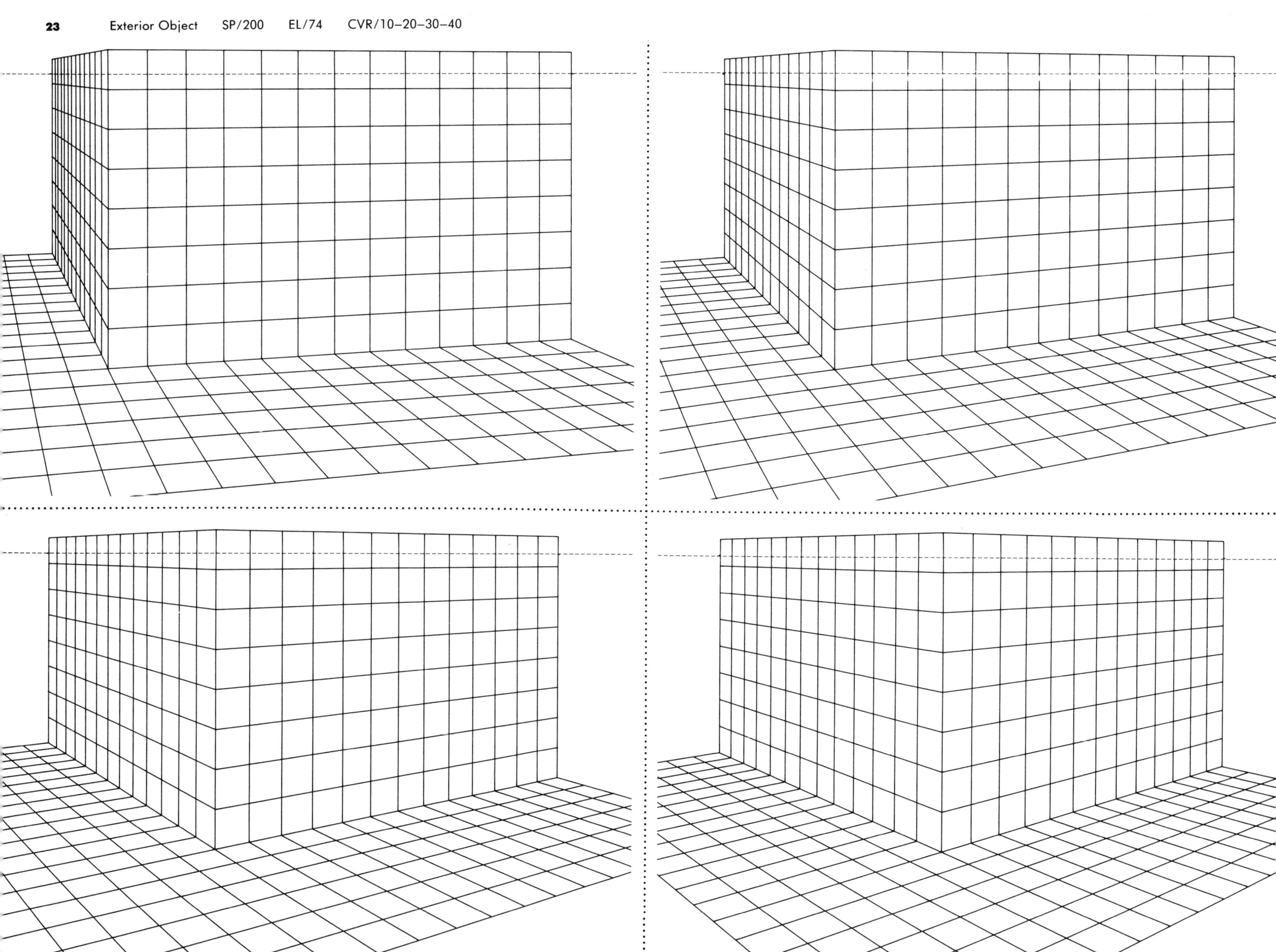

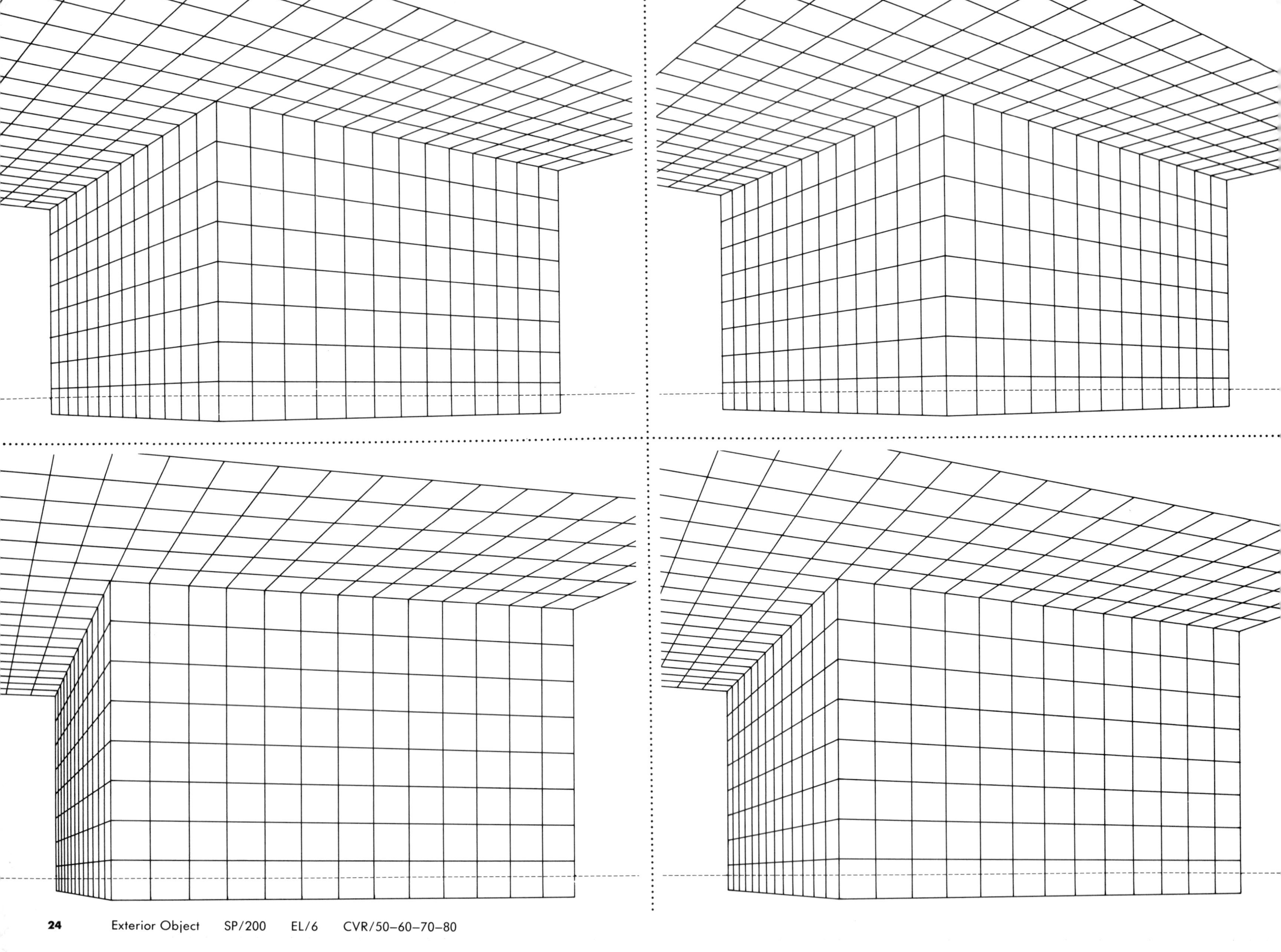

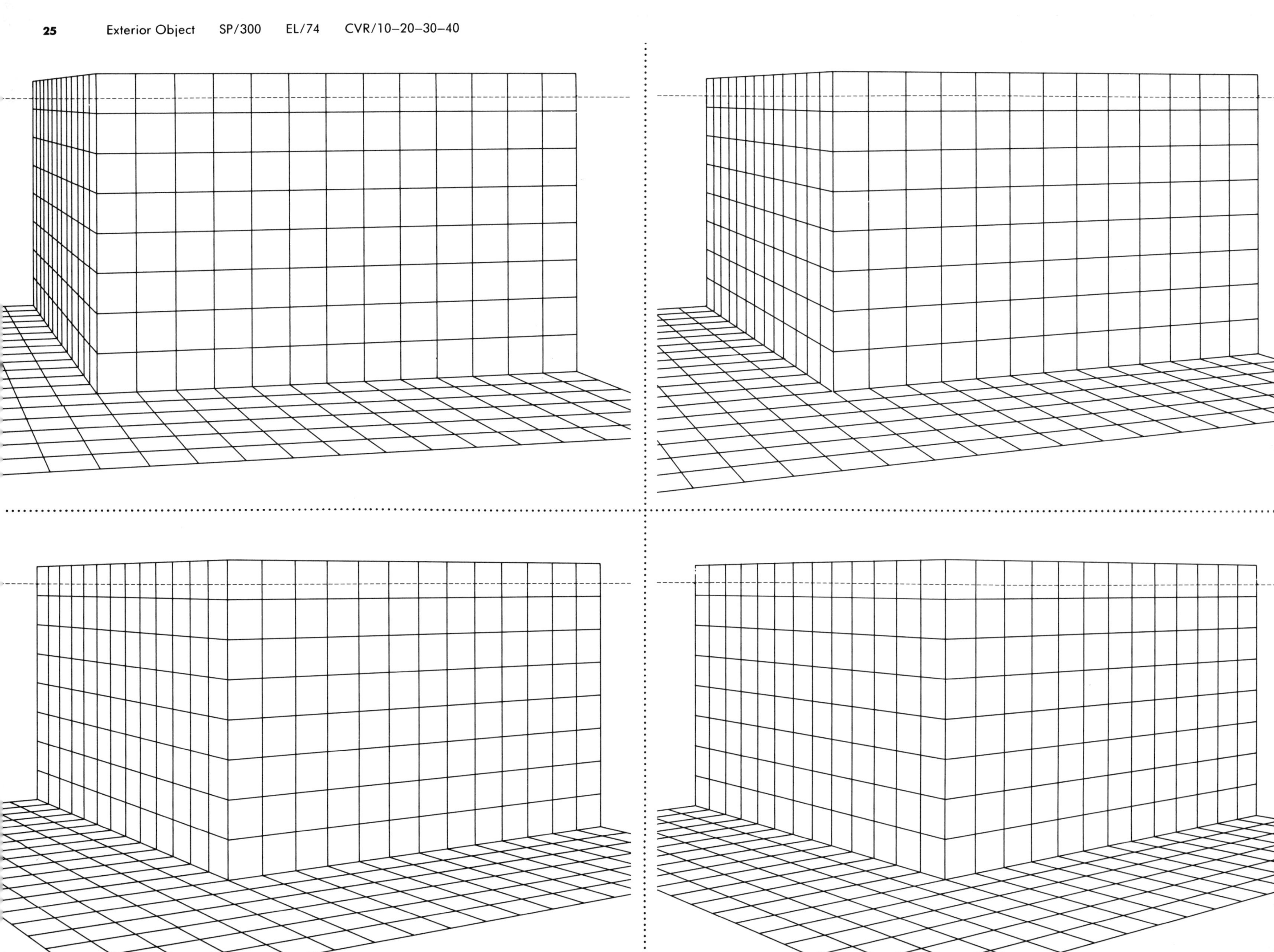

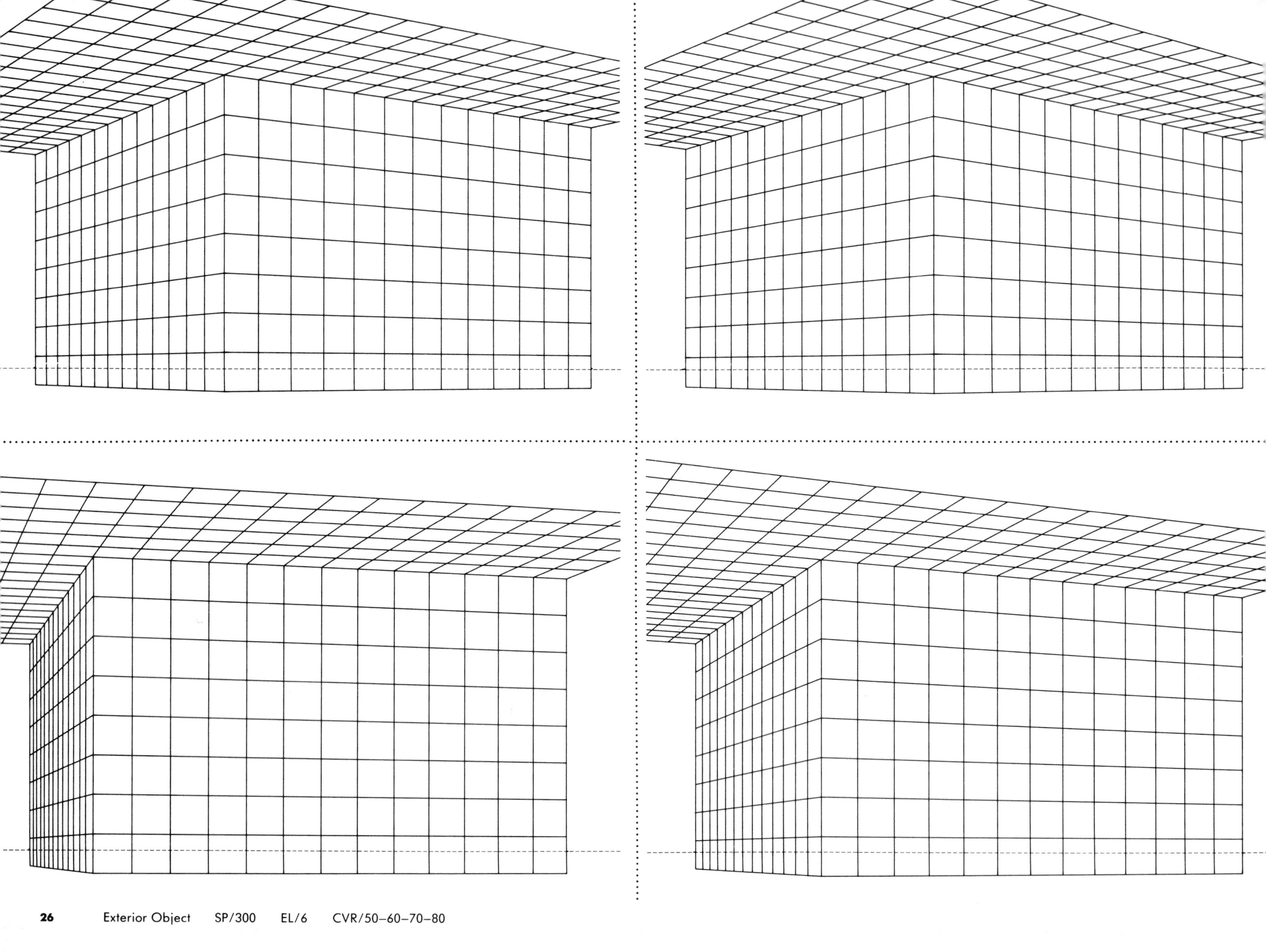

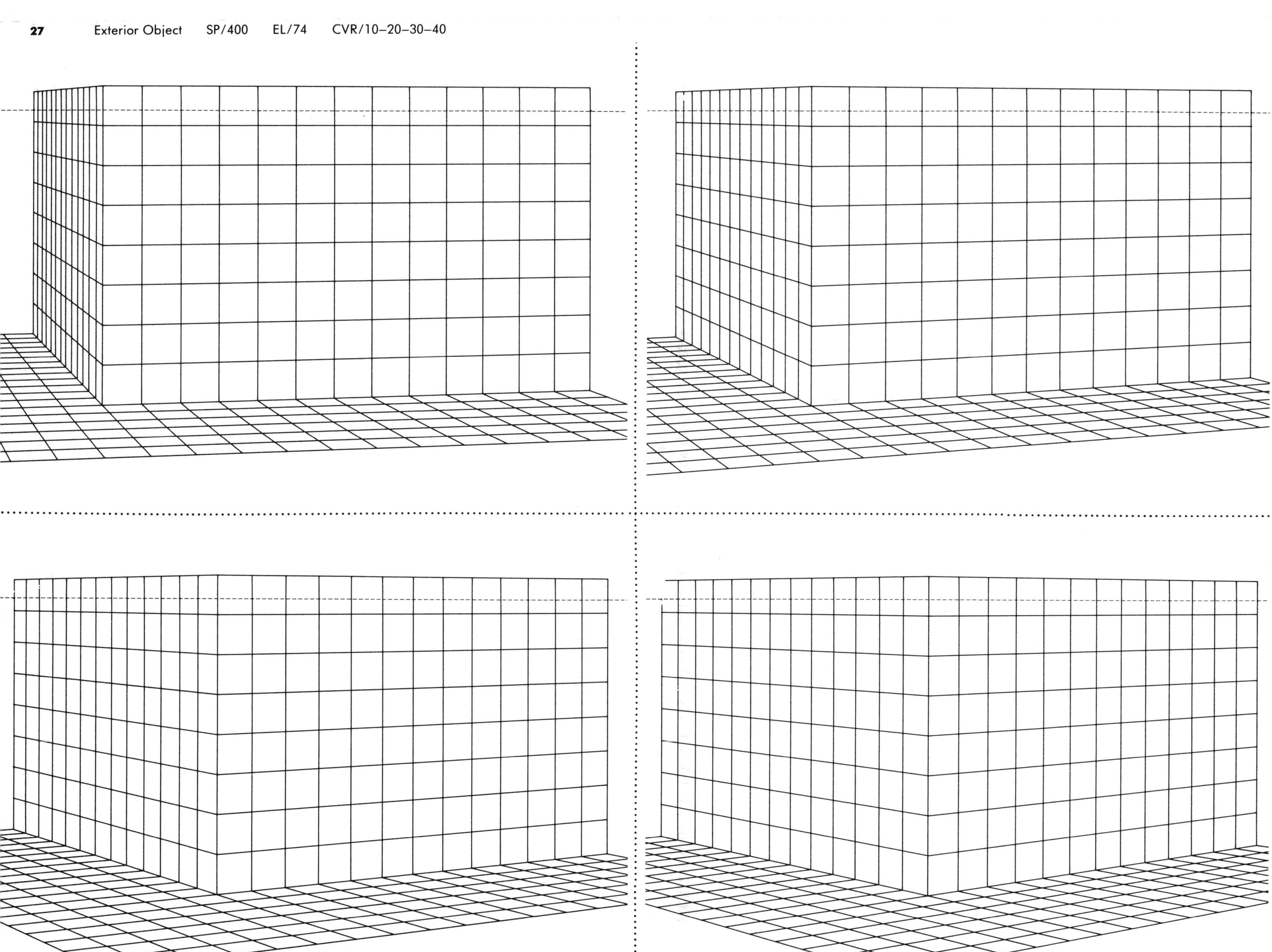

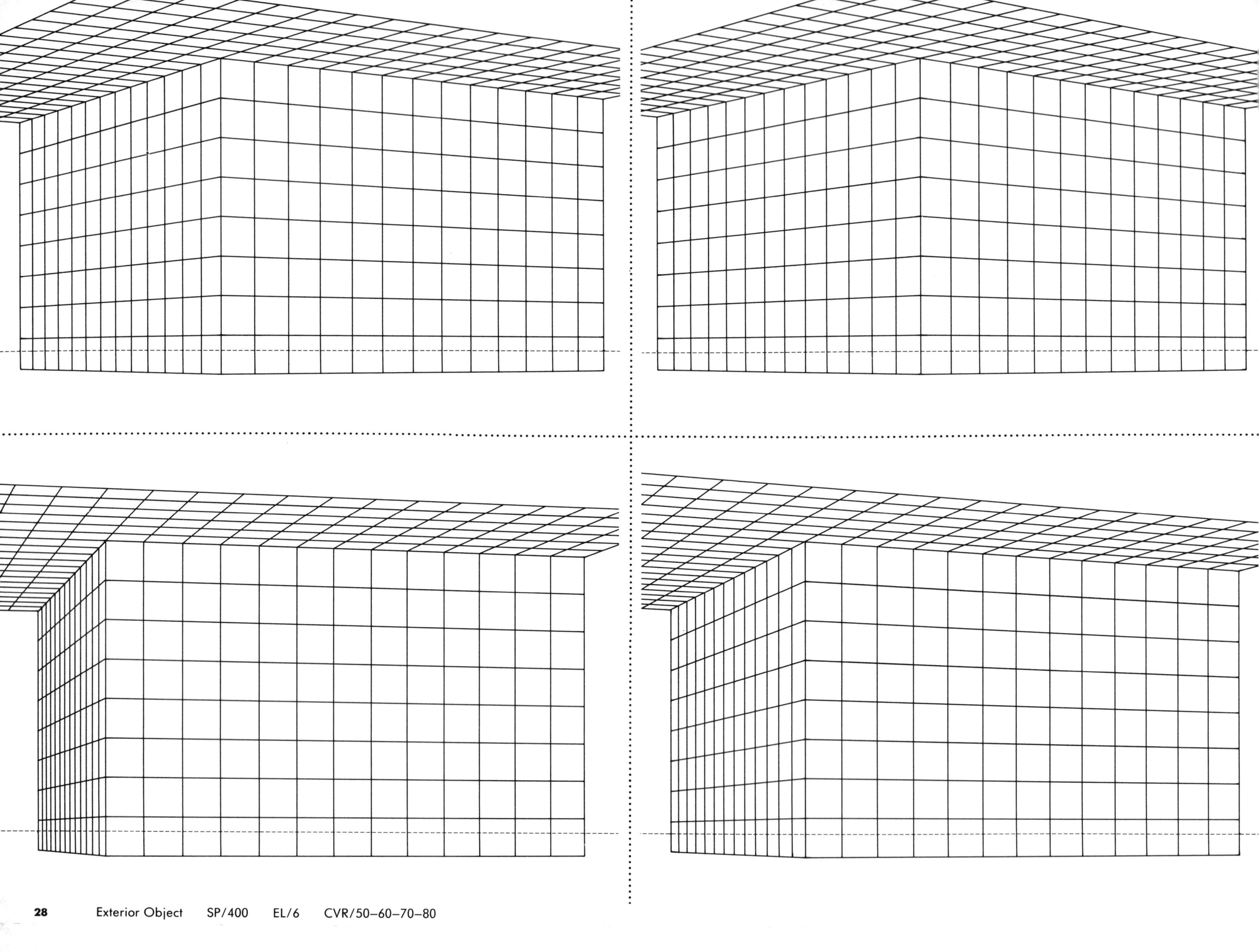

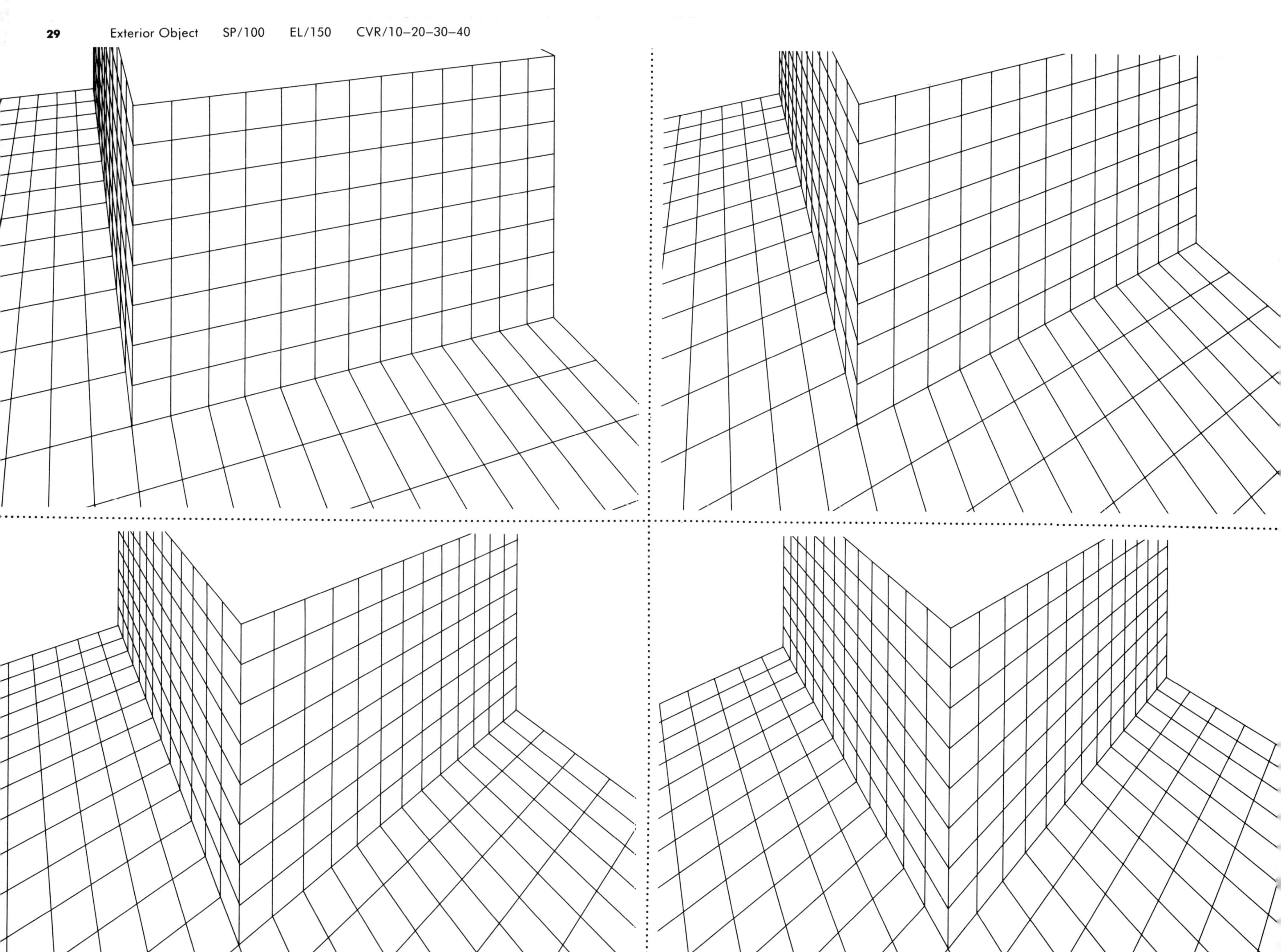

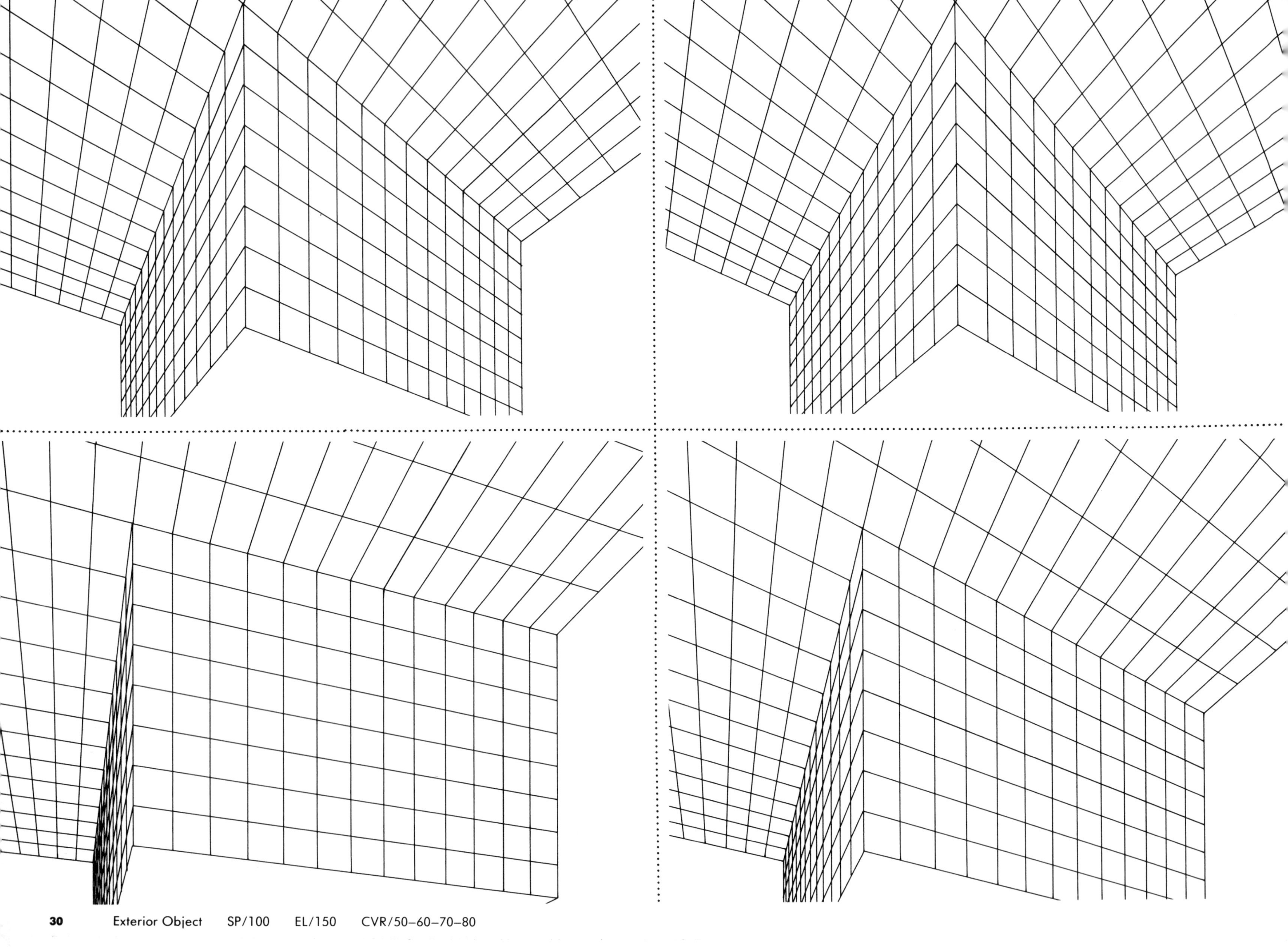

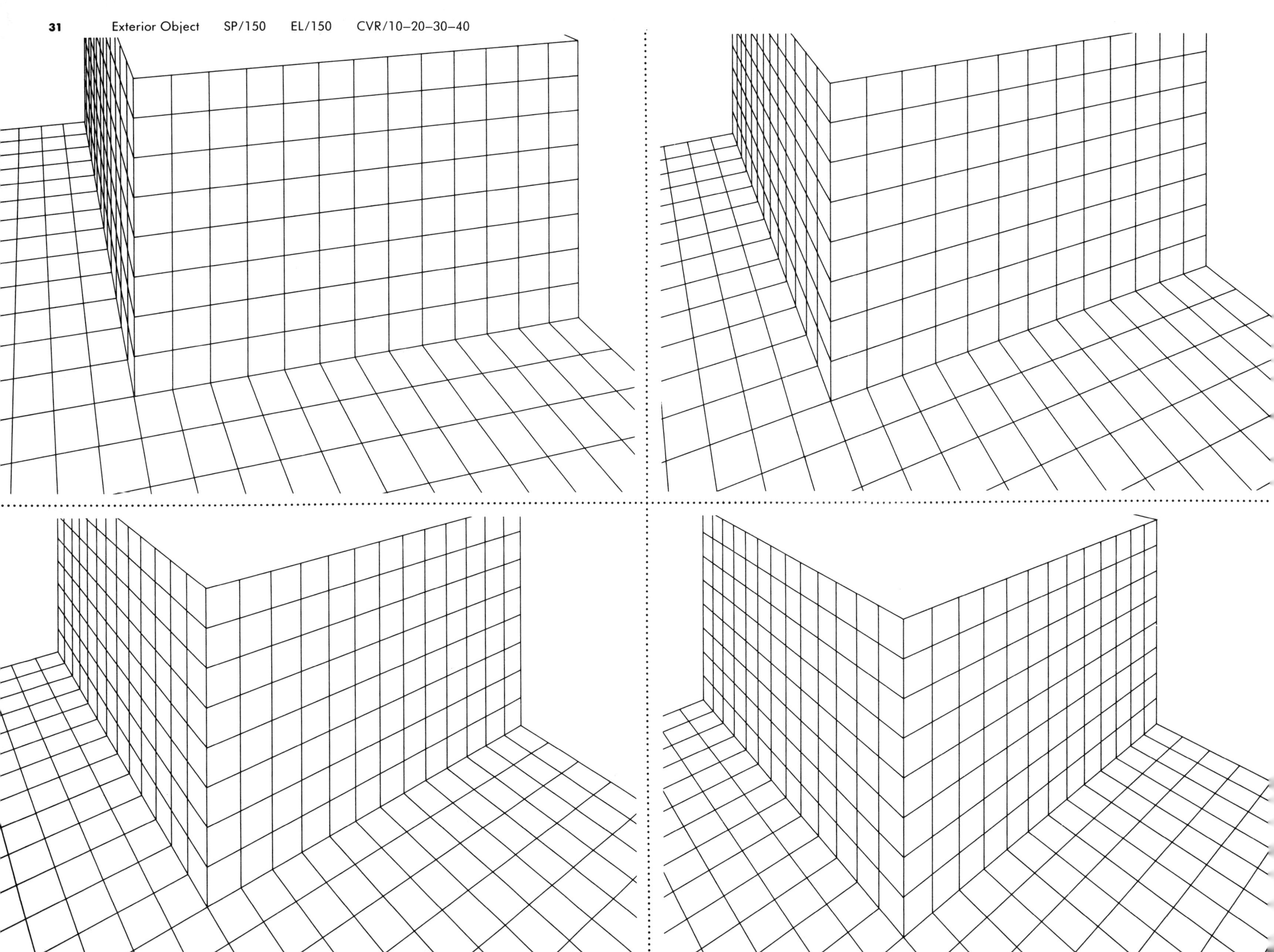

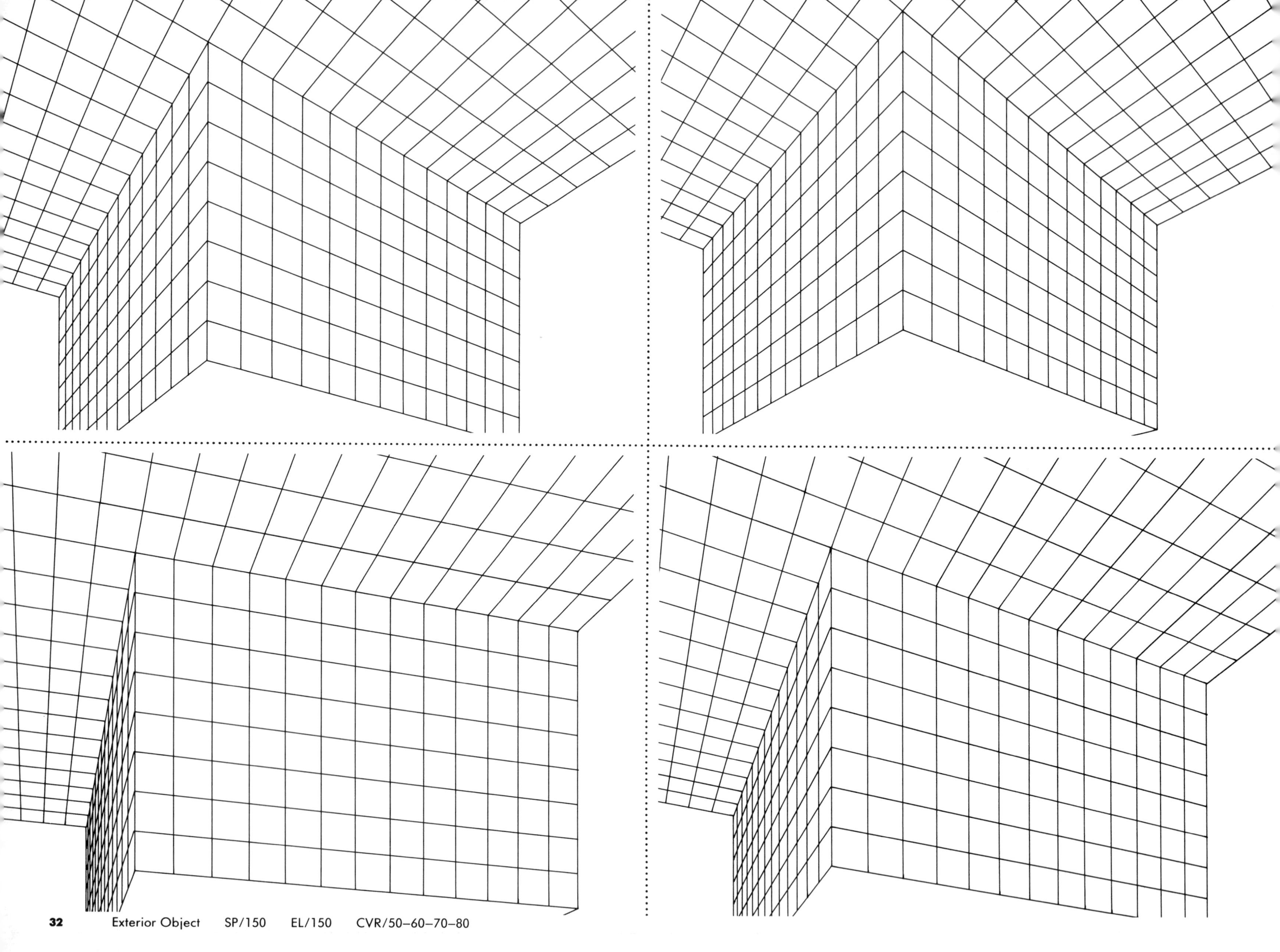



Exterior Object SP/150 EL/150 CVR/50–60–70–80

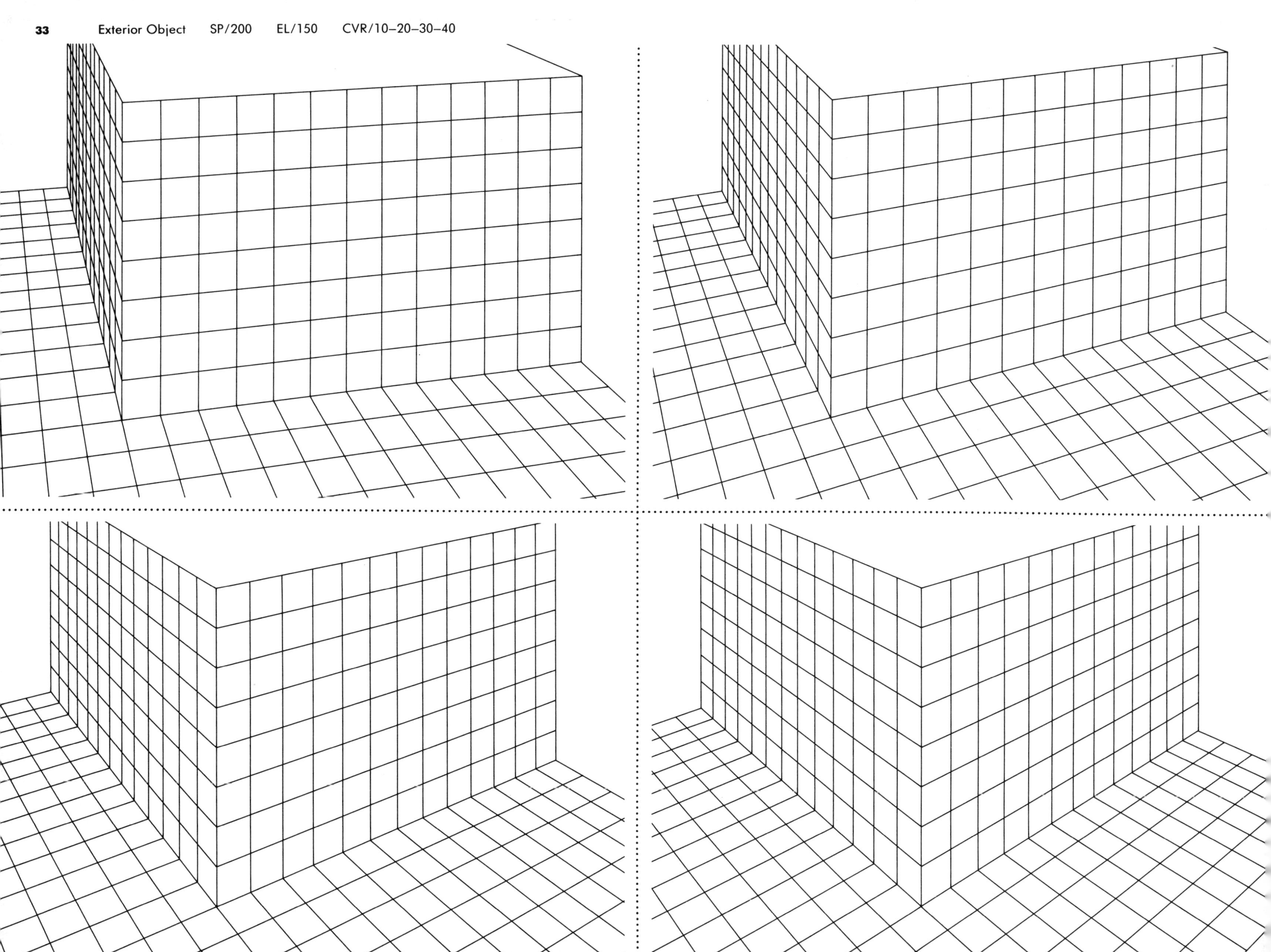

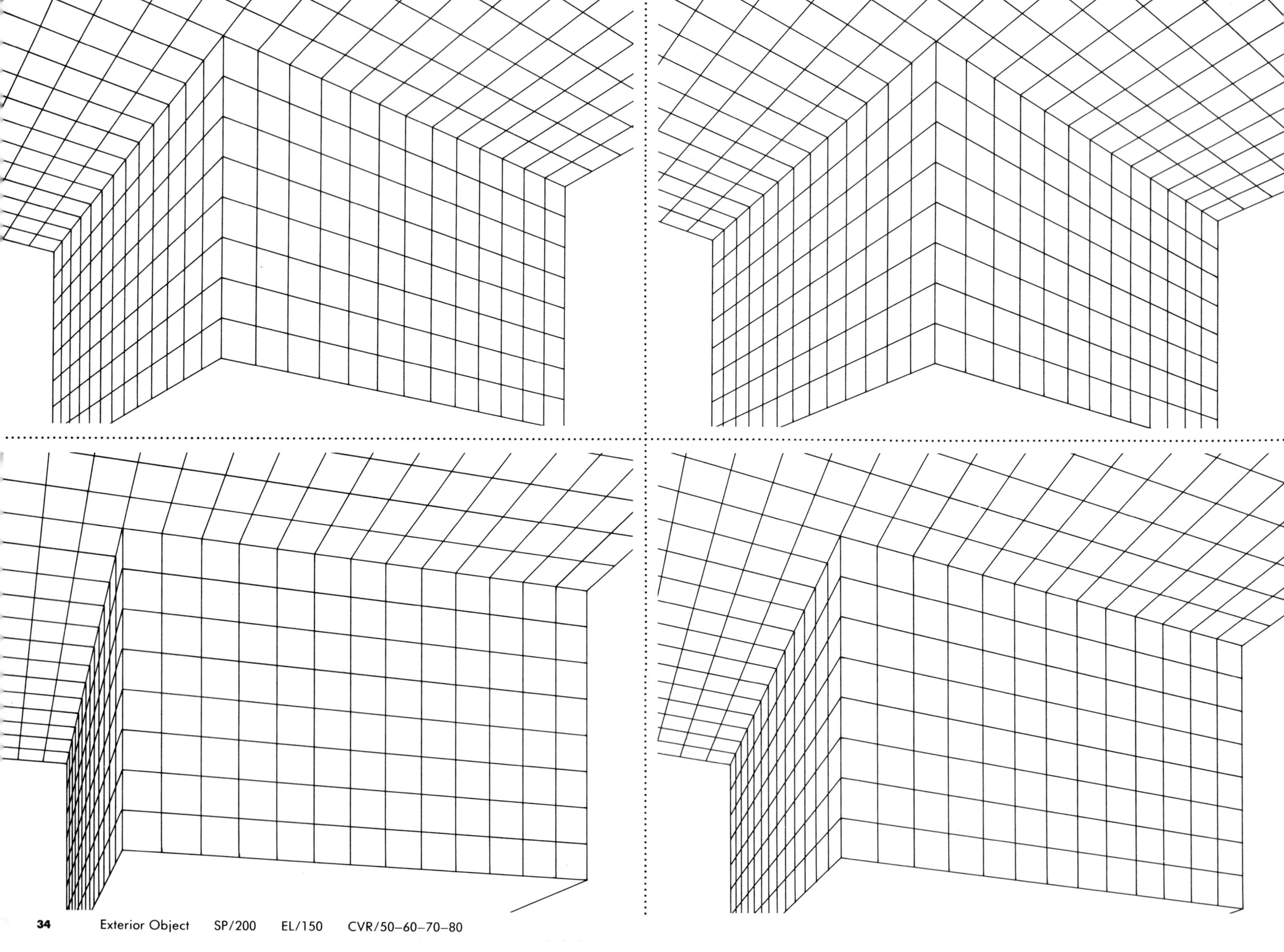

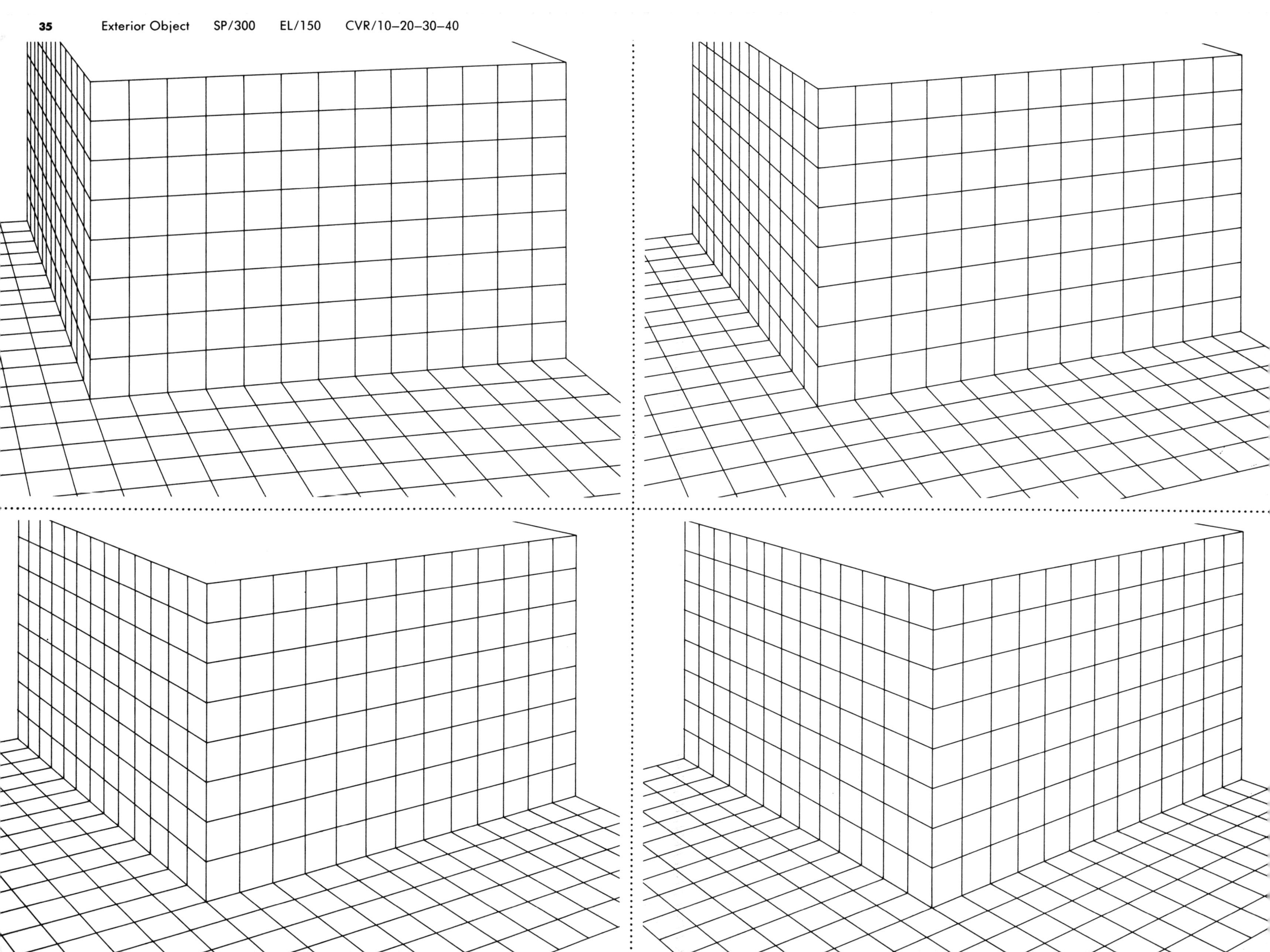

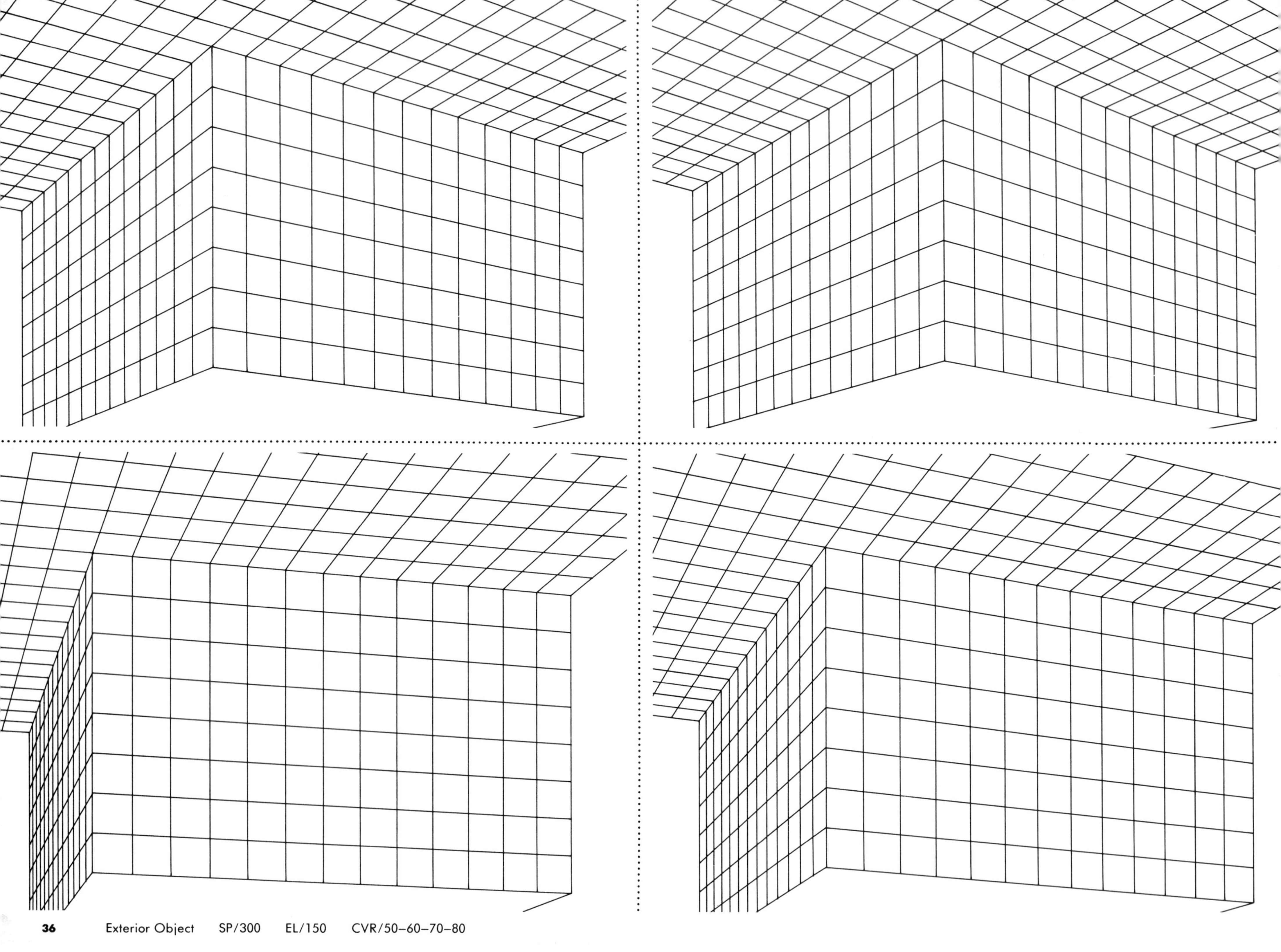

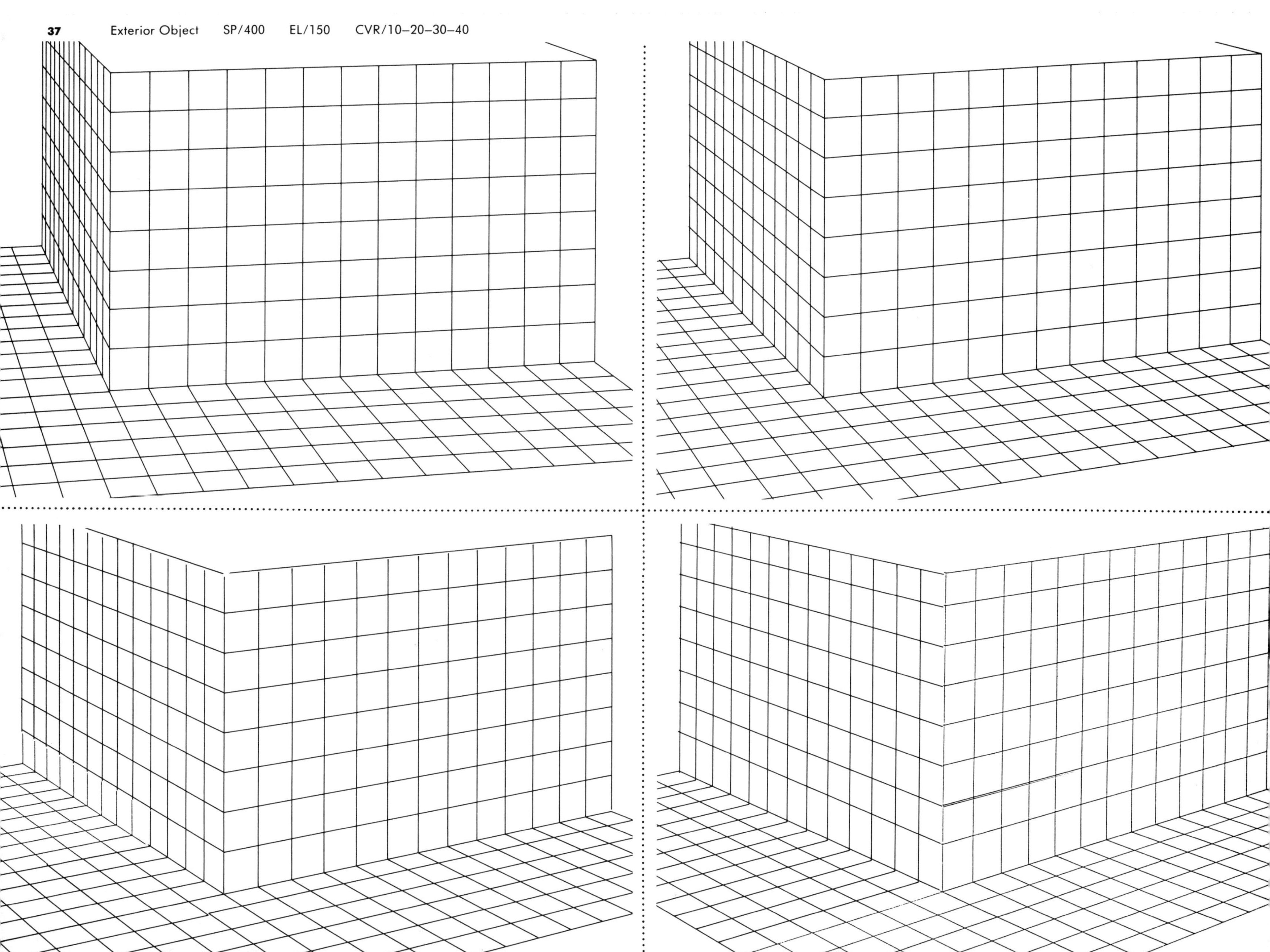

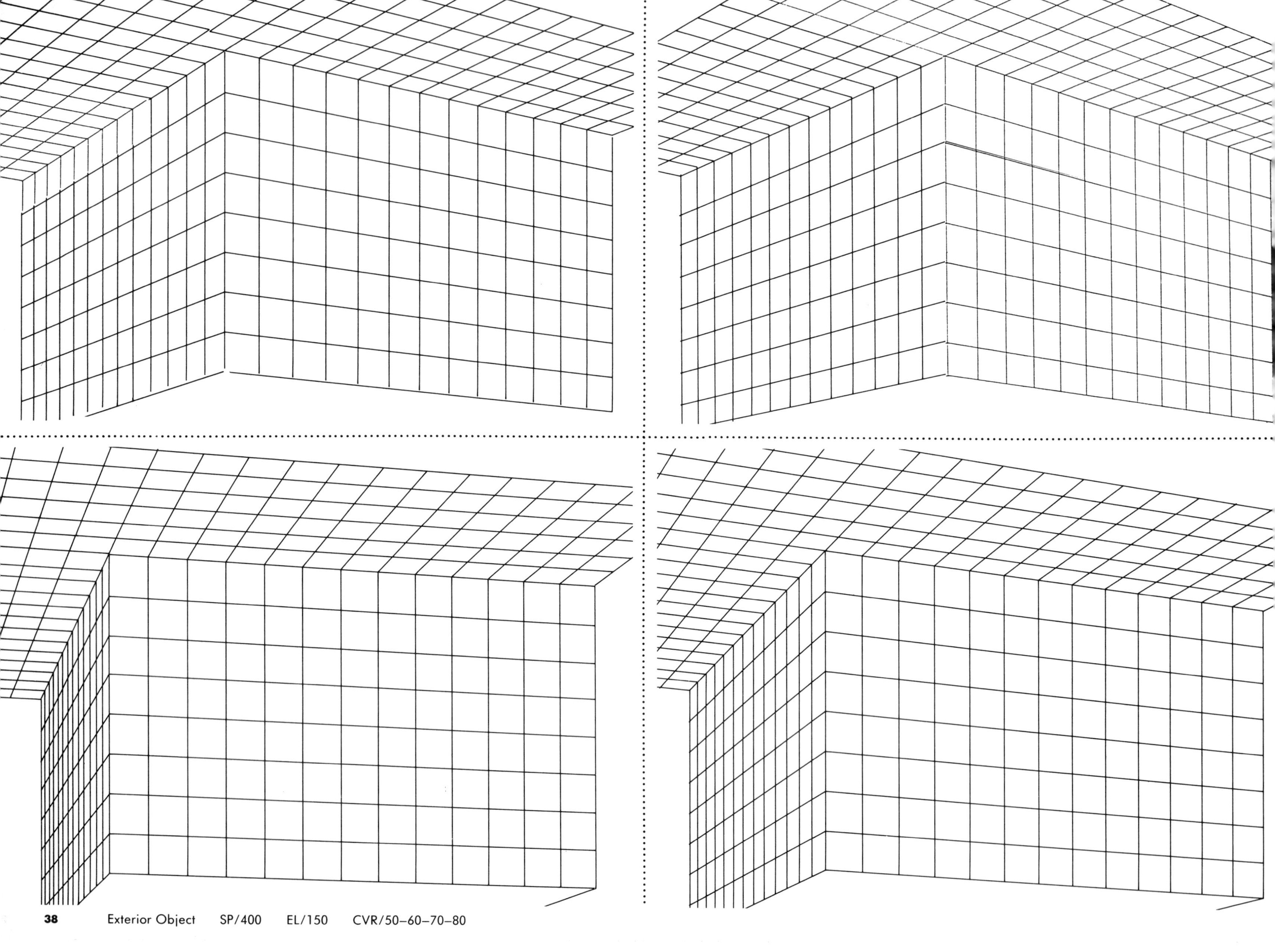

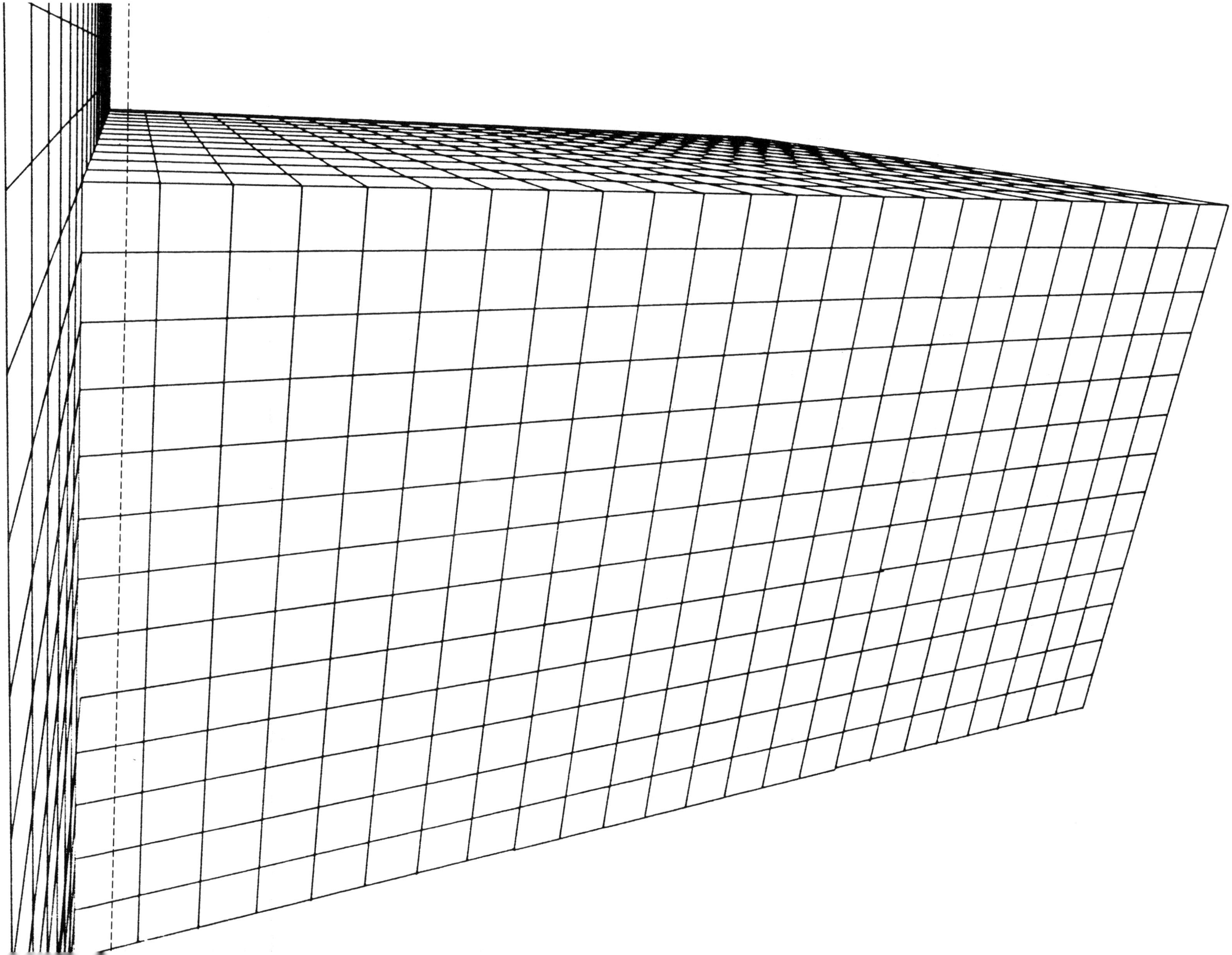

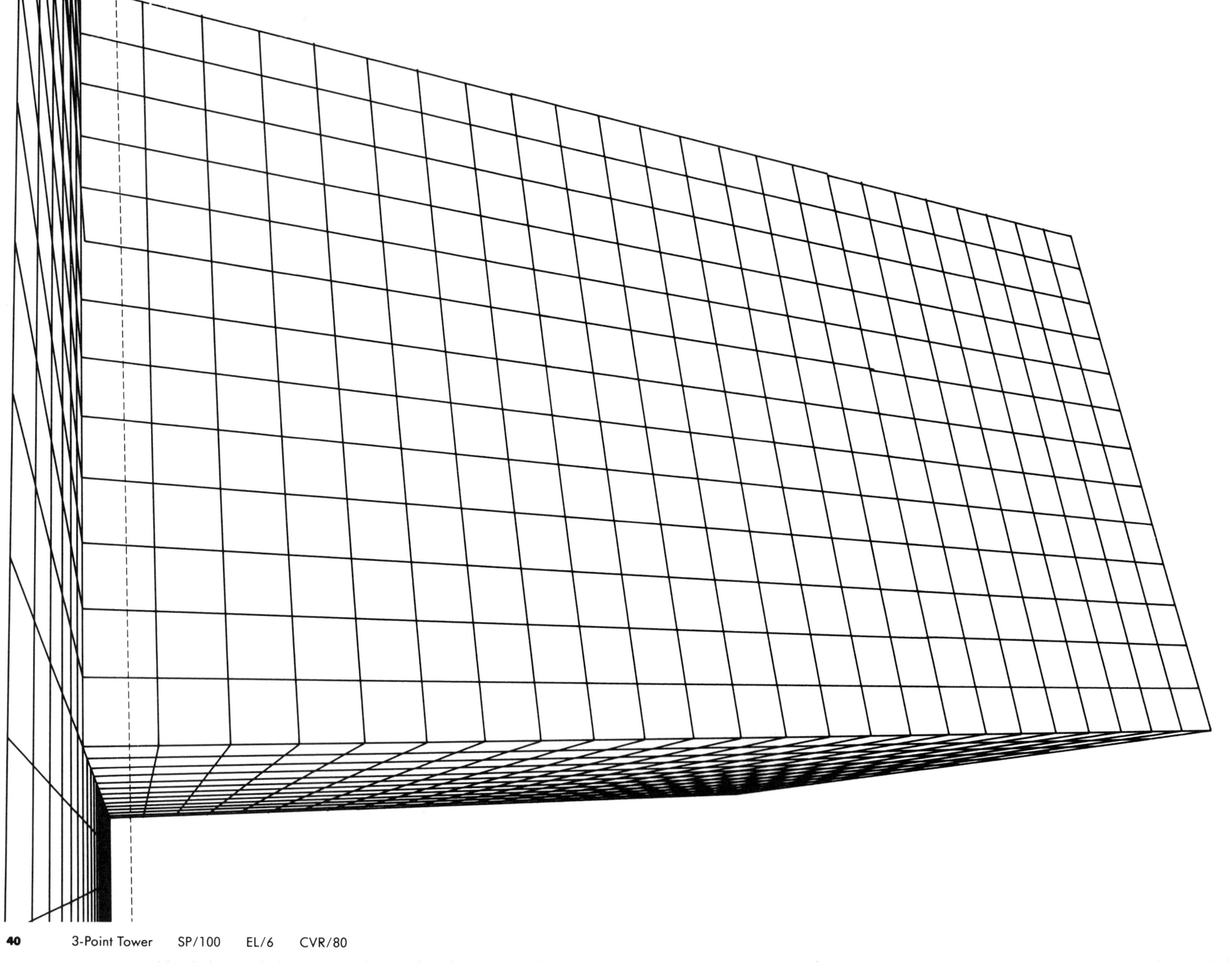

 3-Point Tower SP/100 EL/6 CVR/80

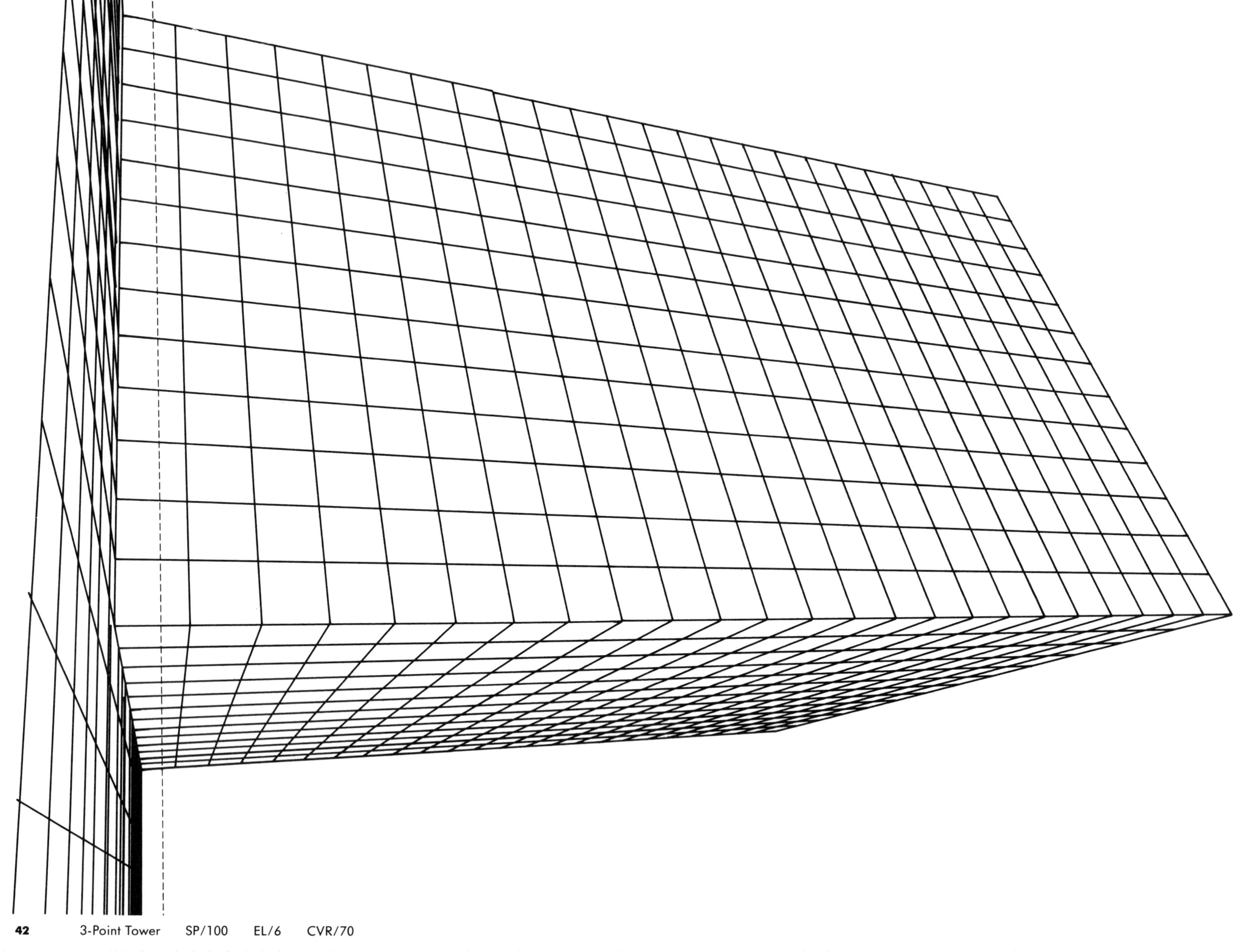



3-Point Tower SP/100 EL/6 CVR/70

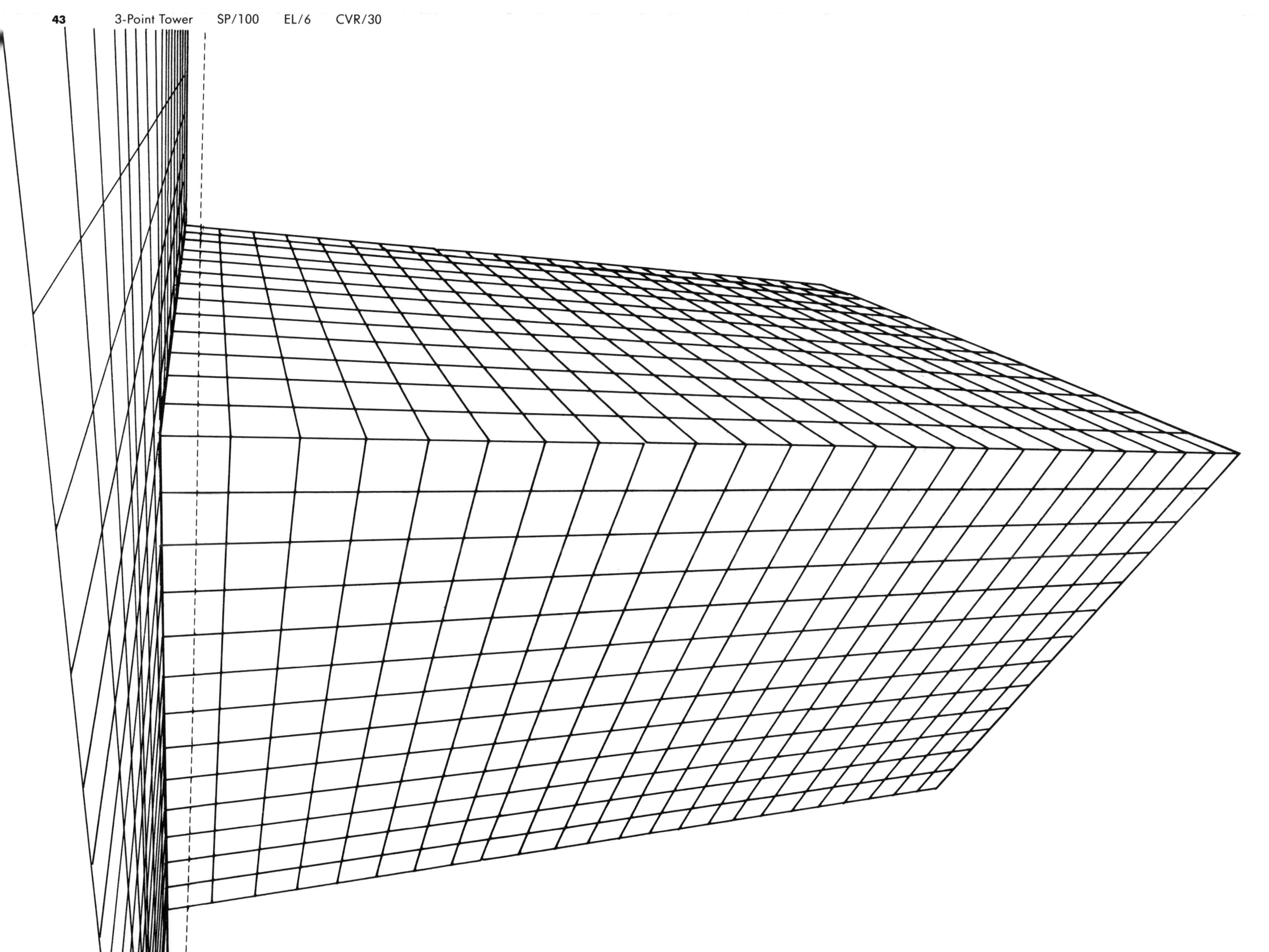

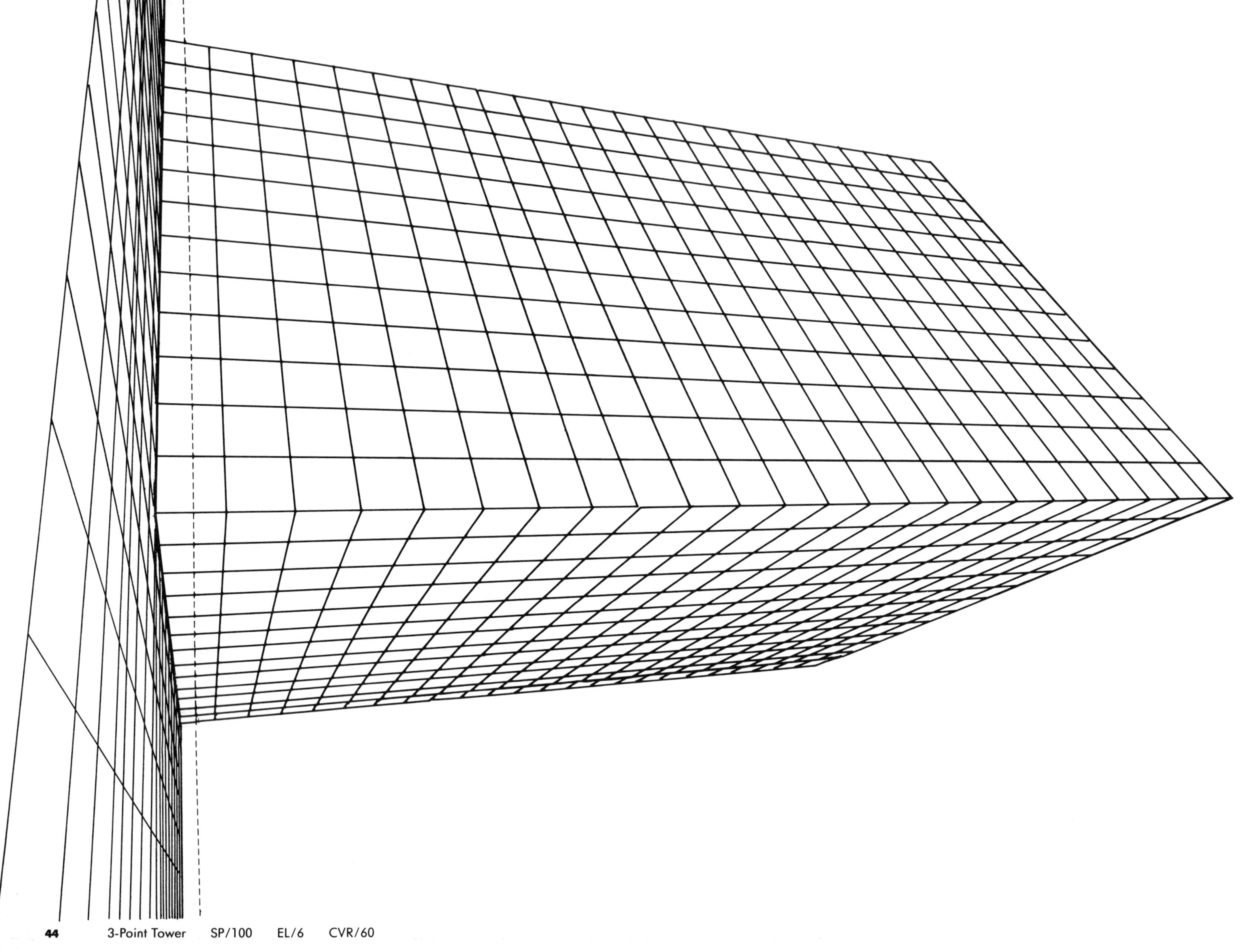

 3-Point Tower SP/100 EL/6 CVR/60

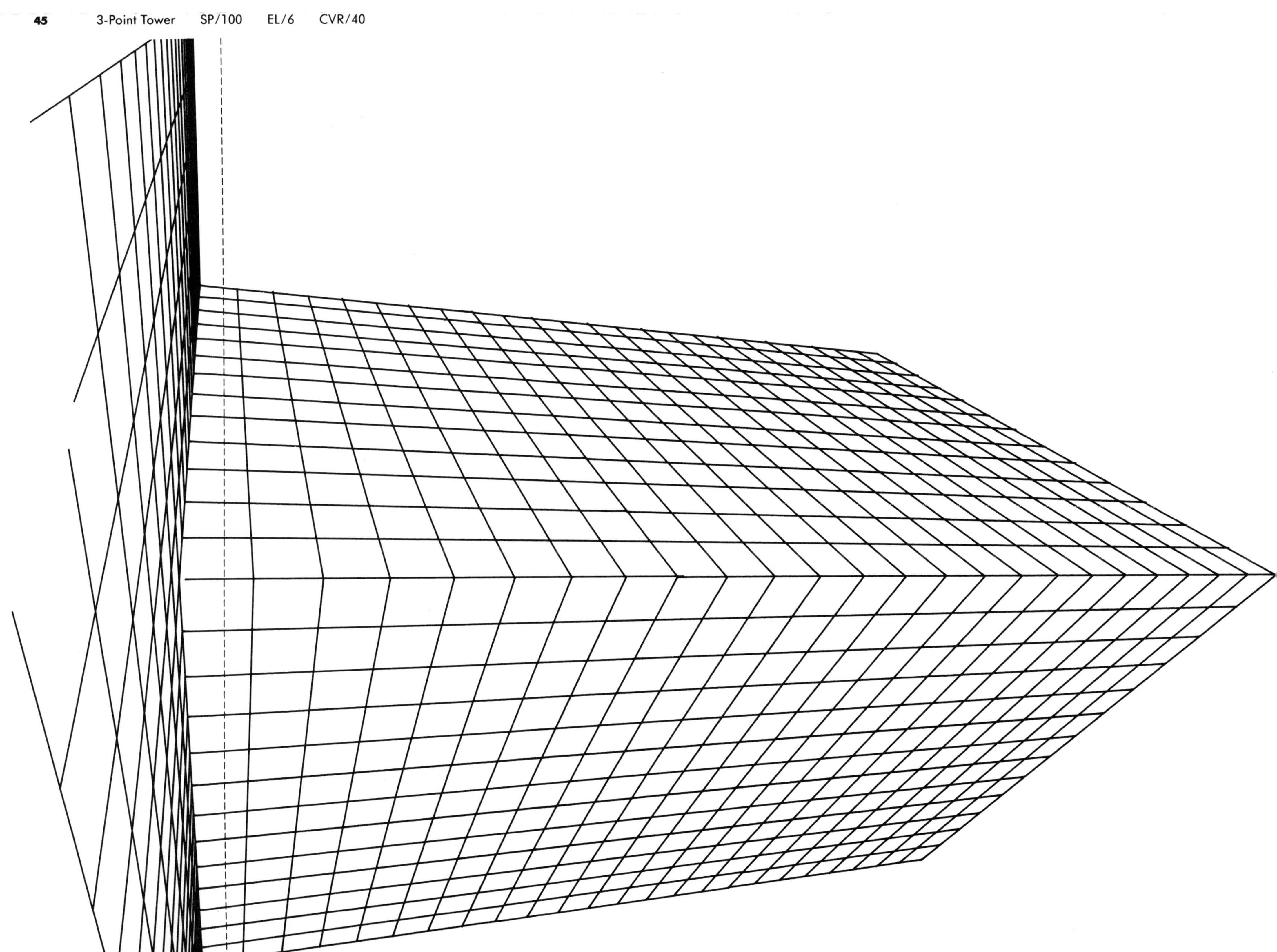

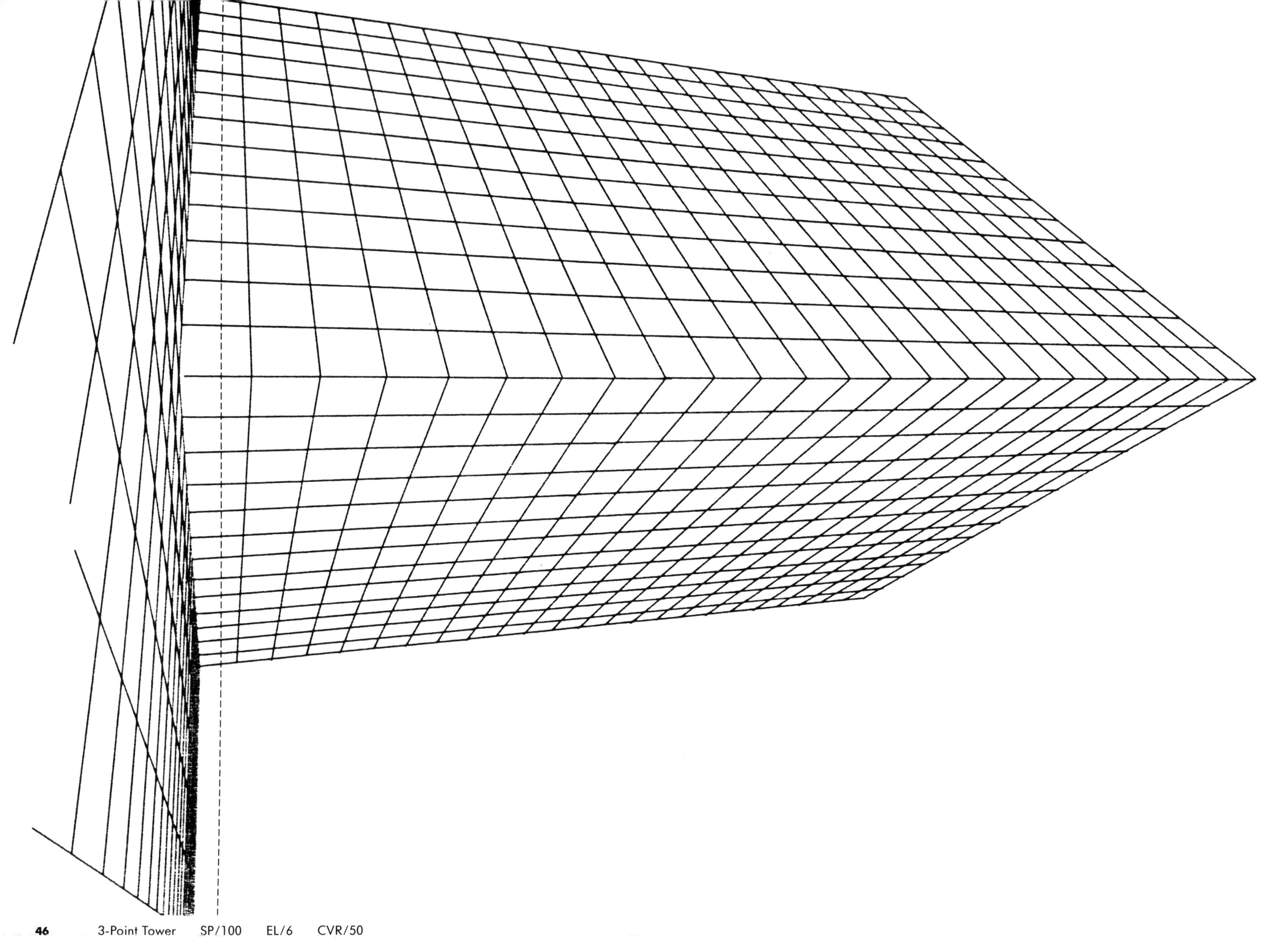

 3-Point Tower SP/100 EL/6 CVR/50

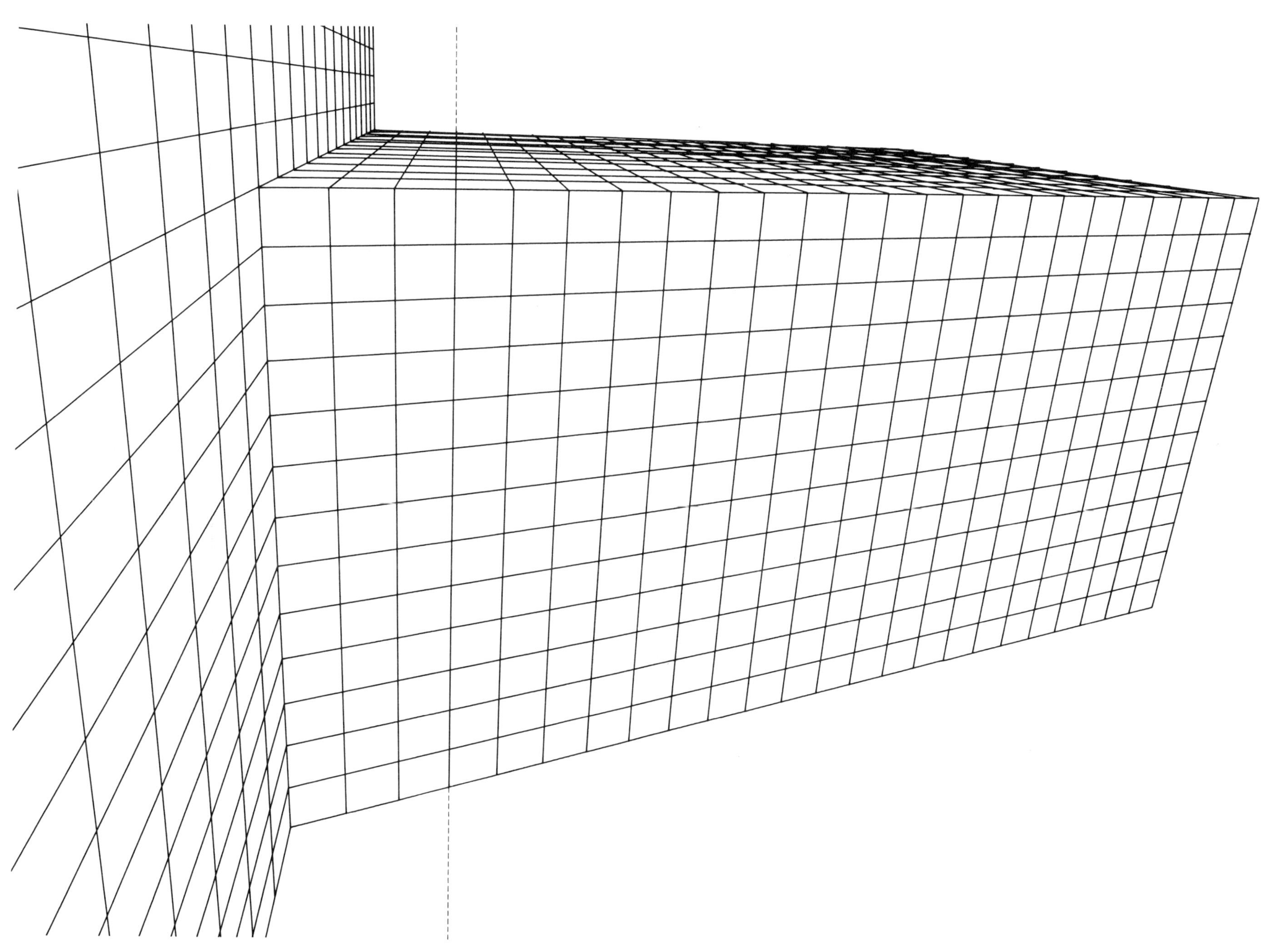

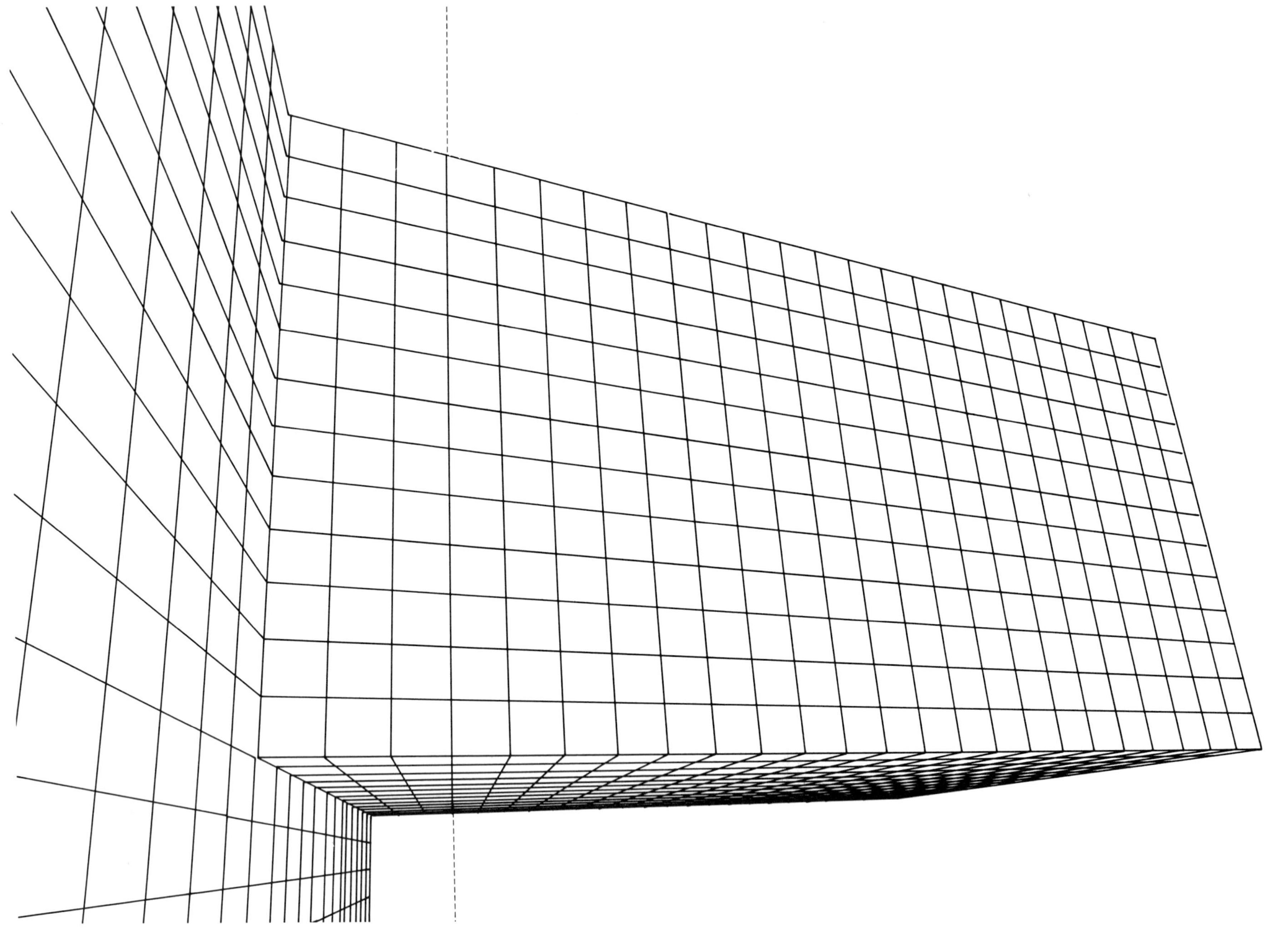



3-Point Tower SP/100 EL/30 CVR/80

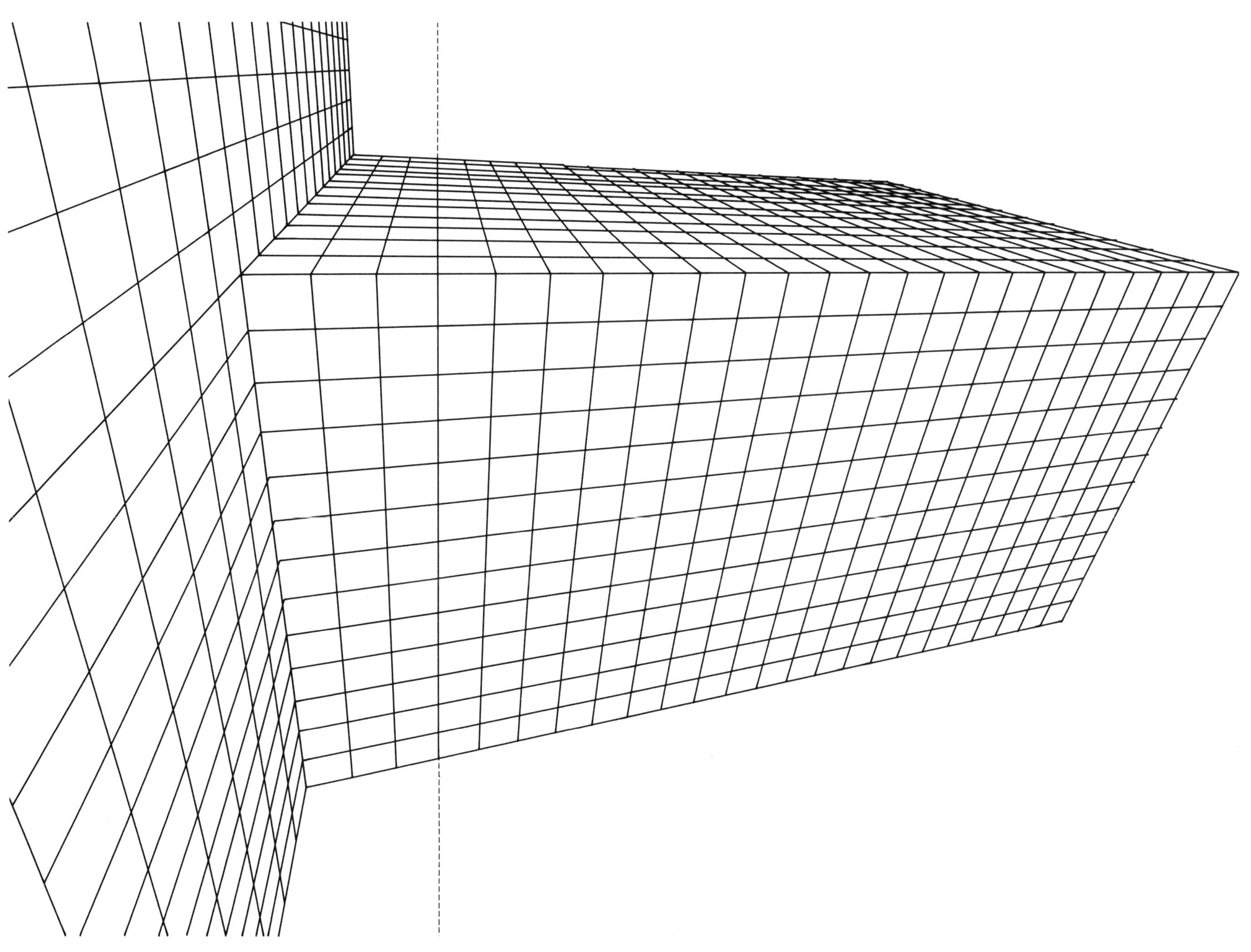

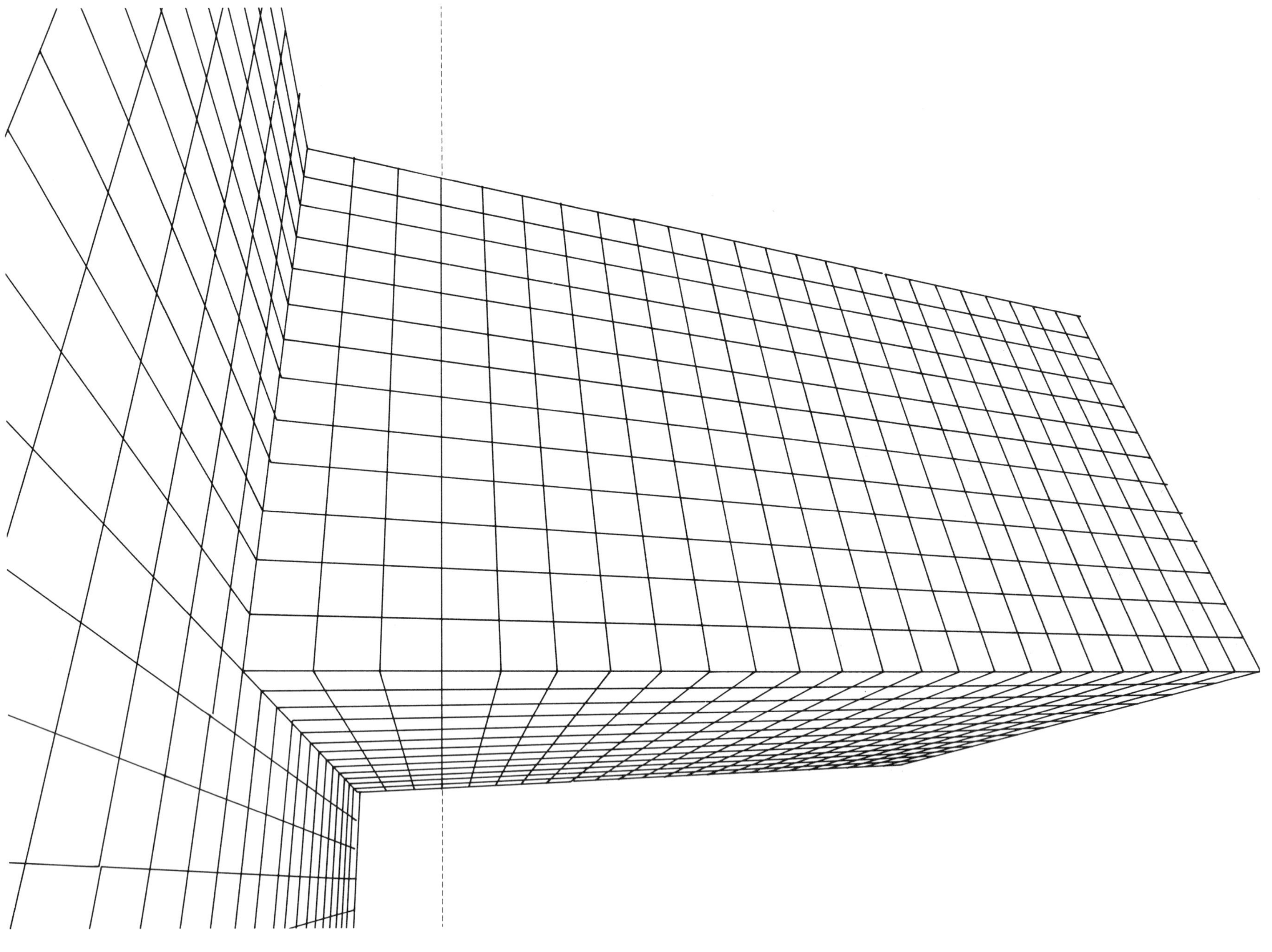

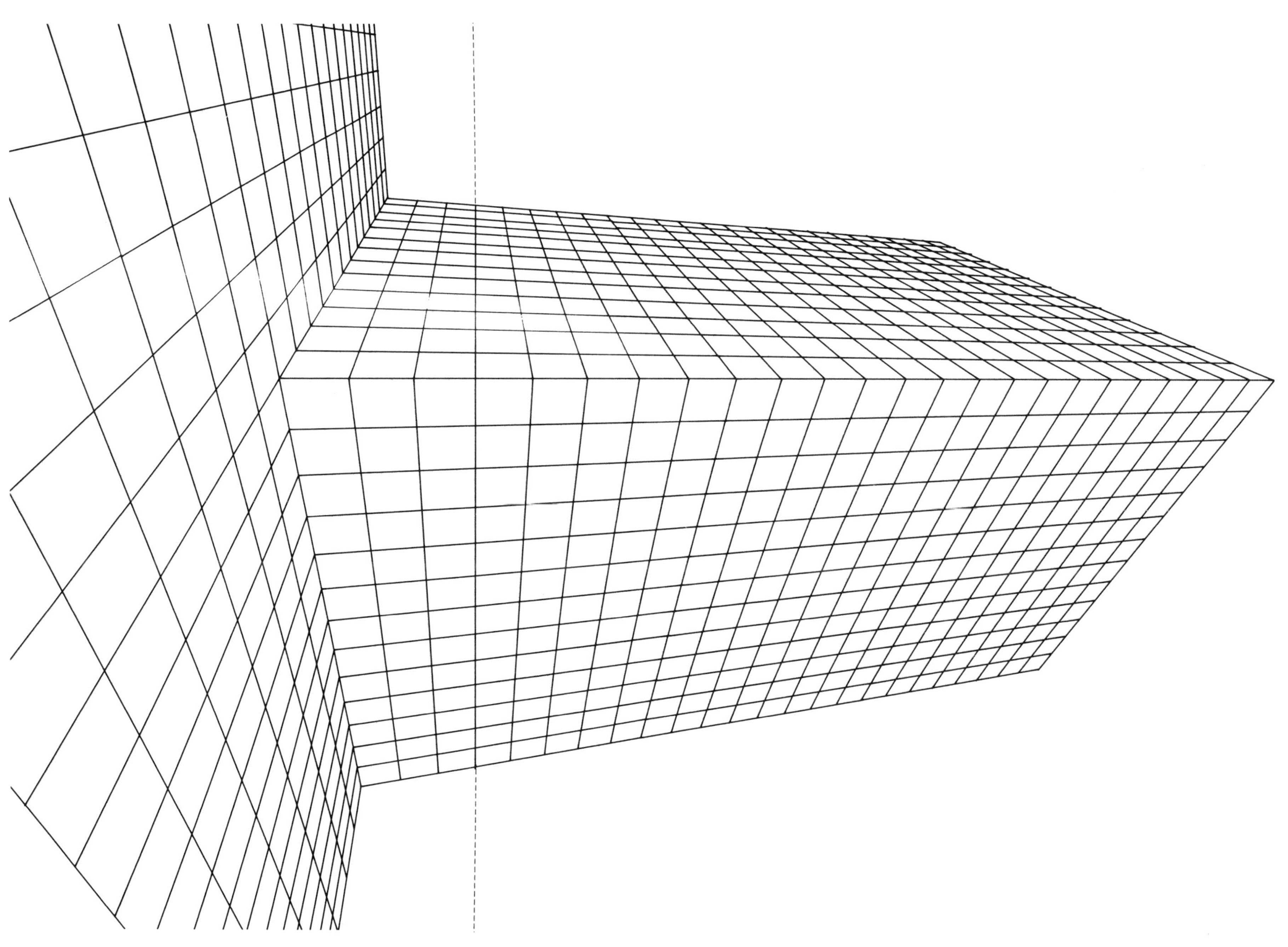

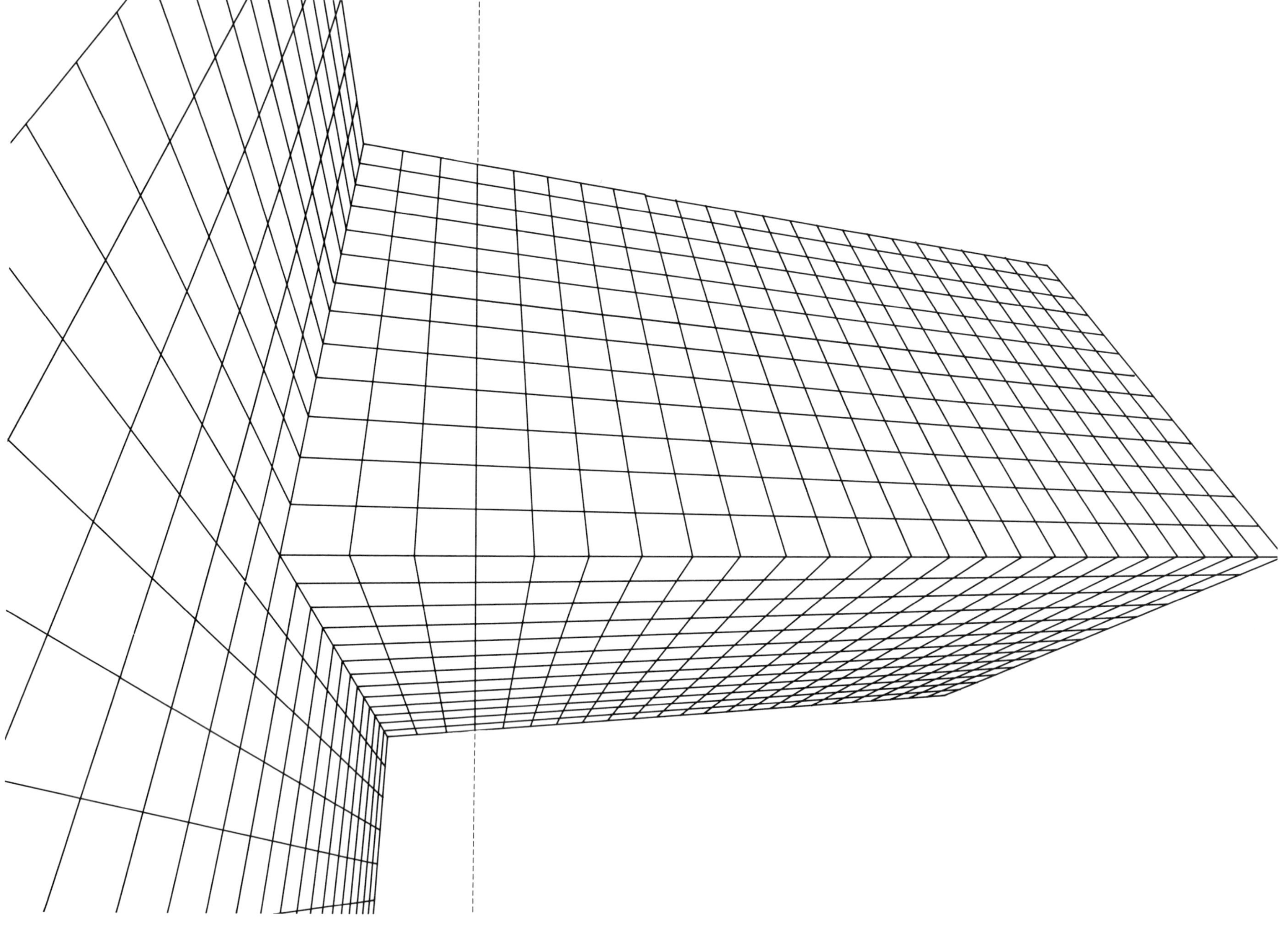

 3-Point Tower SP/100 EL/30 CVR/60

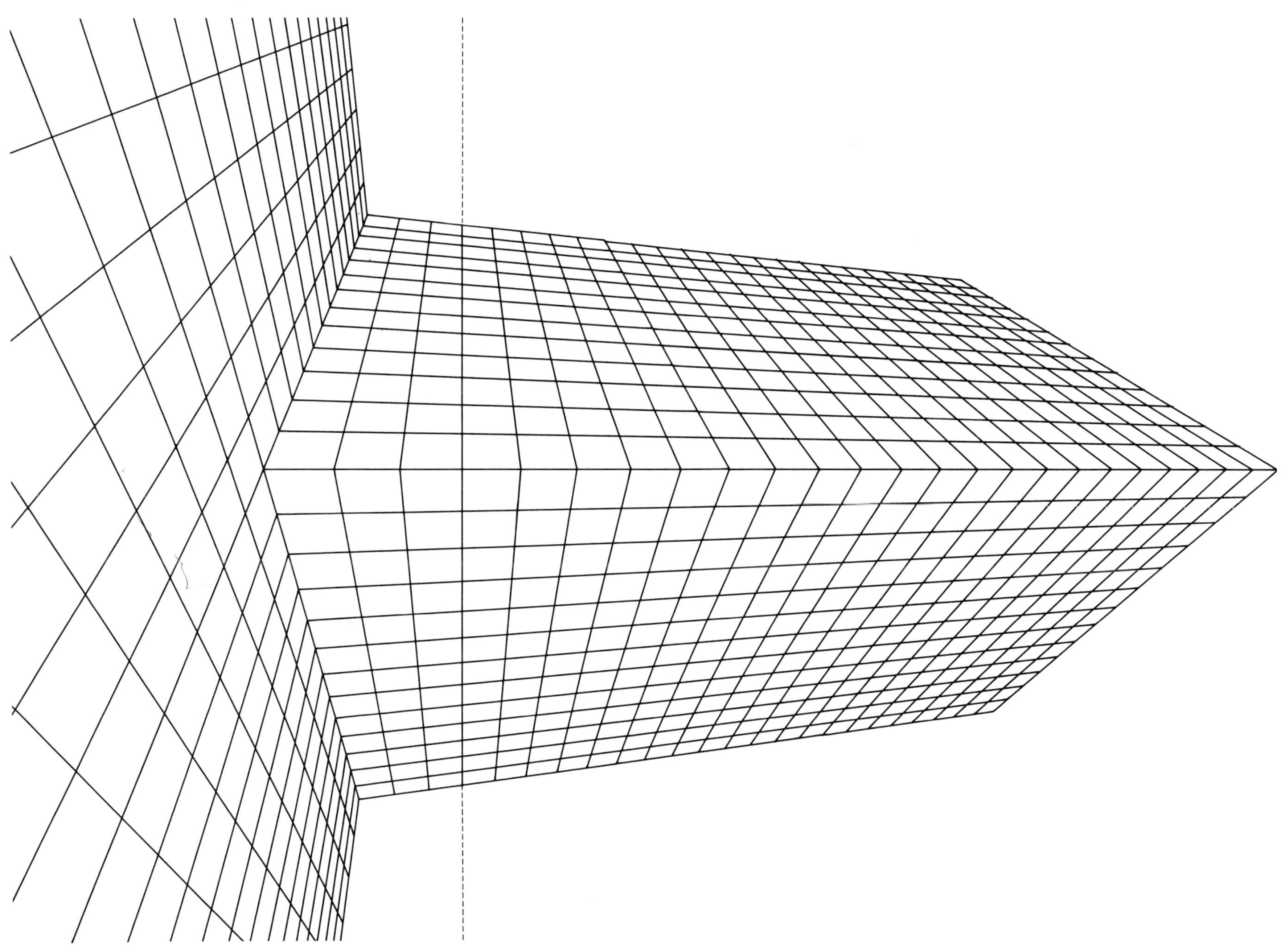

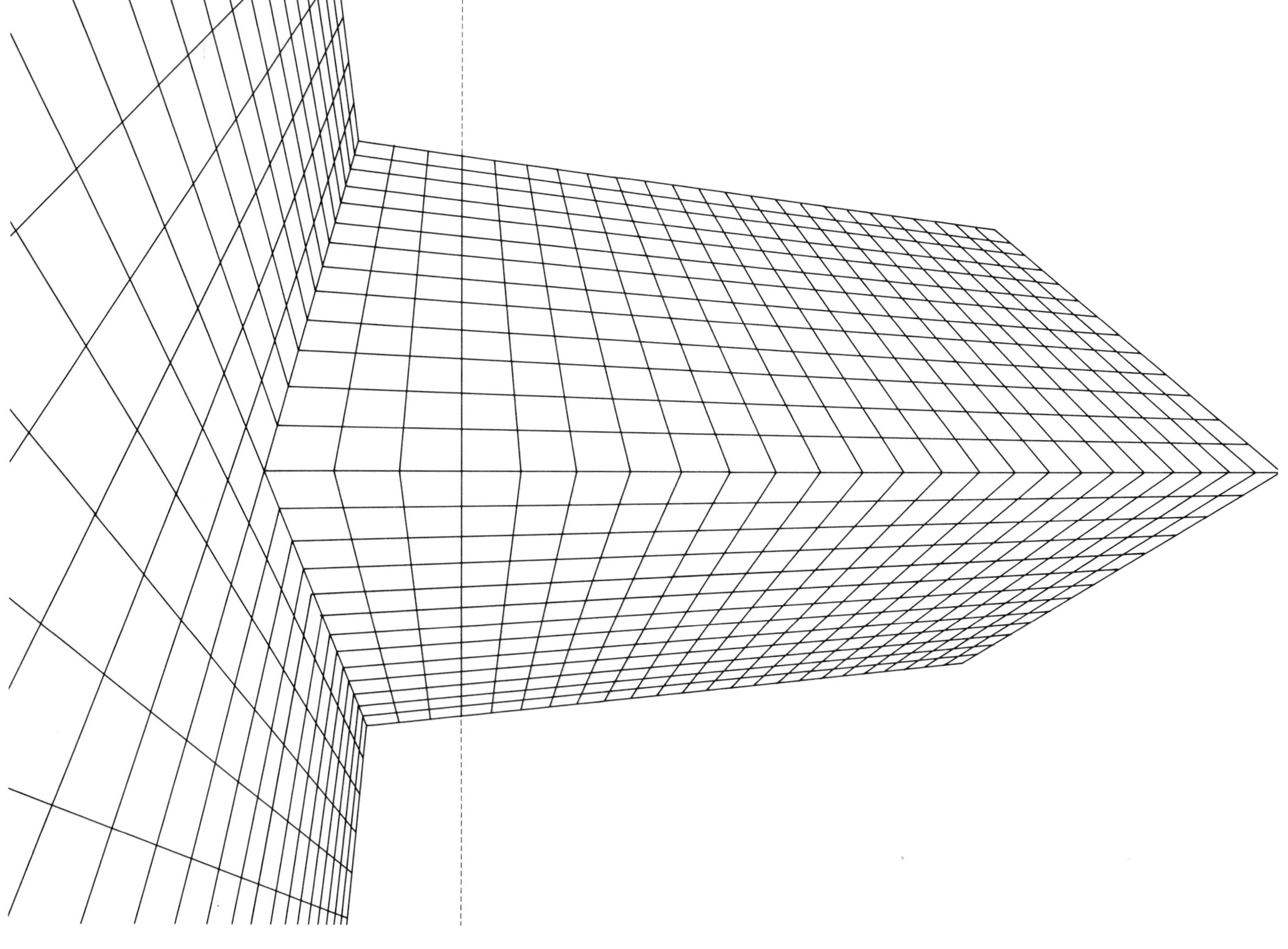

54 3-Point Tower SP/100 EL/30 CVR/50

# GRIDS FOR SKETCHING

## Interior Room

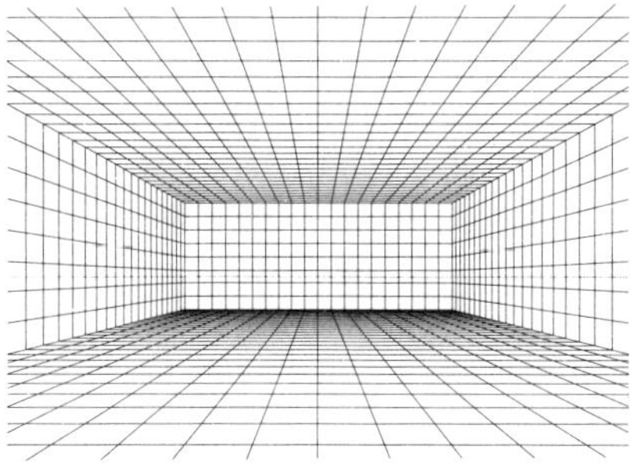

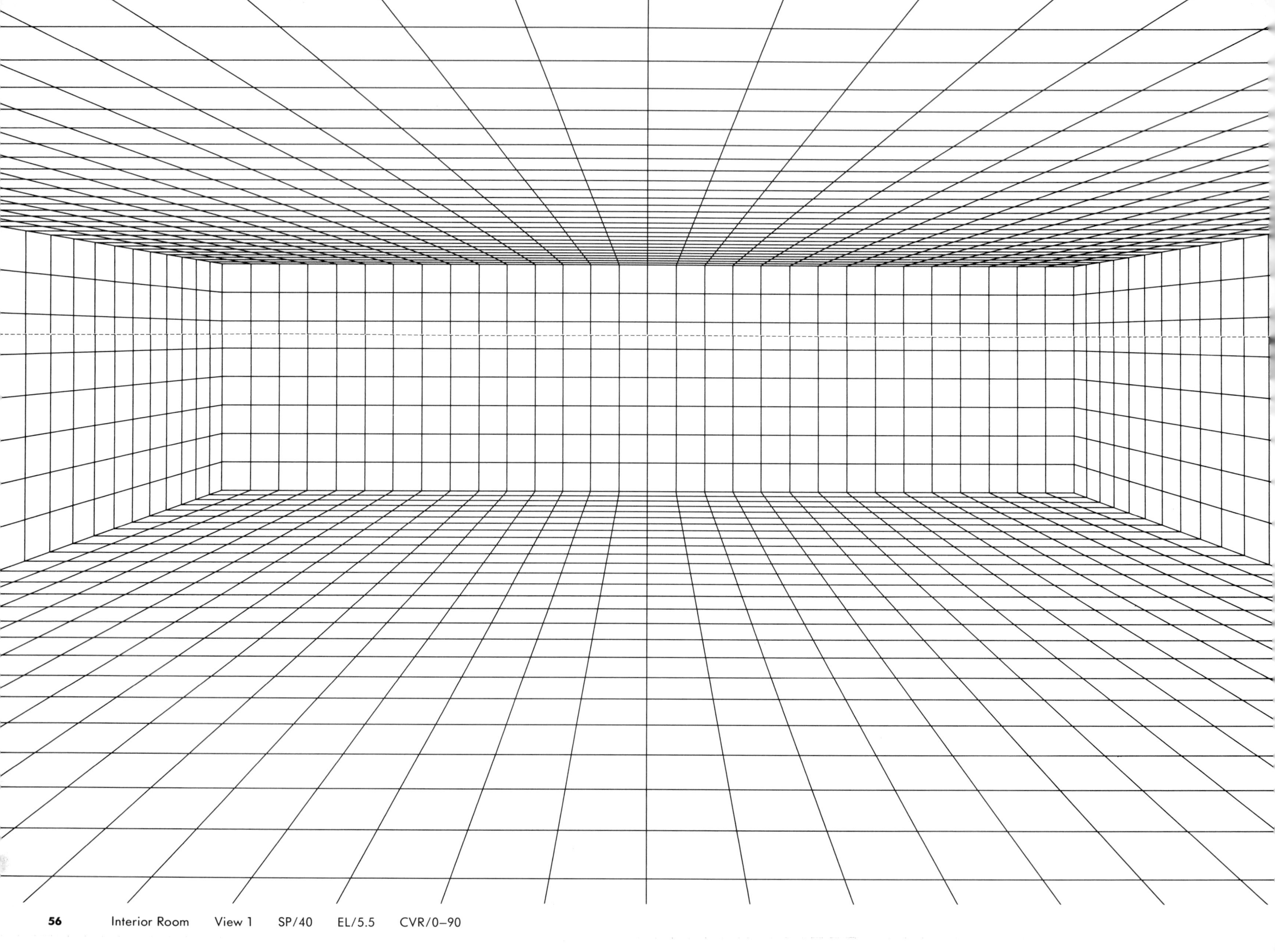

 Interior Room View 1 SP/40 EL/5.5 CVR/0–90

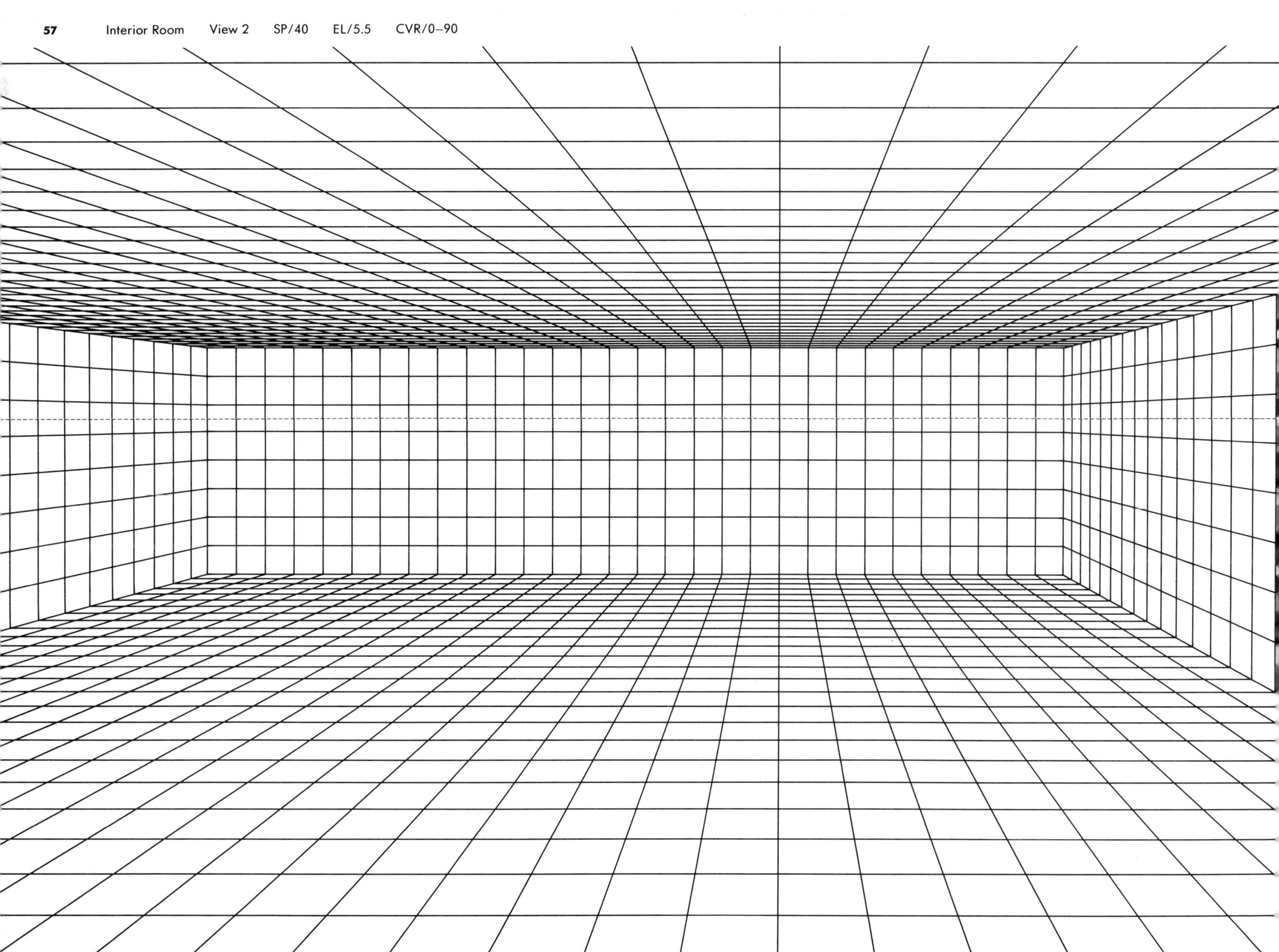

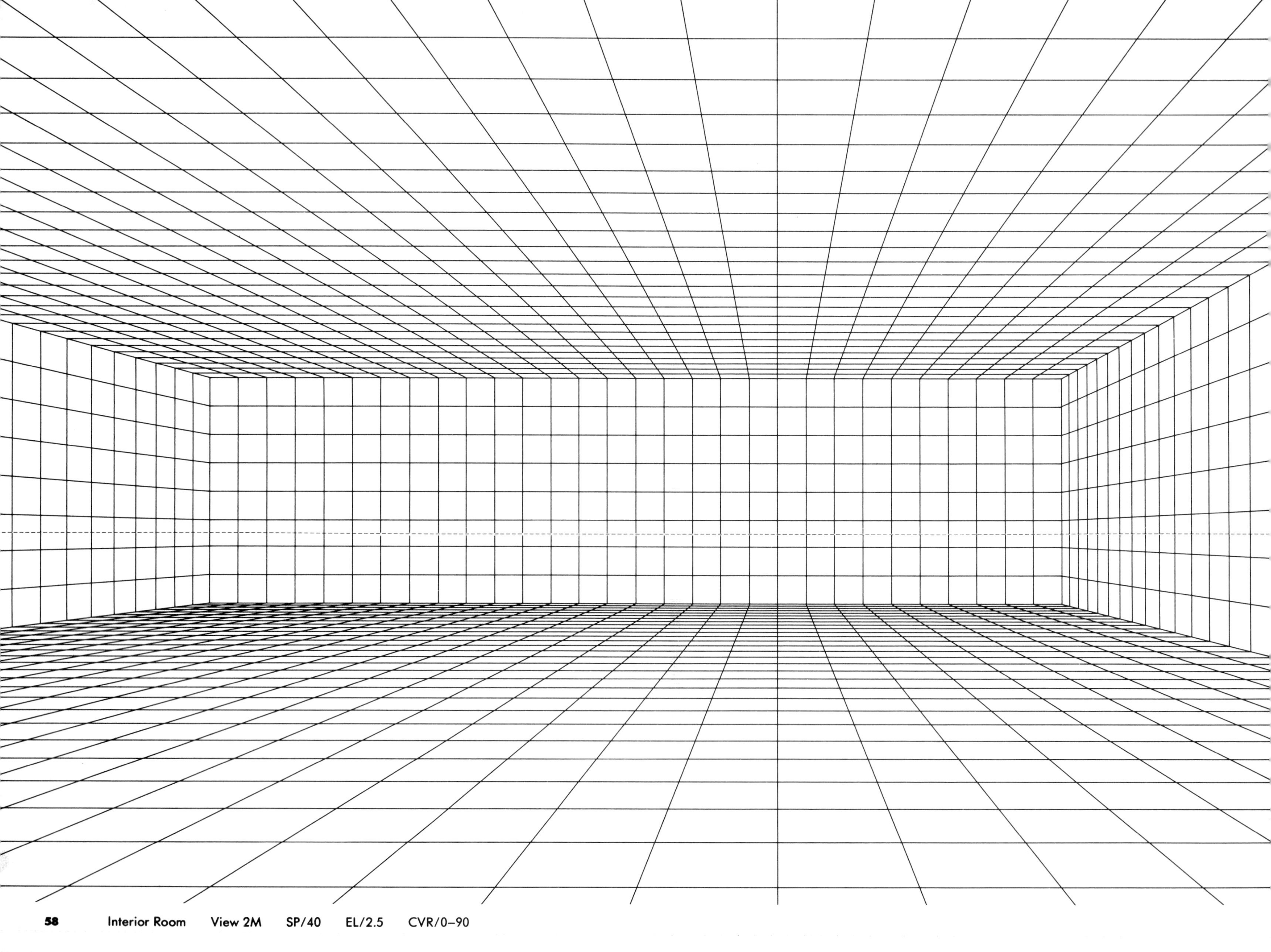

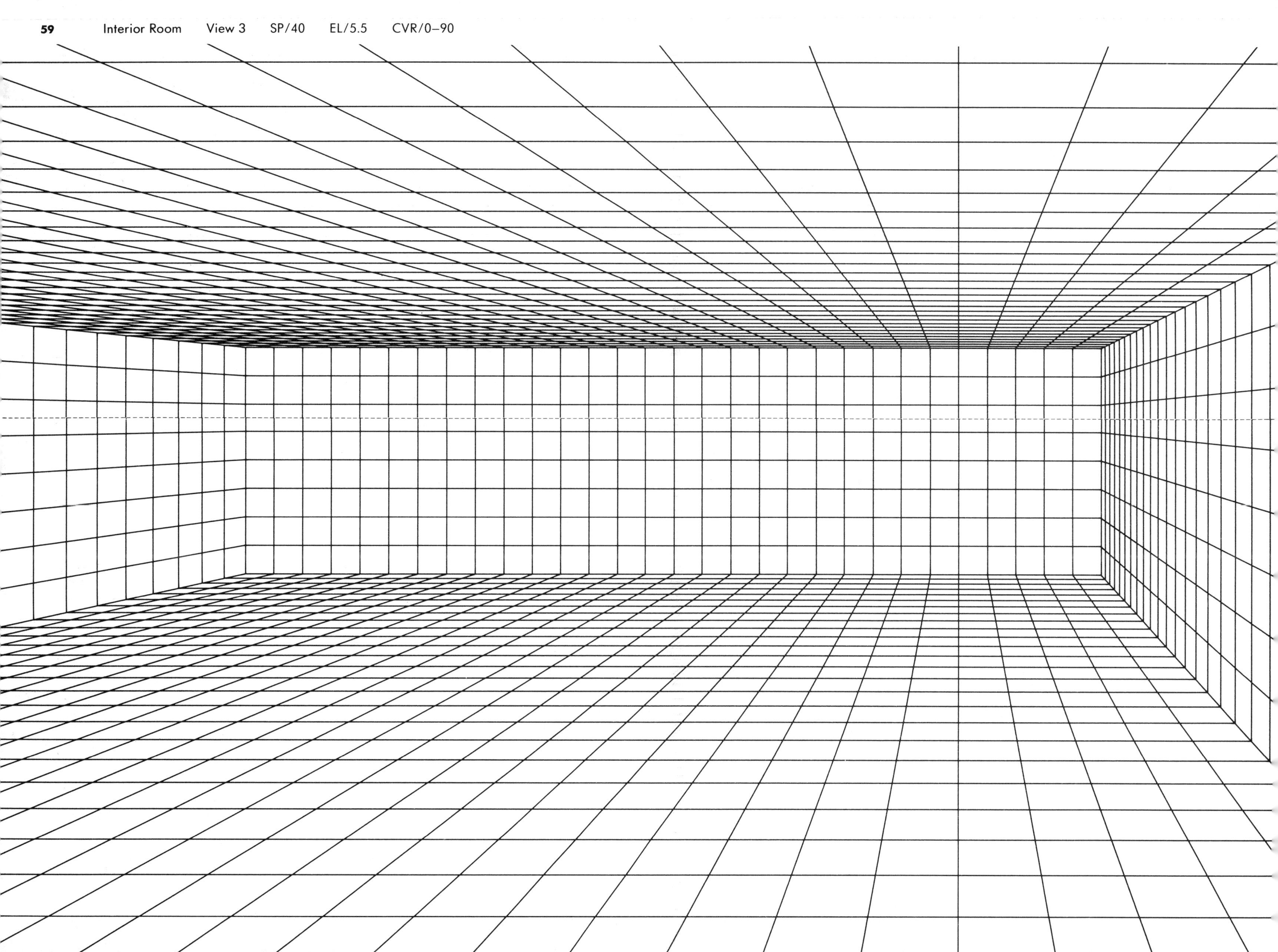

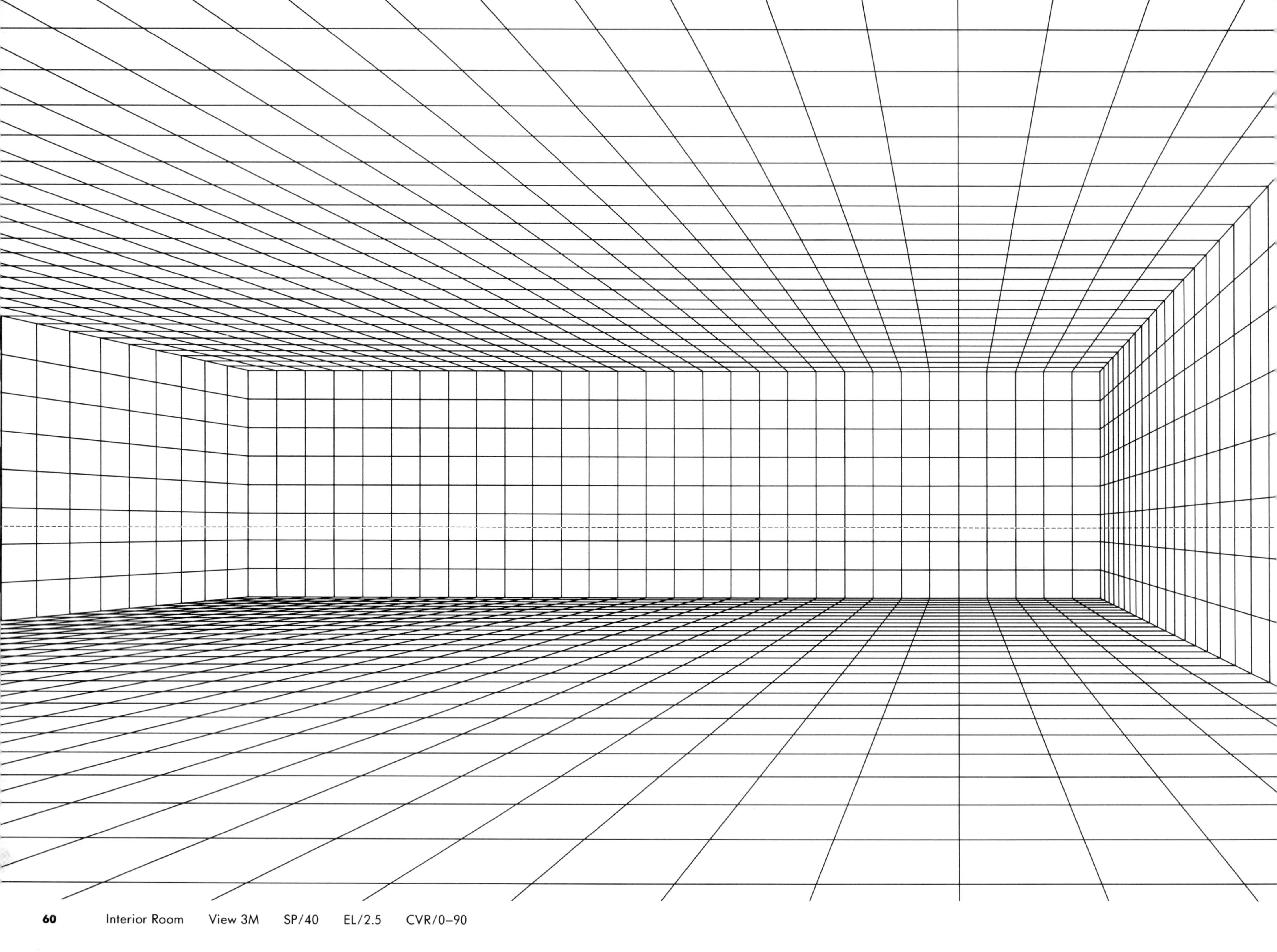

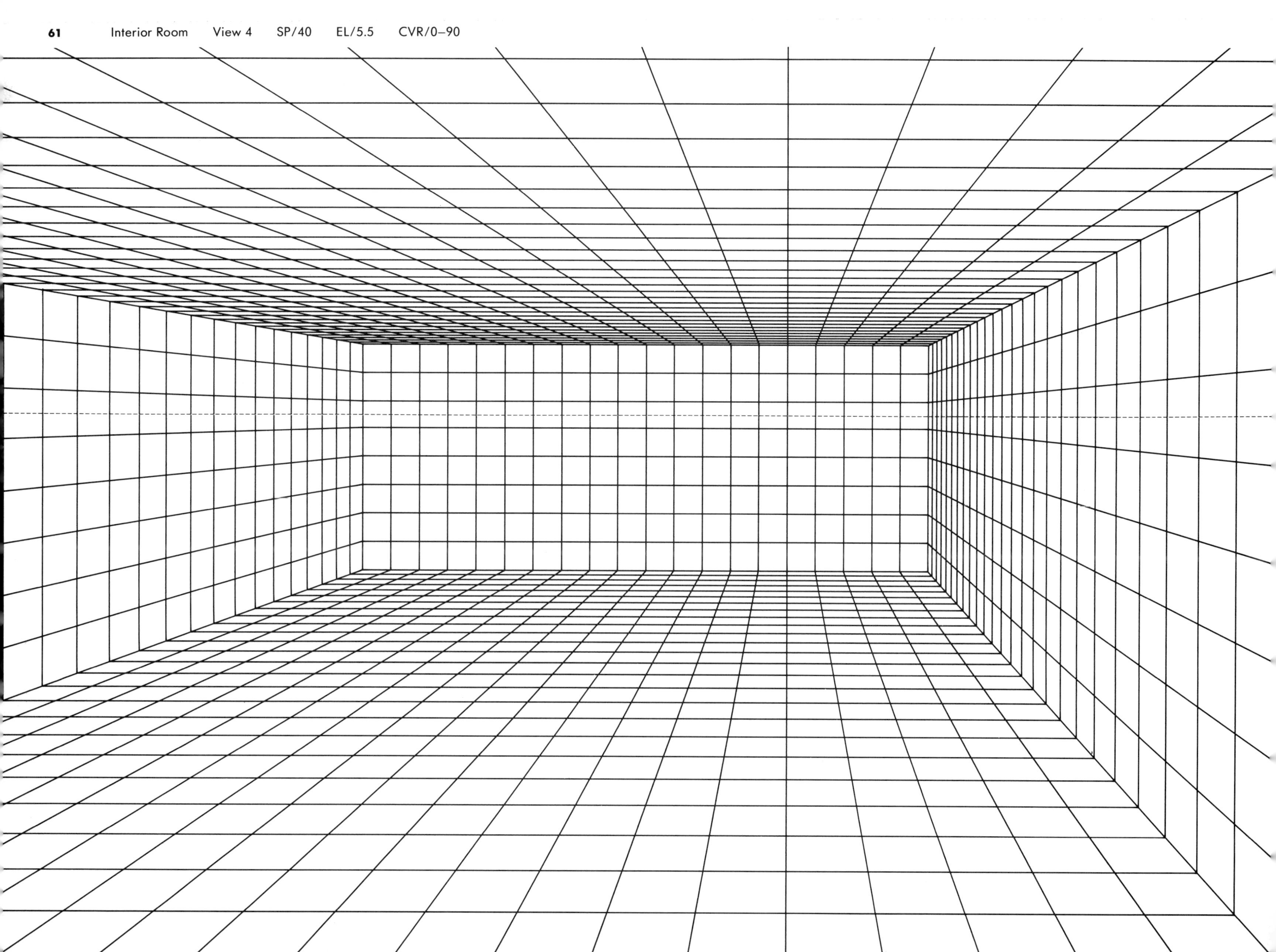

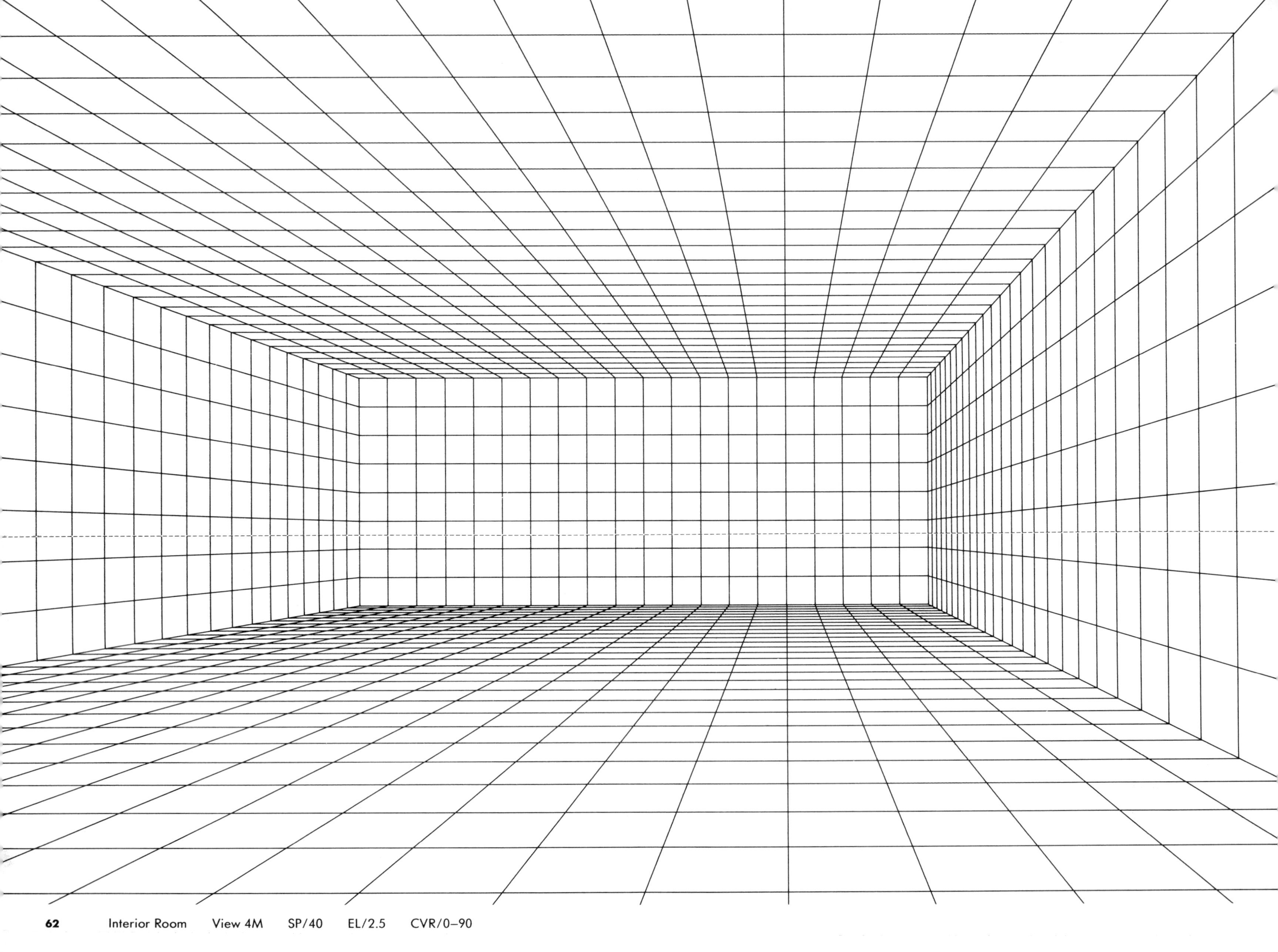



Interior Room View 4M SP/40 EL/2.5 CVR/0–90

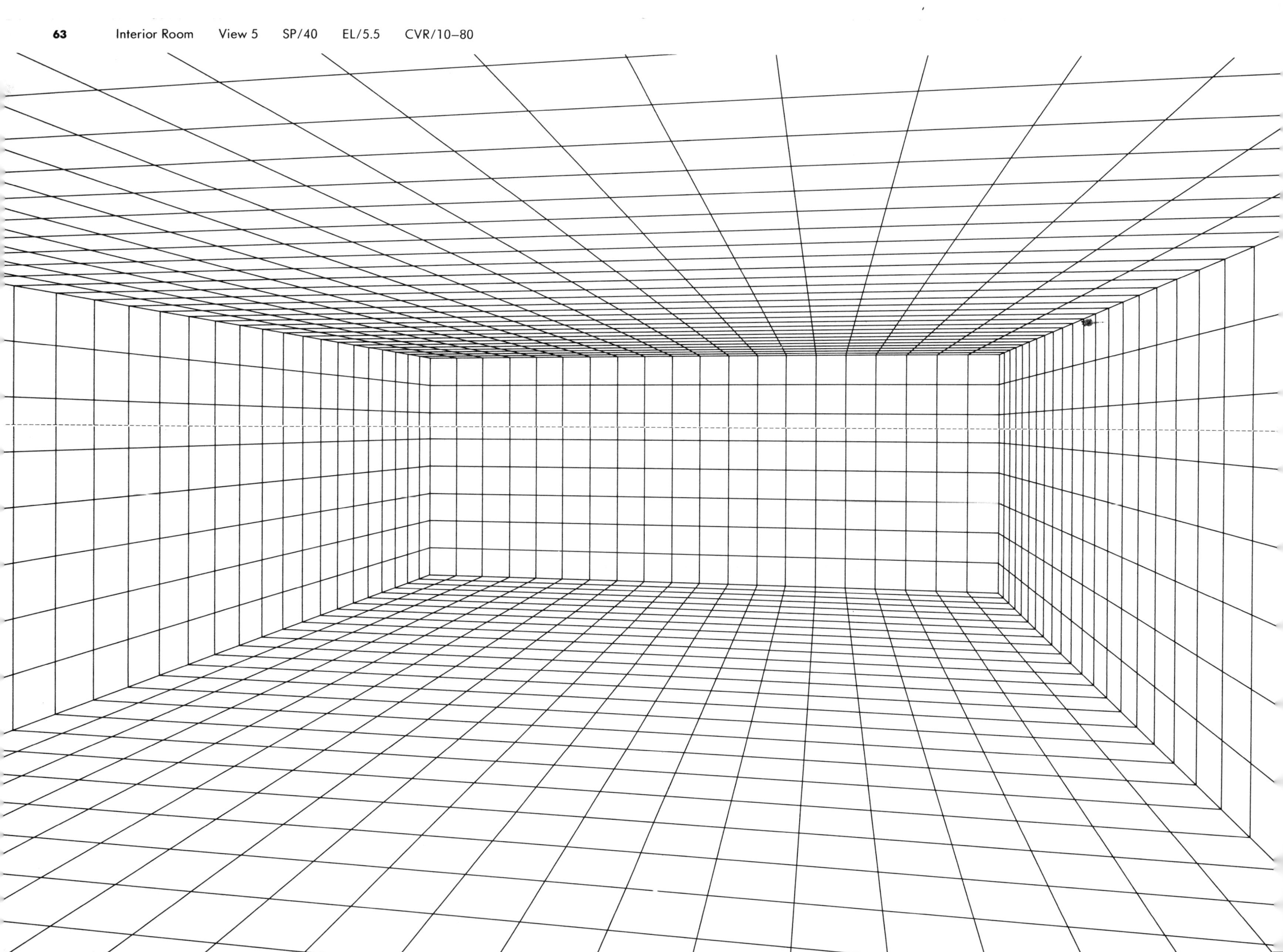

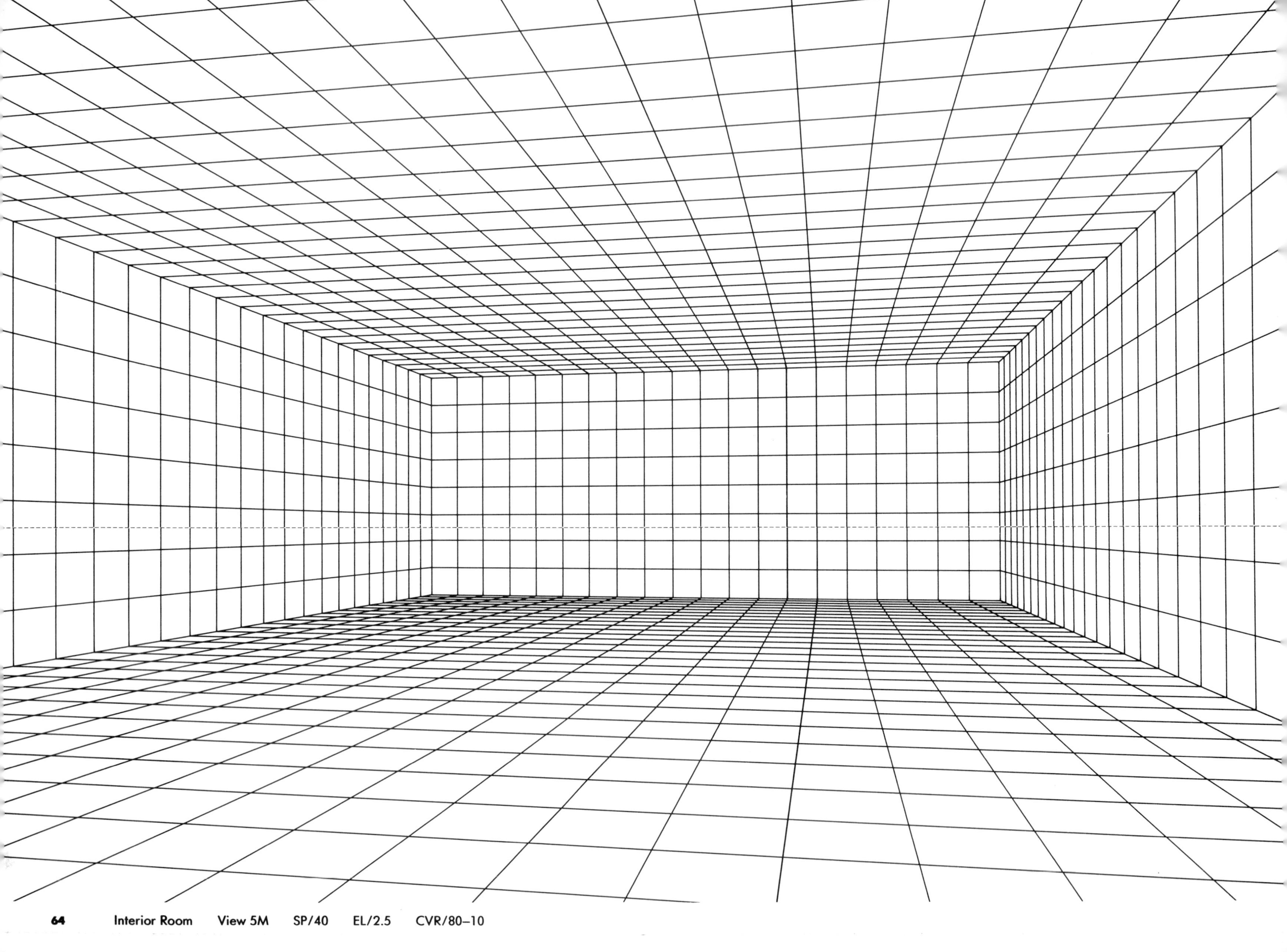

 Interior Room View 5M SP/40 EL/2.5 CVR/80–10

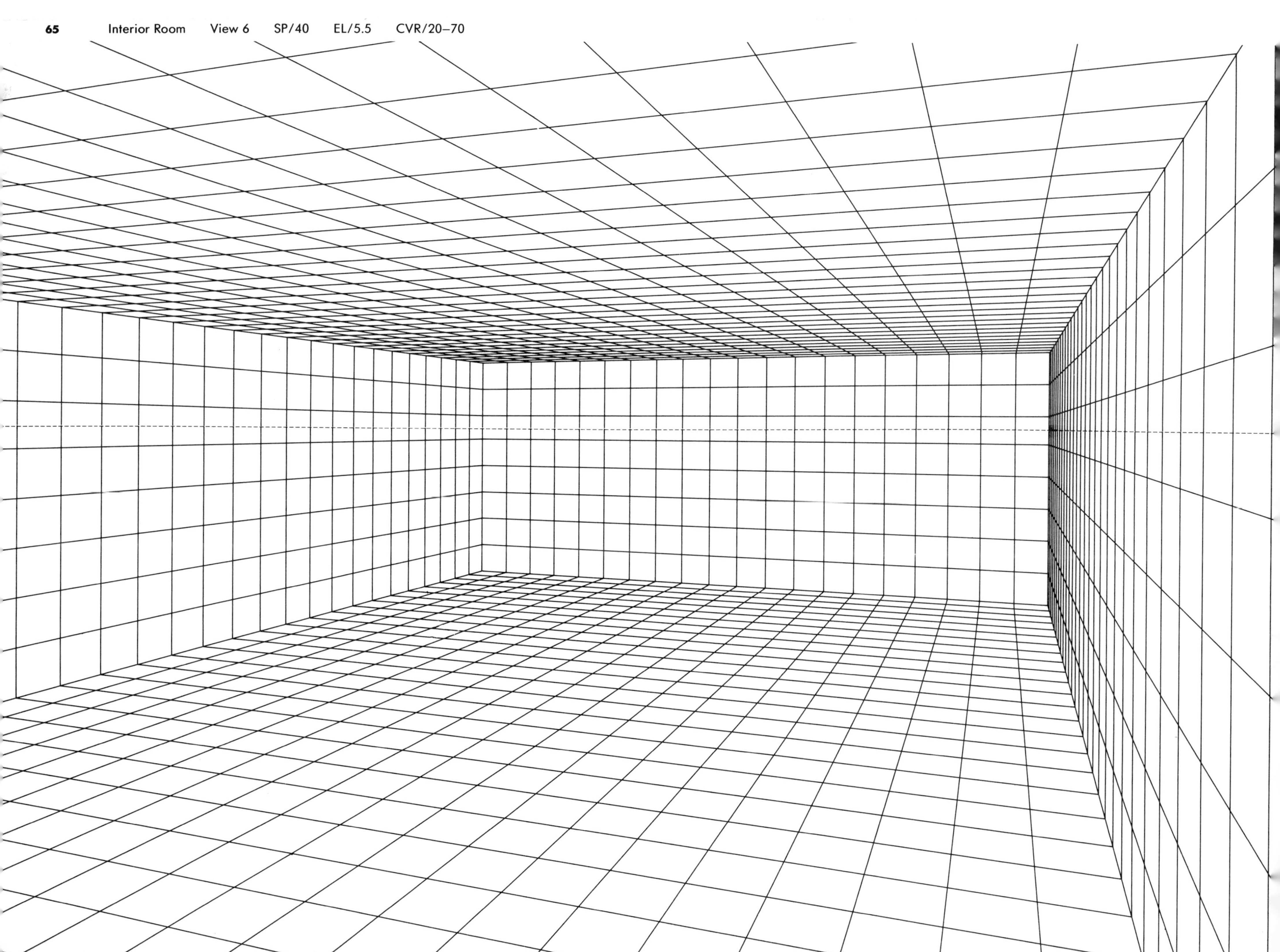

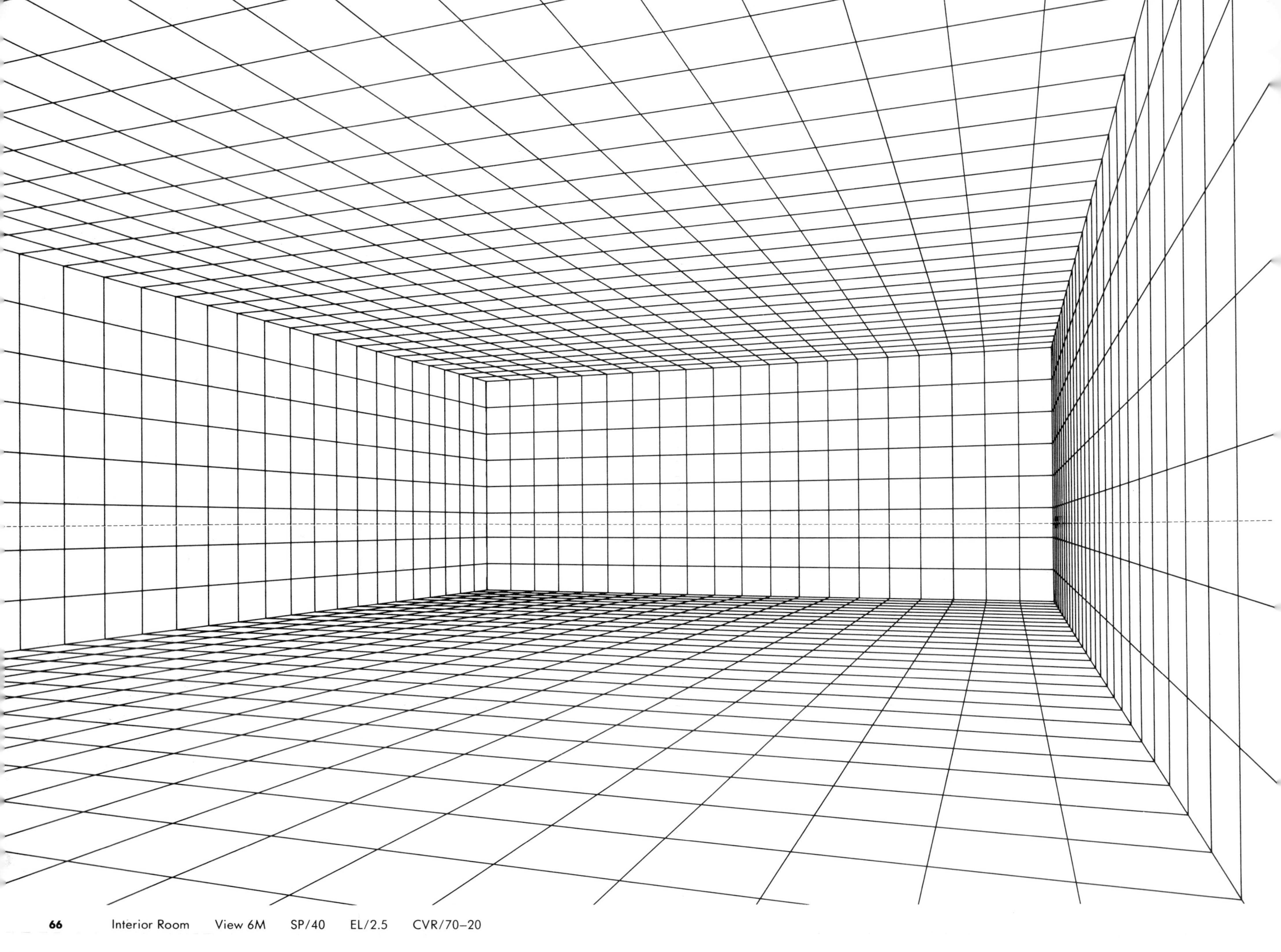

 Interior Room View 6M SP/40 EL/2.5 CVR/70–20

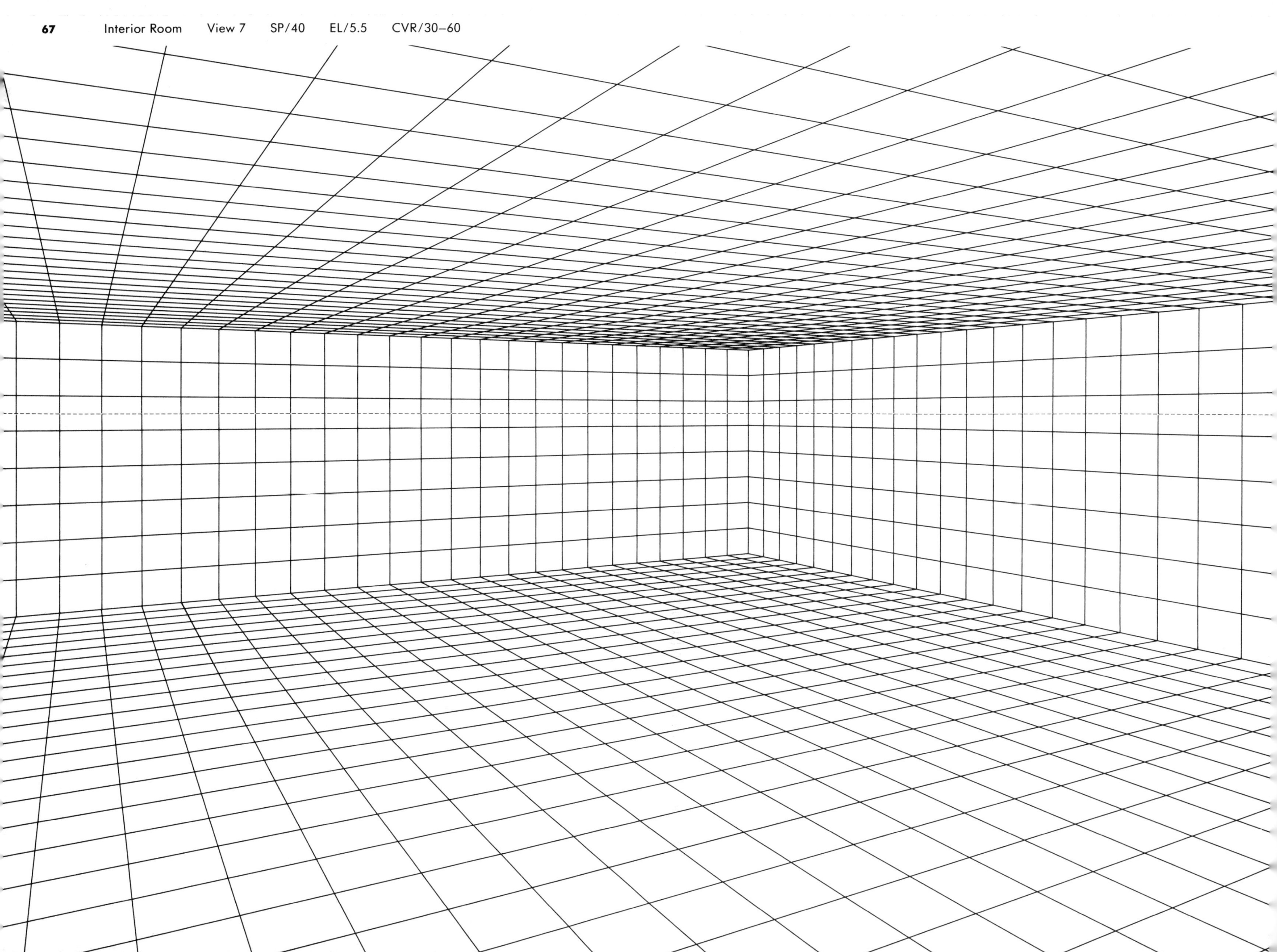

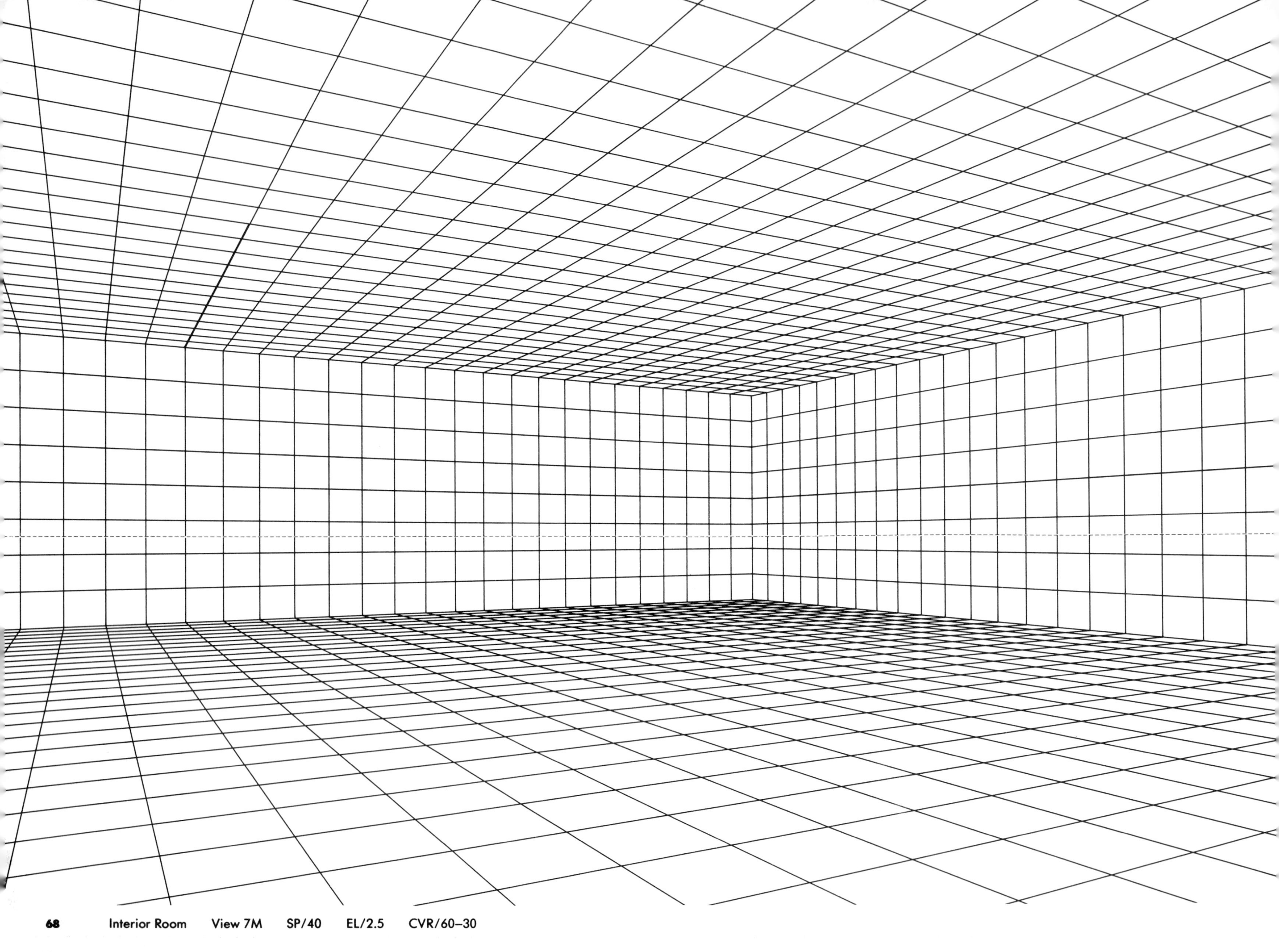

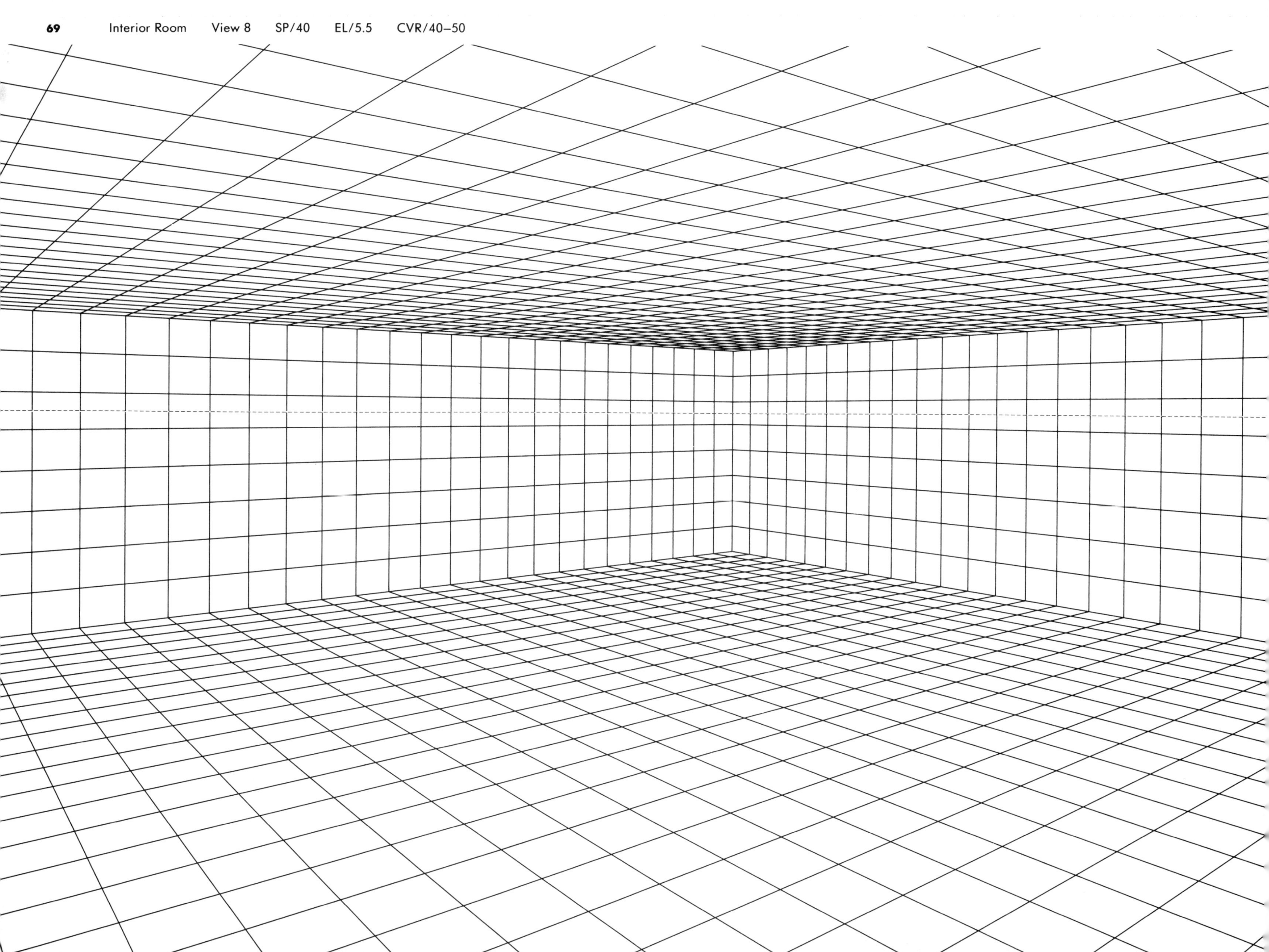

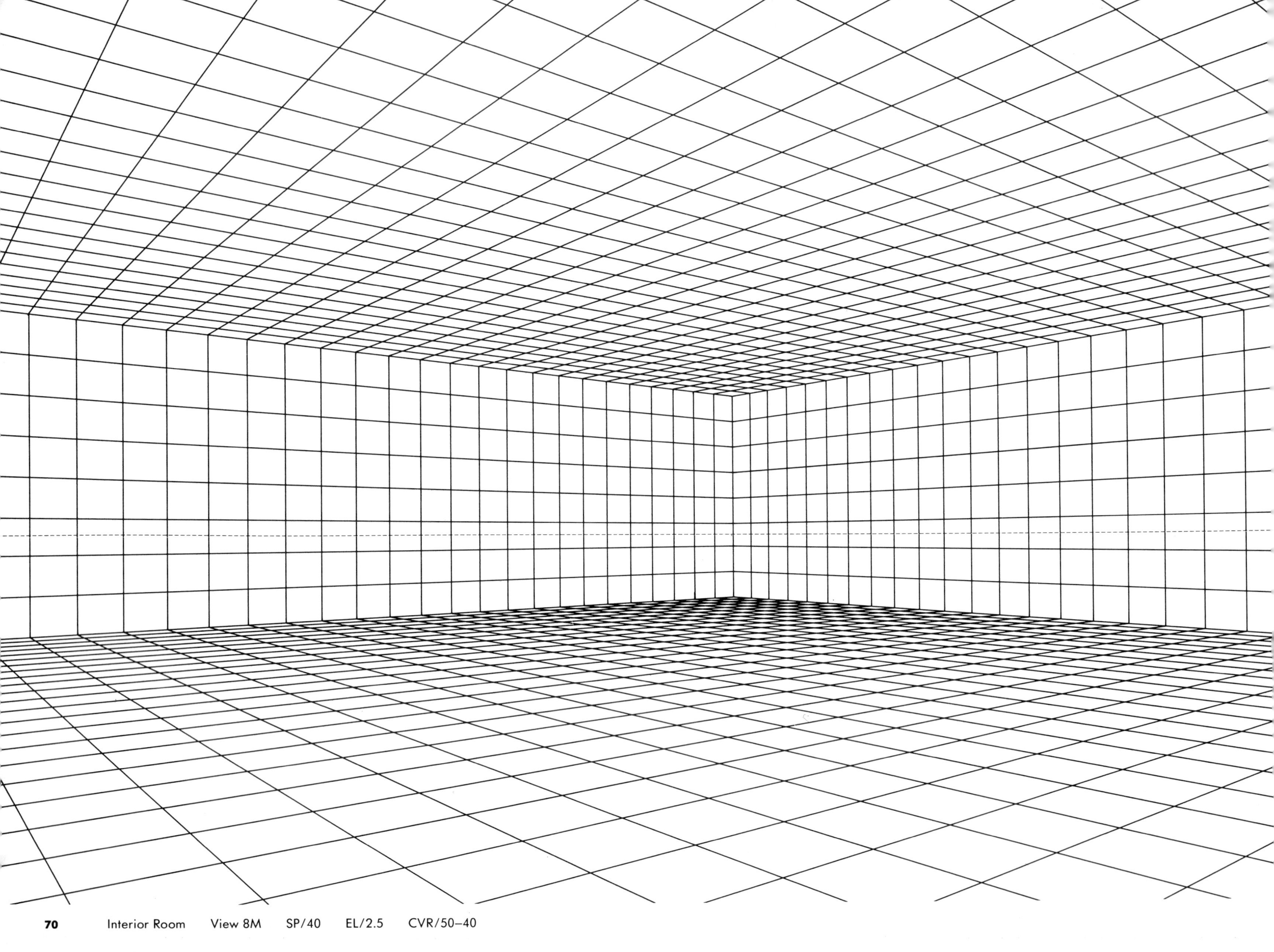

 Interior Room View 8M SP/40 EL/2.5 CVR/50–40

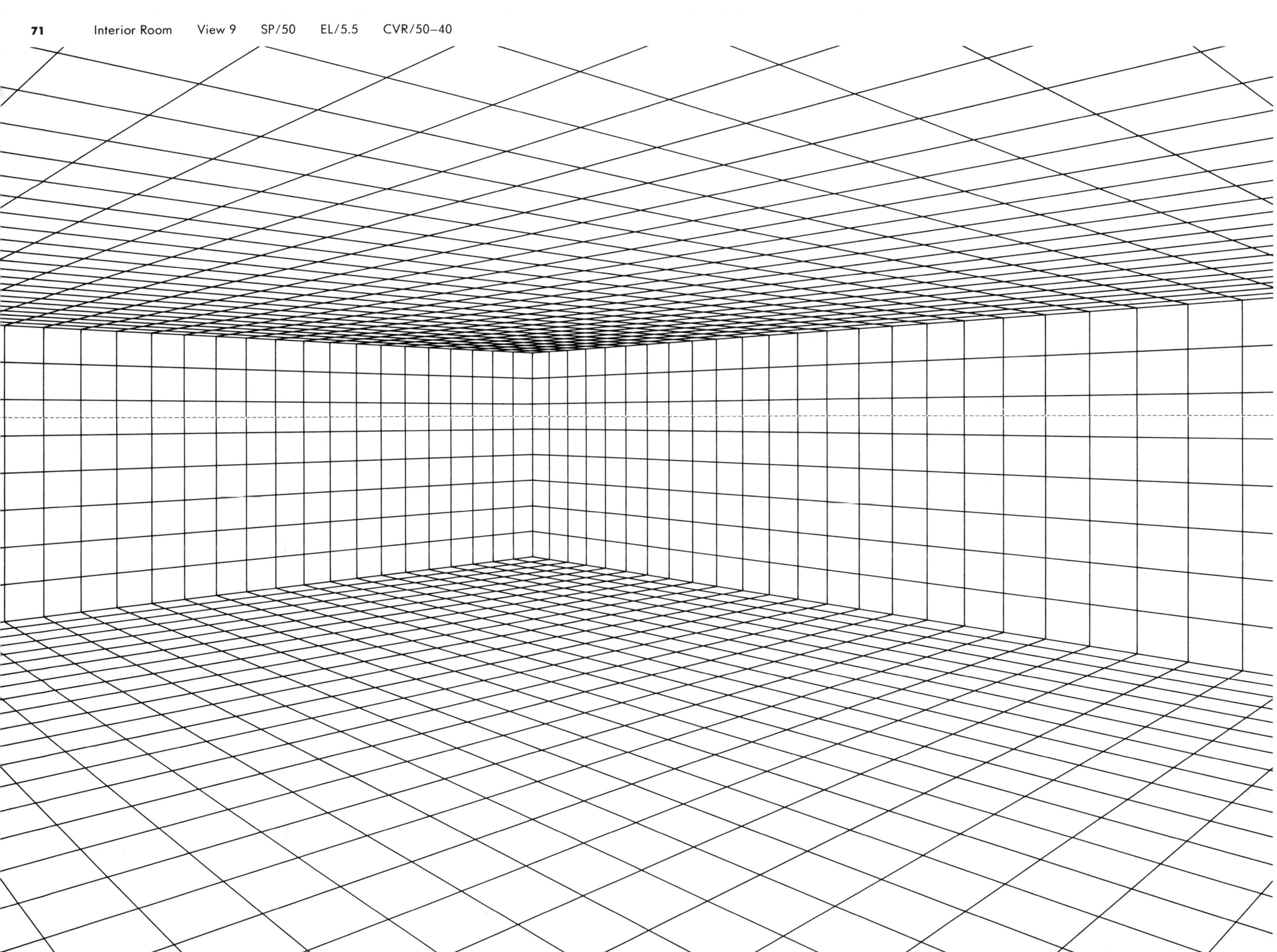

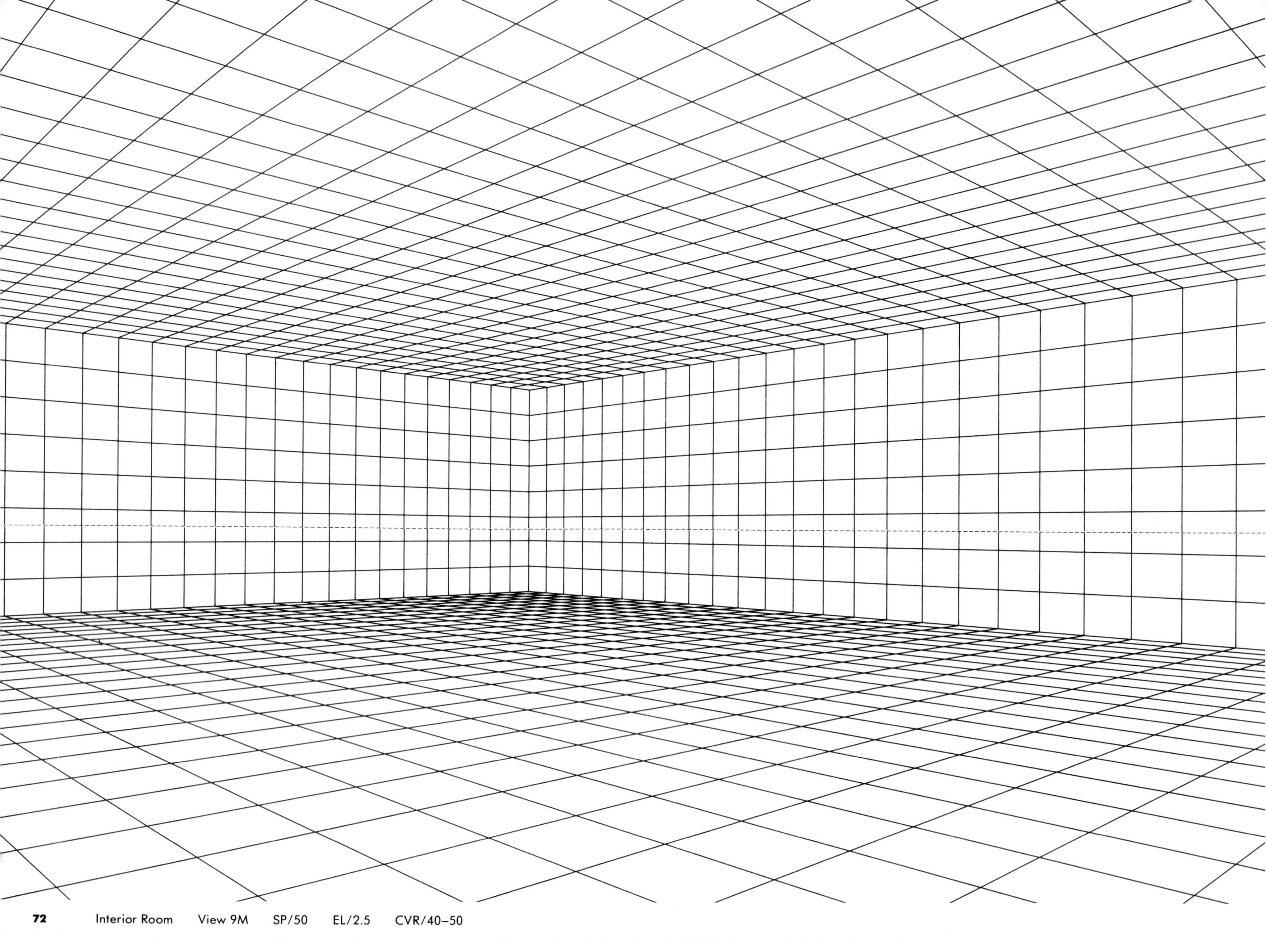

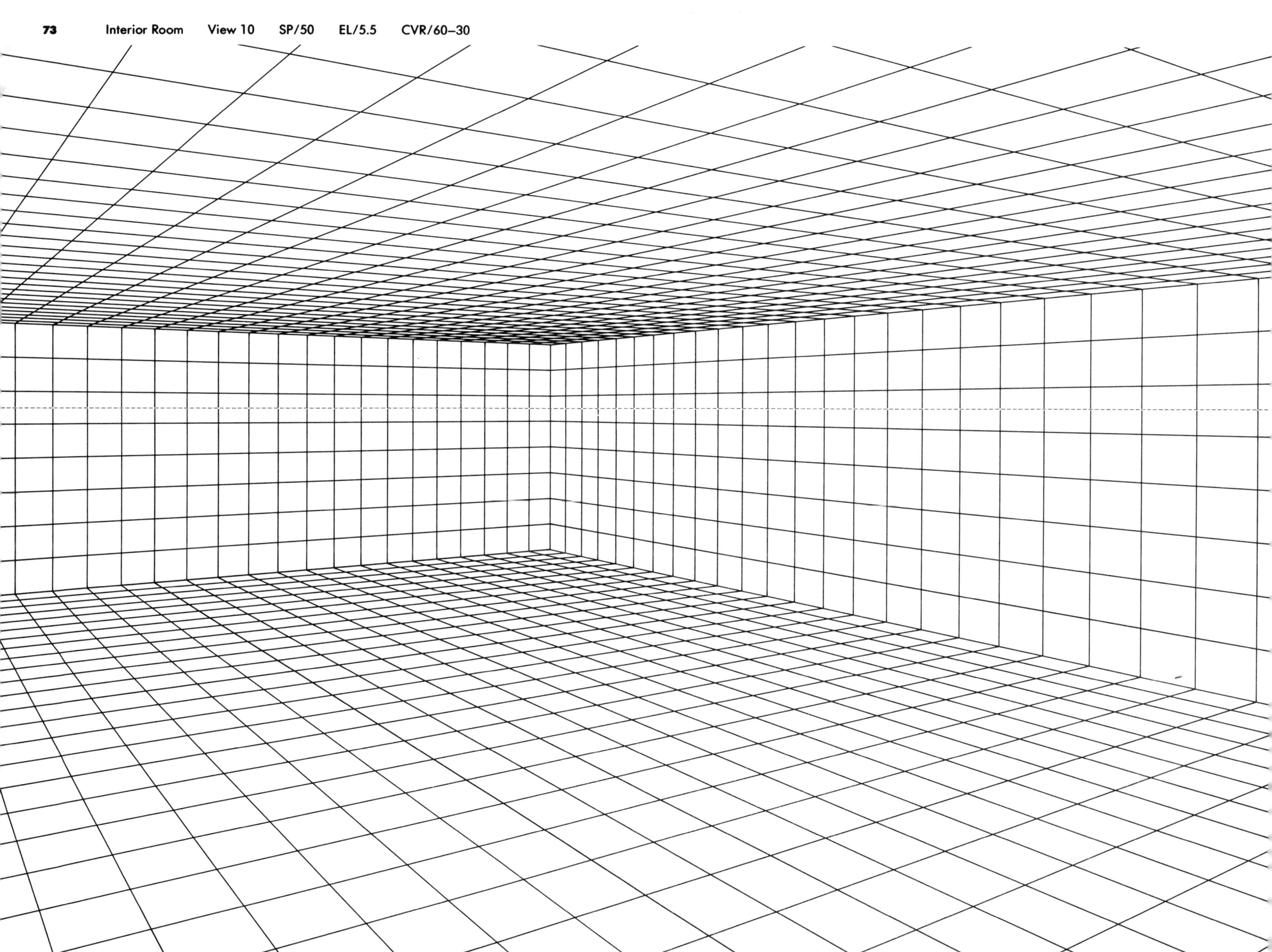

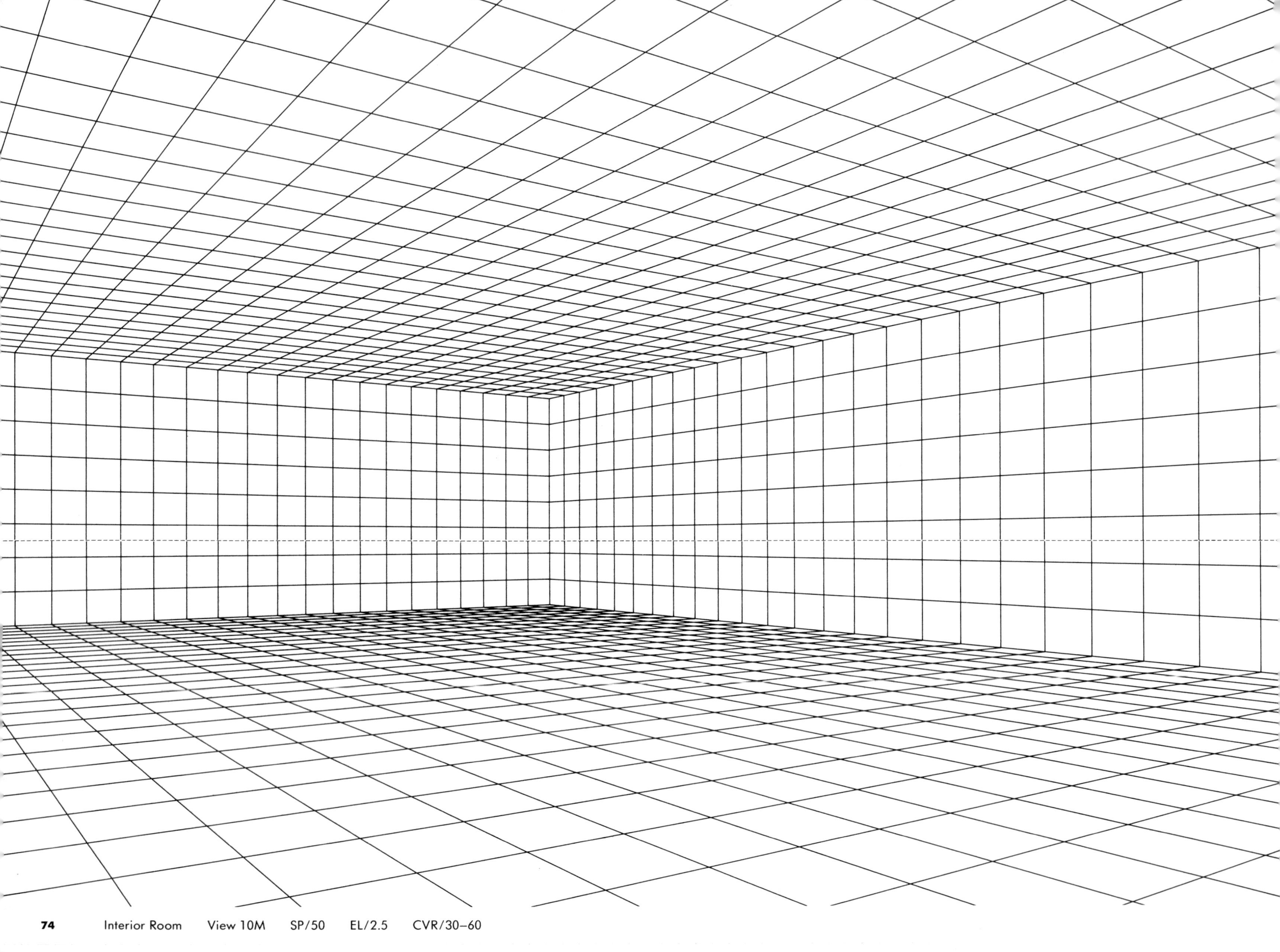

 Interior Room View 10M SP/50 EL/2.5 CVR/30–60

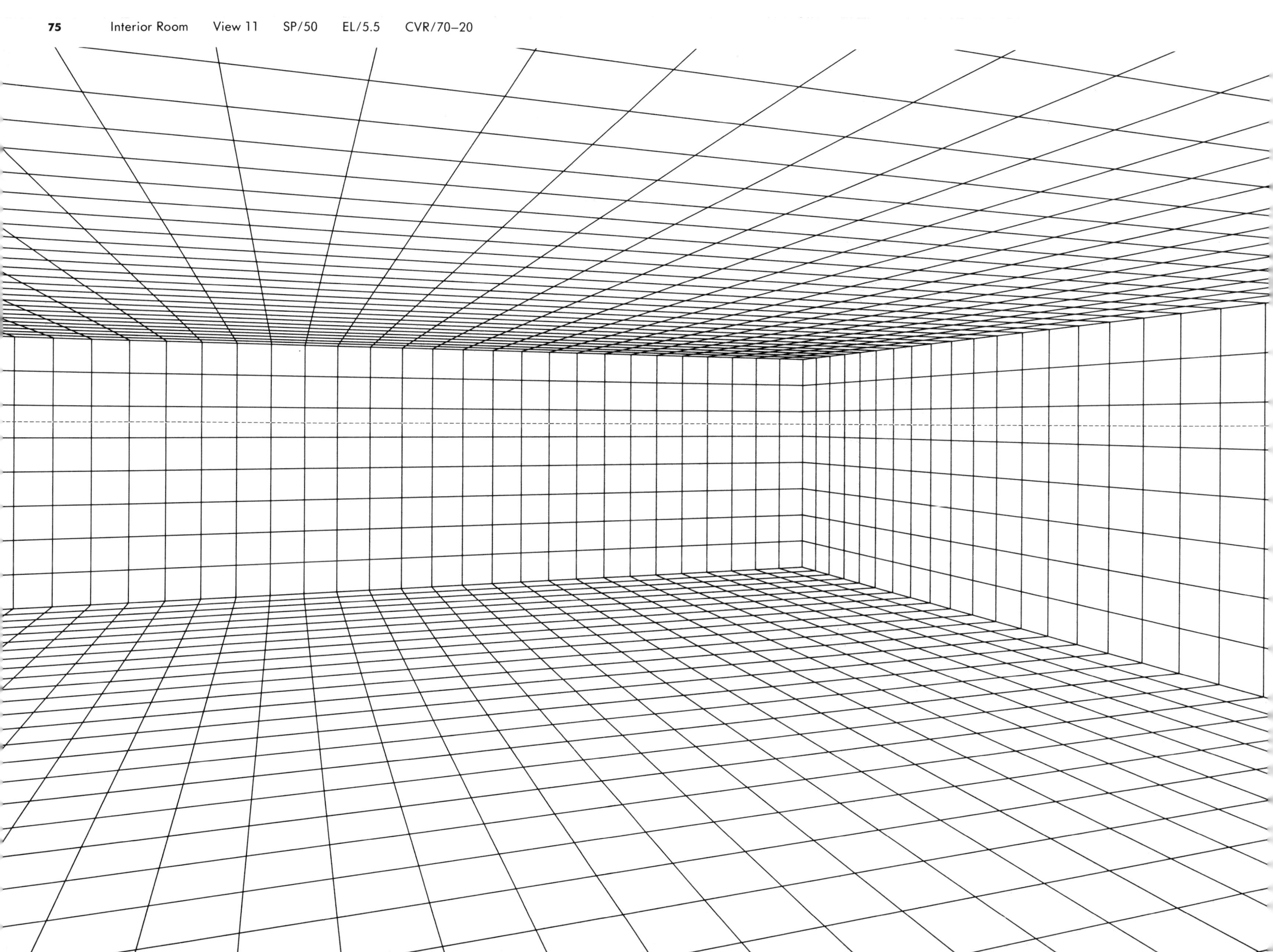

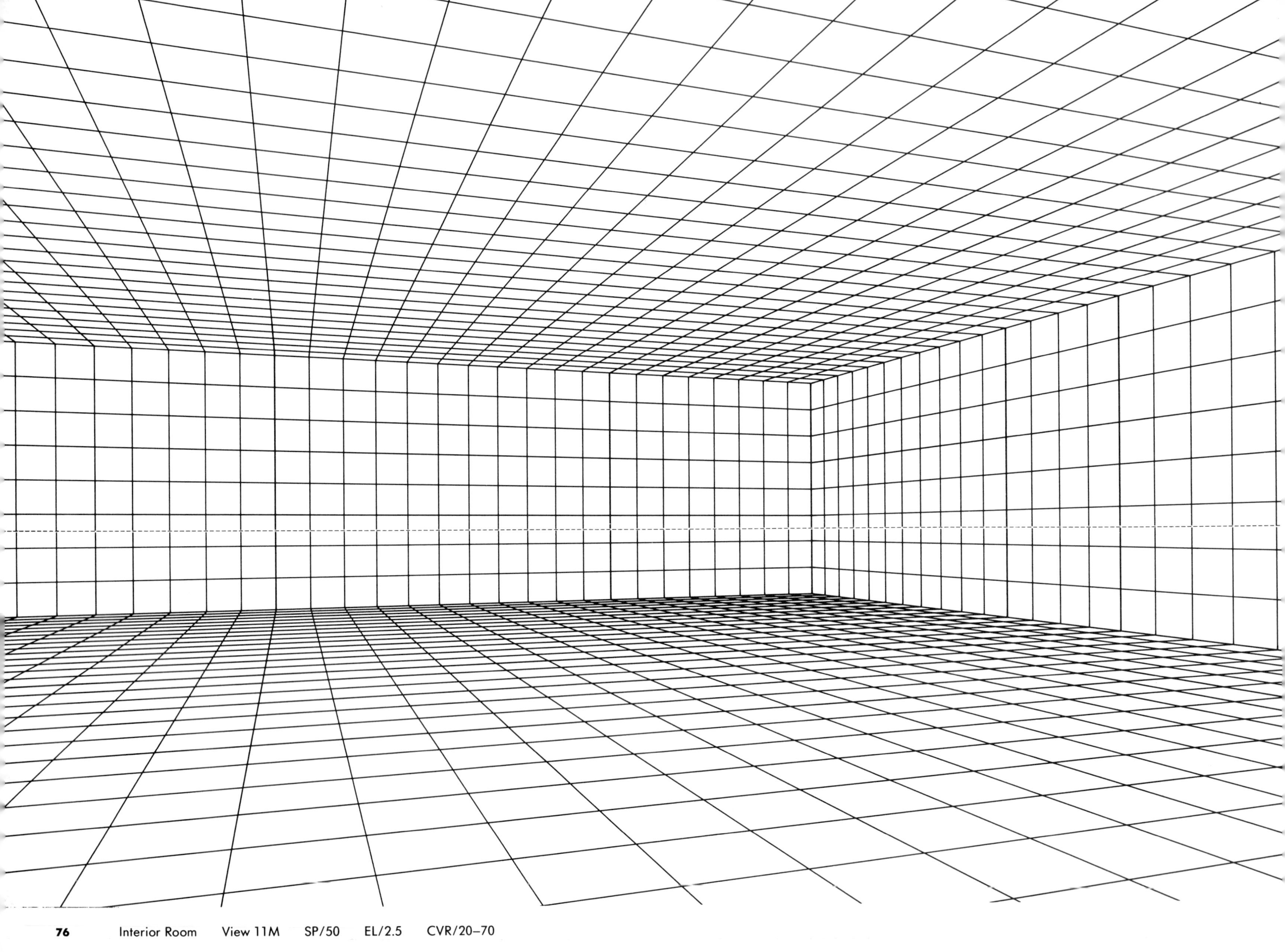

 Interior Room View 11M SP/50 EL/2.5 CVR/20–70

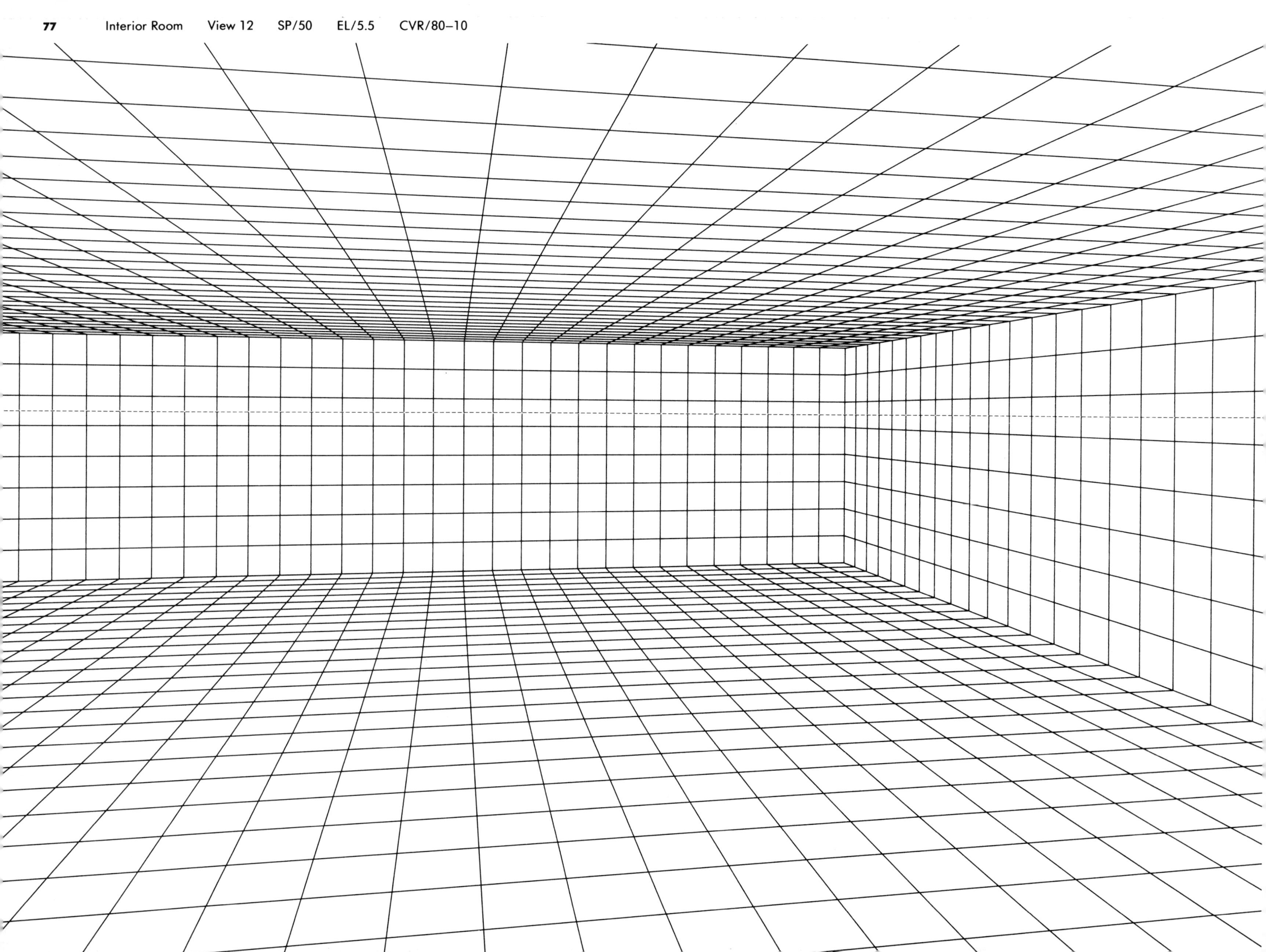

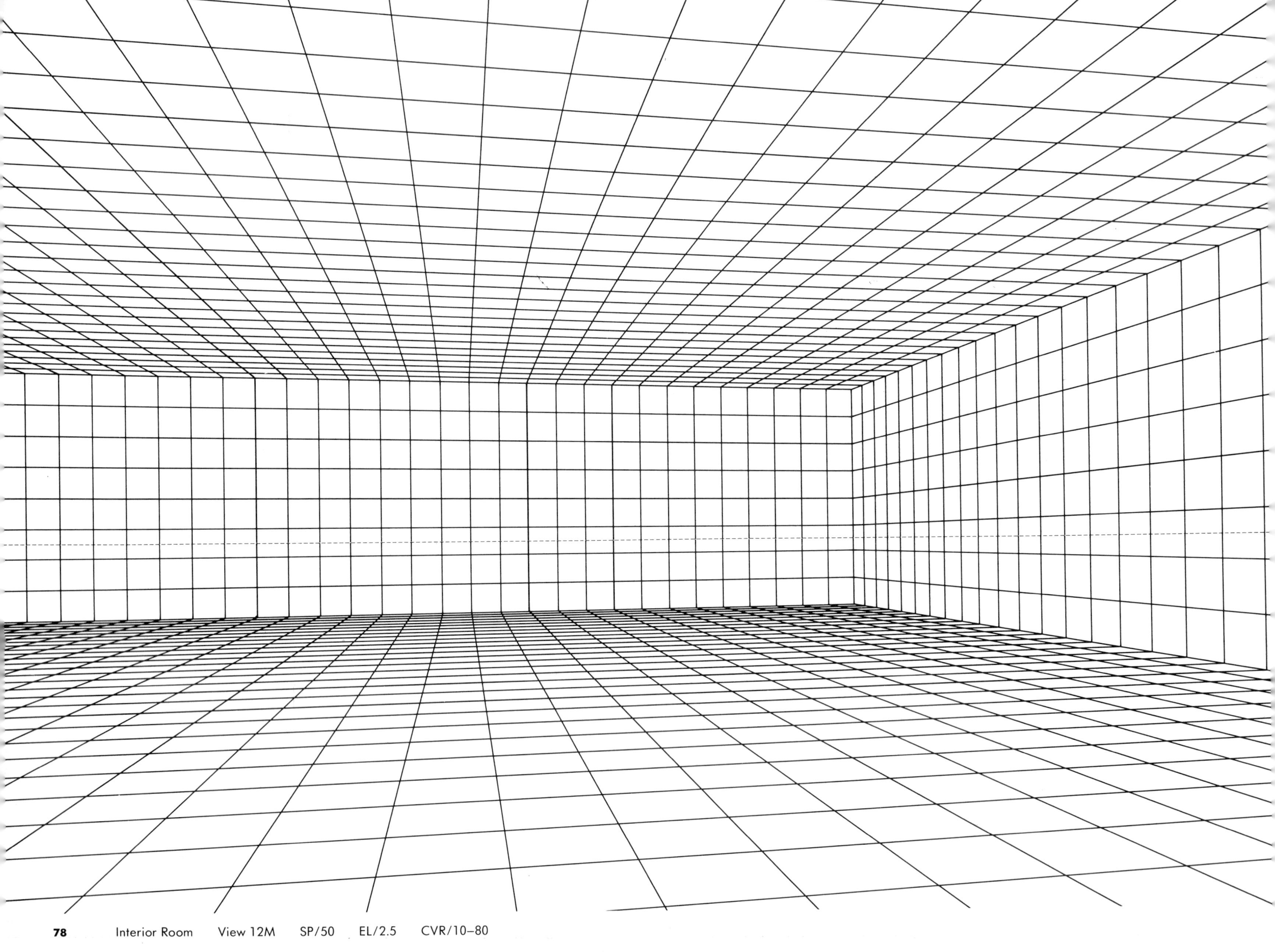

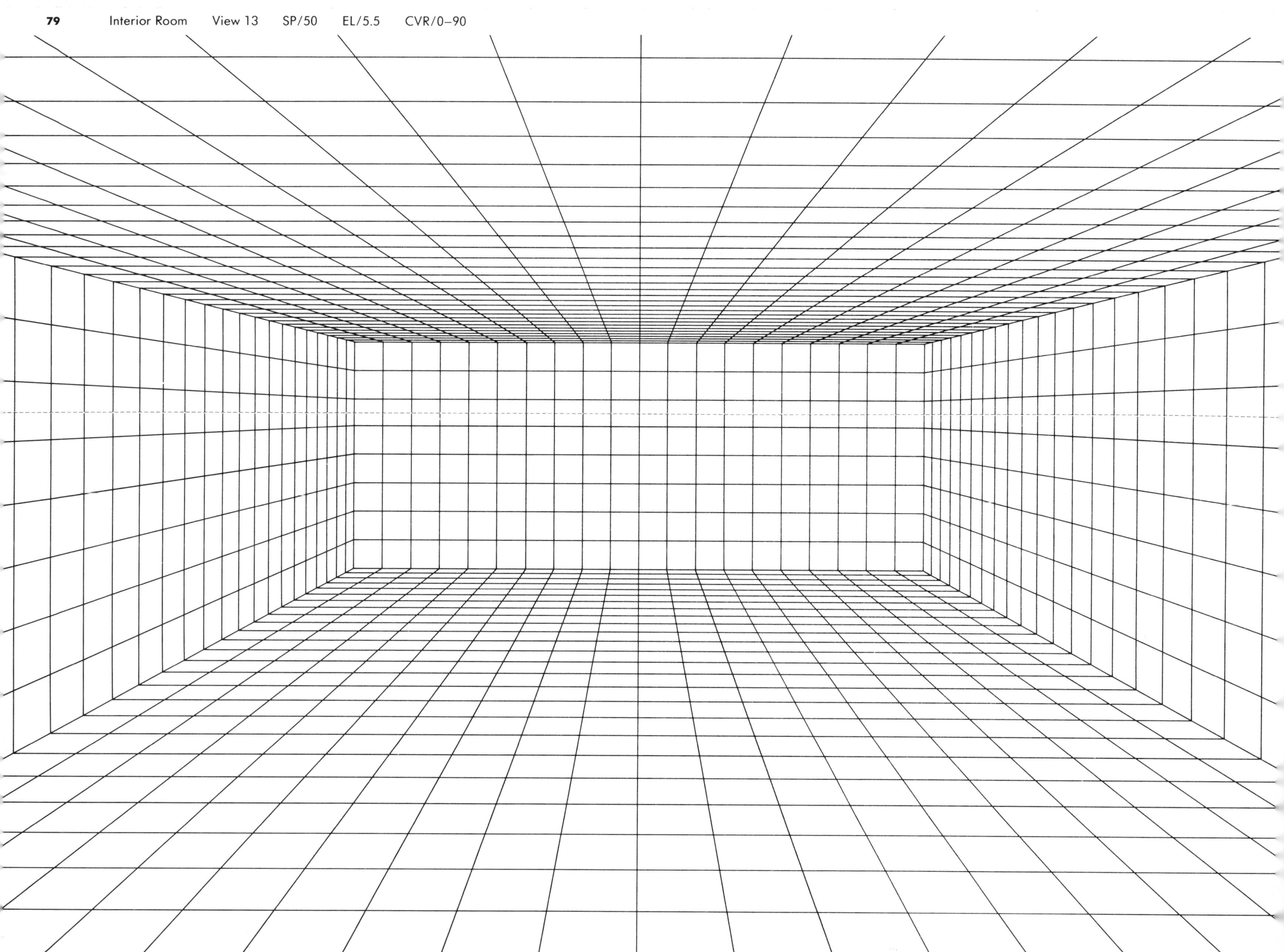

**PART TWO**

# FLOOR AND CEILING GRIDS

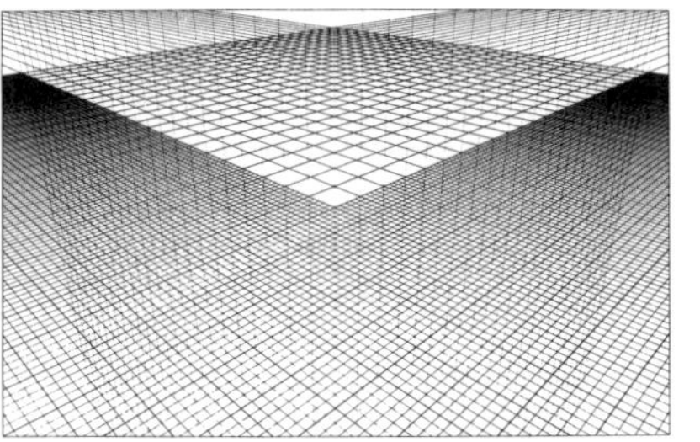

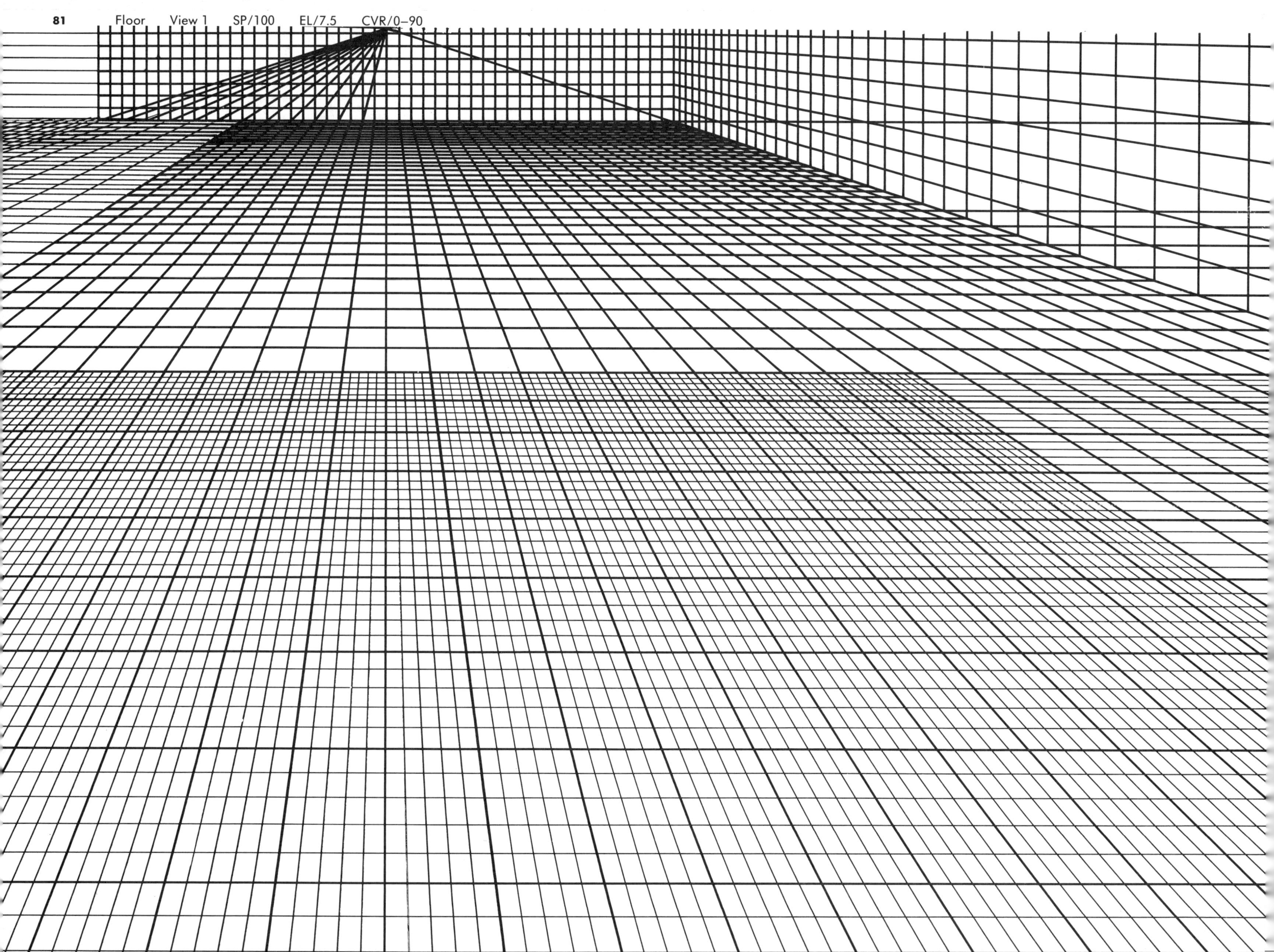
Floor View 1 SP/100 EL/7.5 CVR/0–90

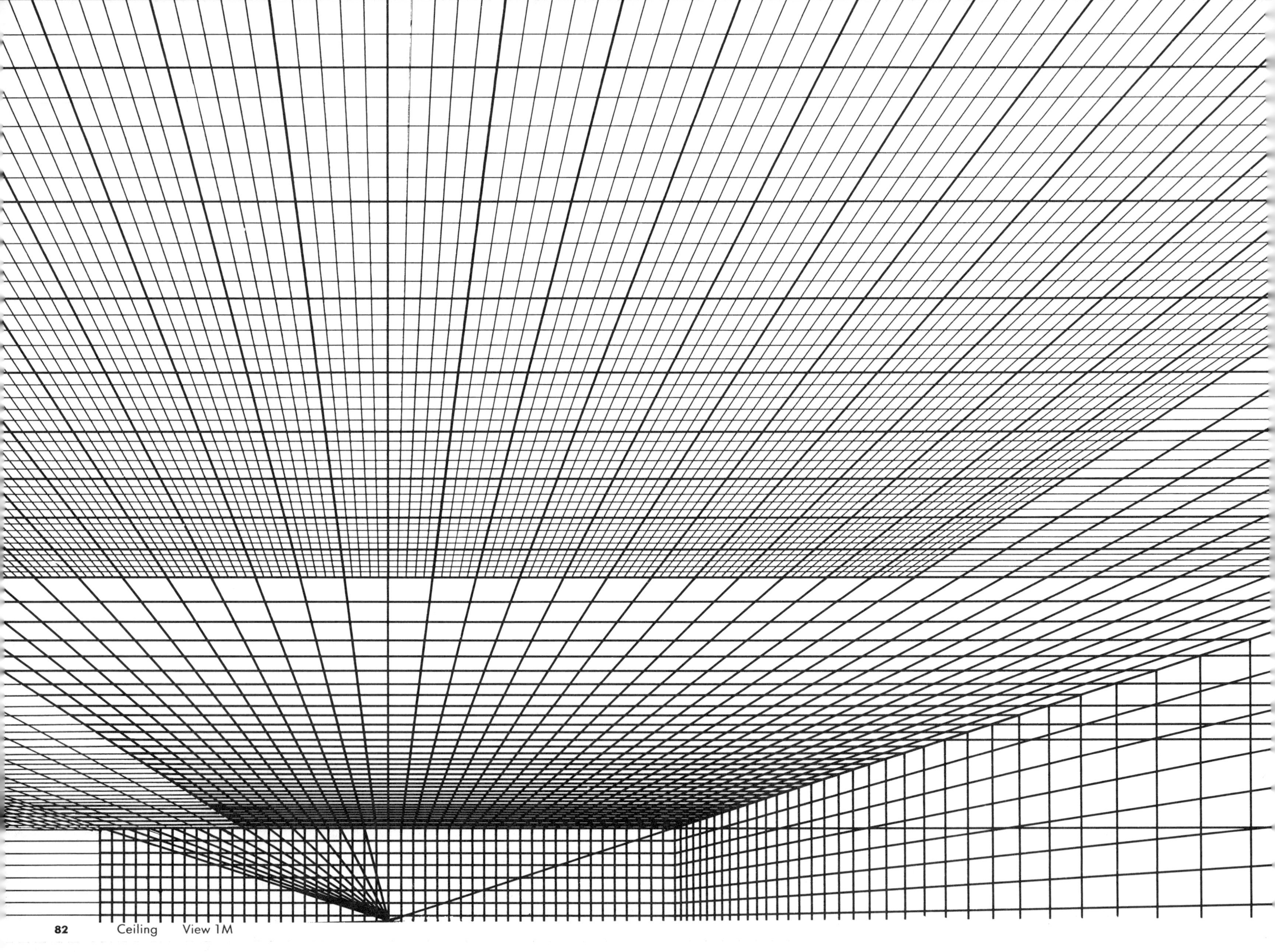

Ceiling View 1M

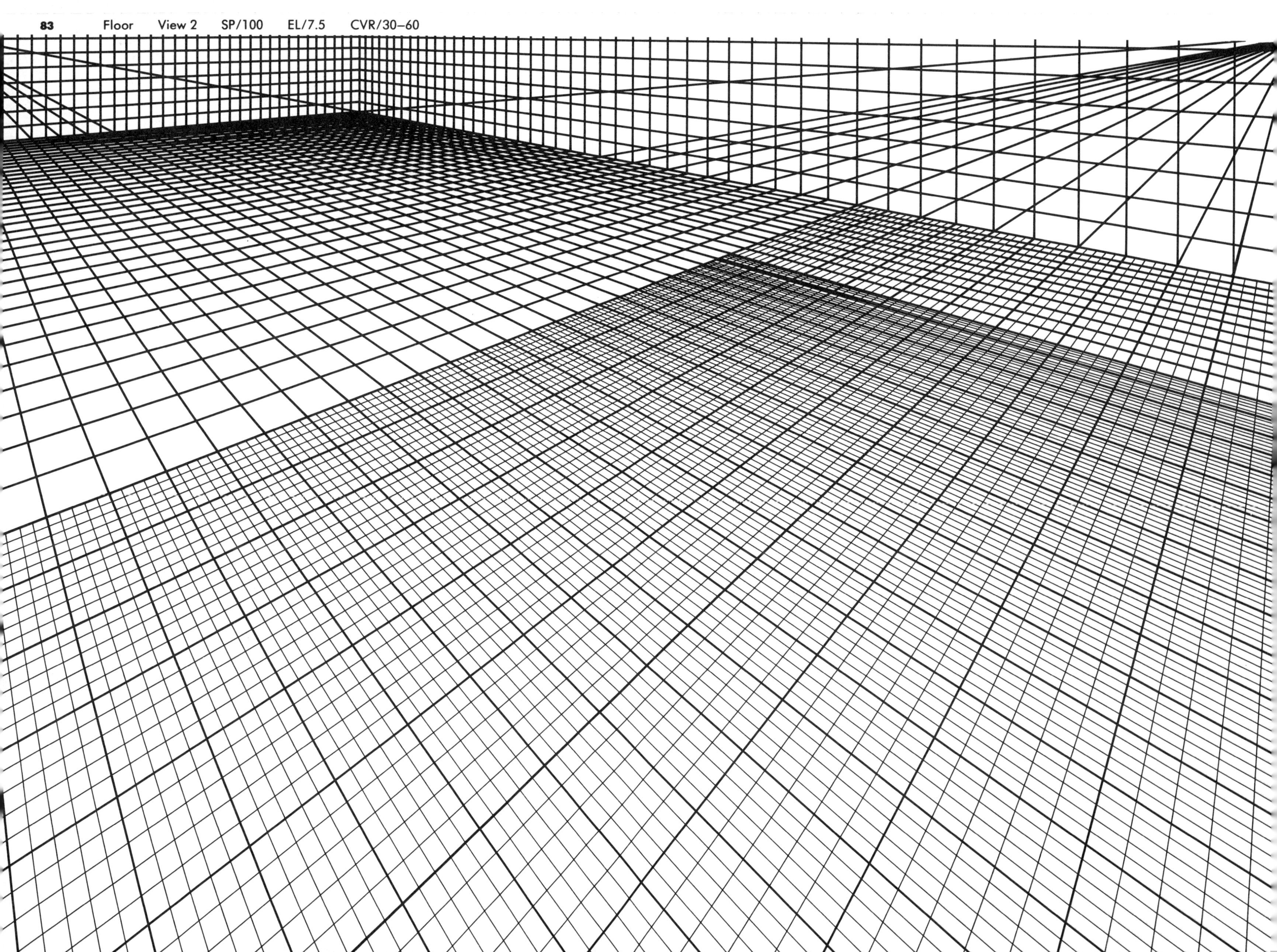

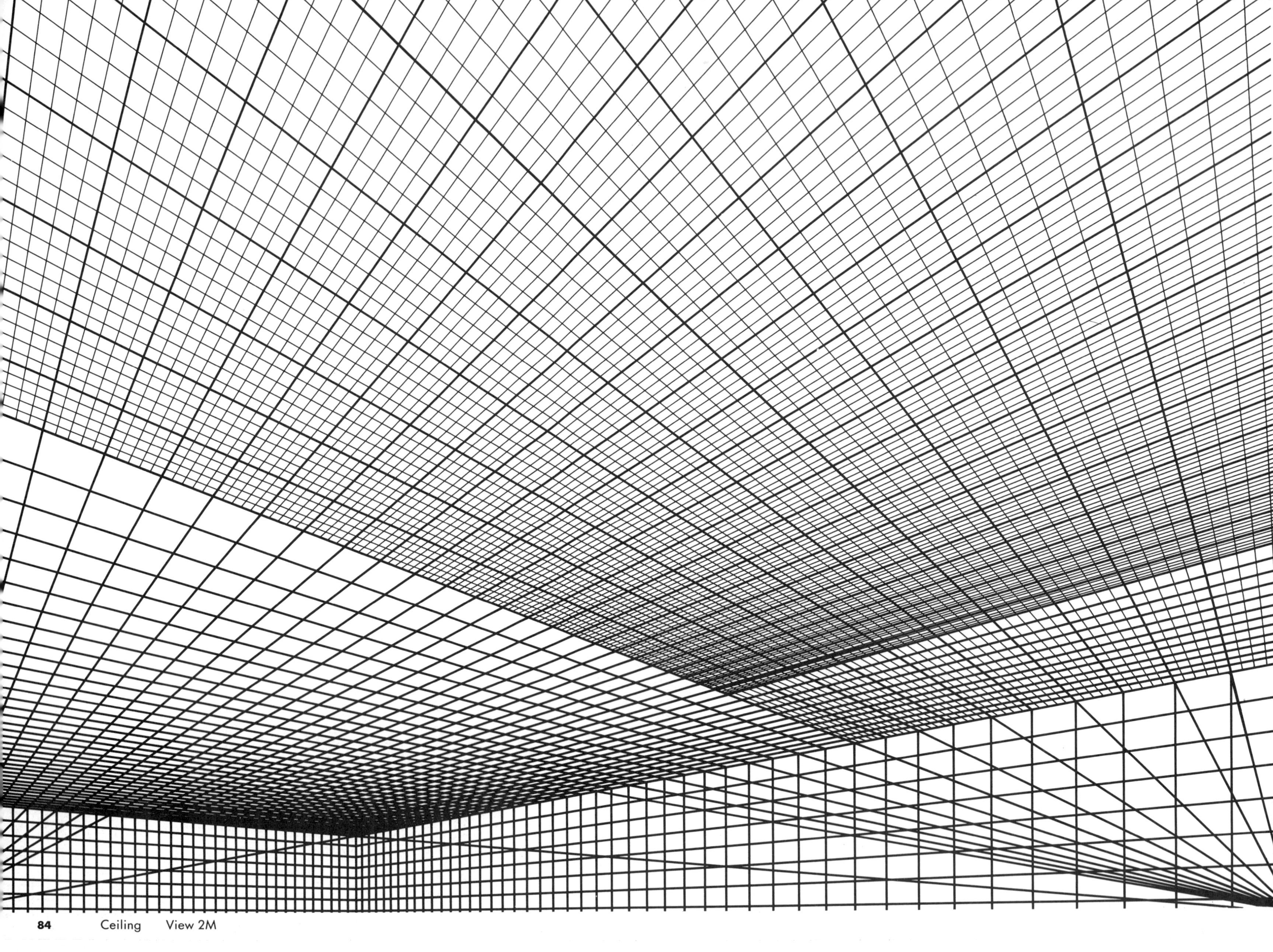

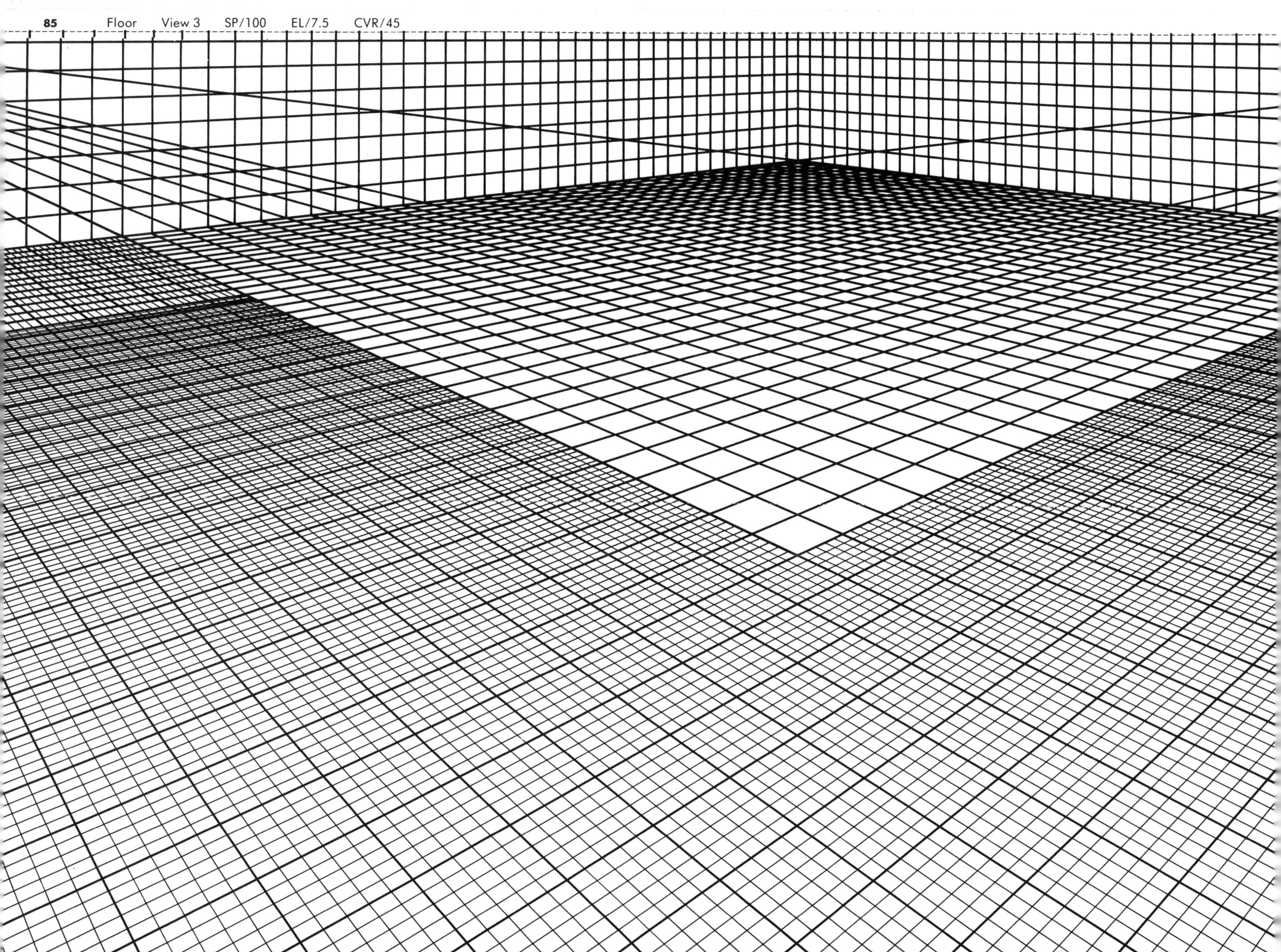

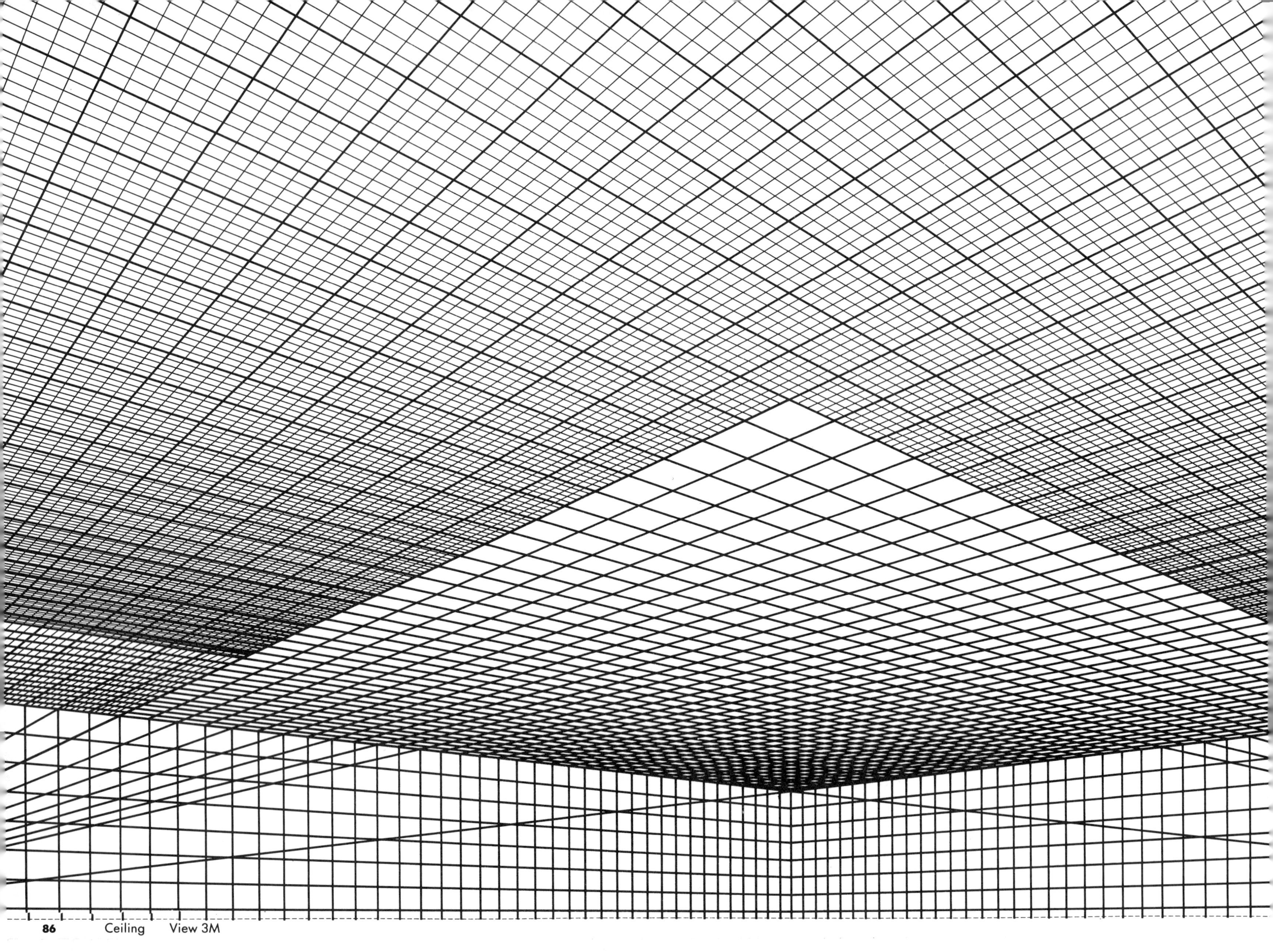

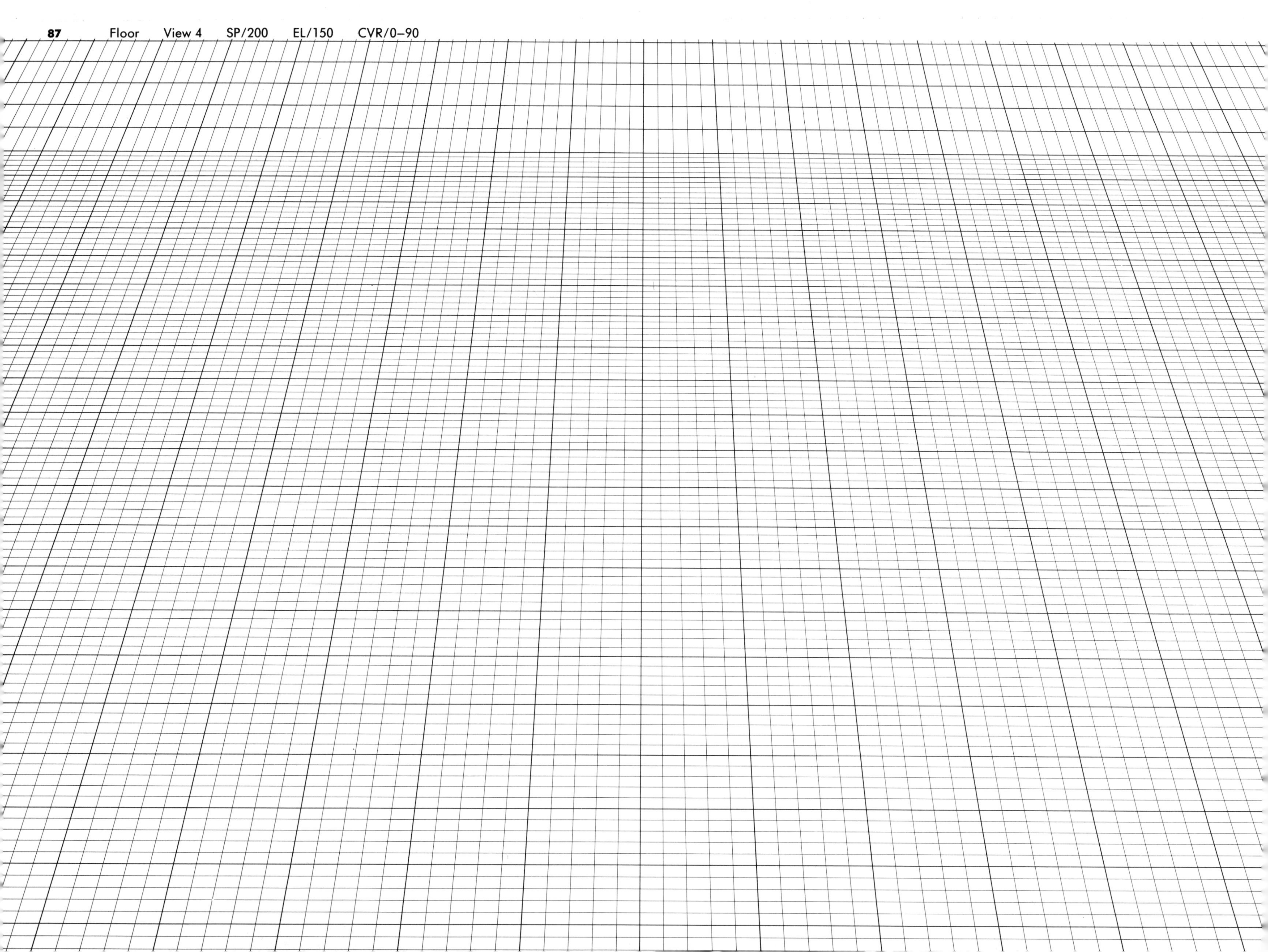

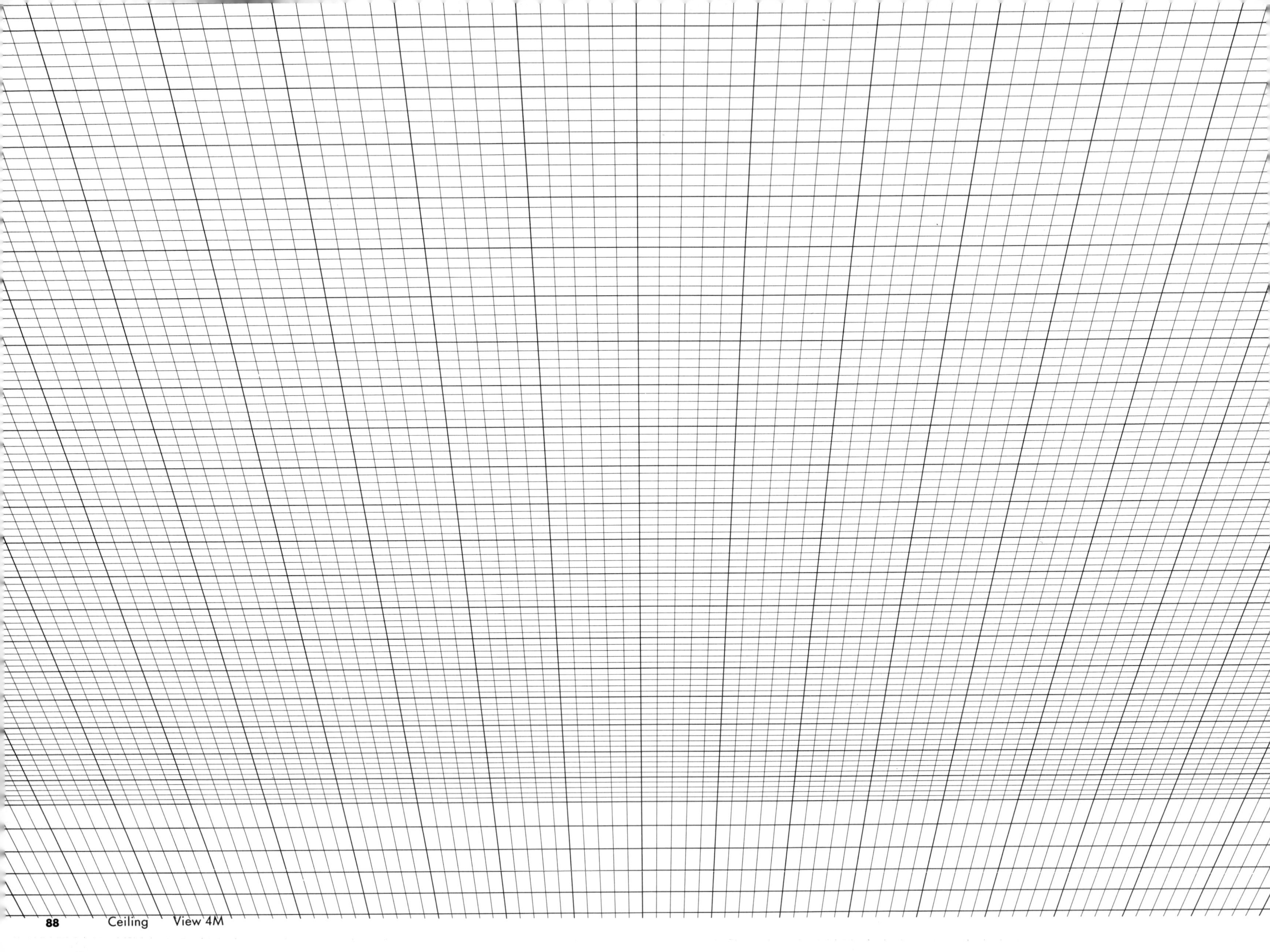

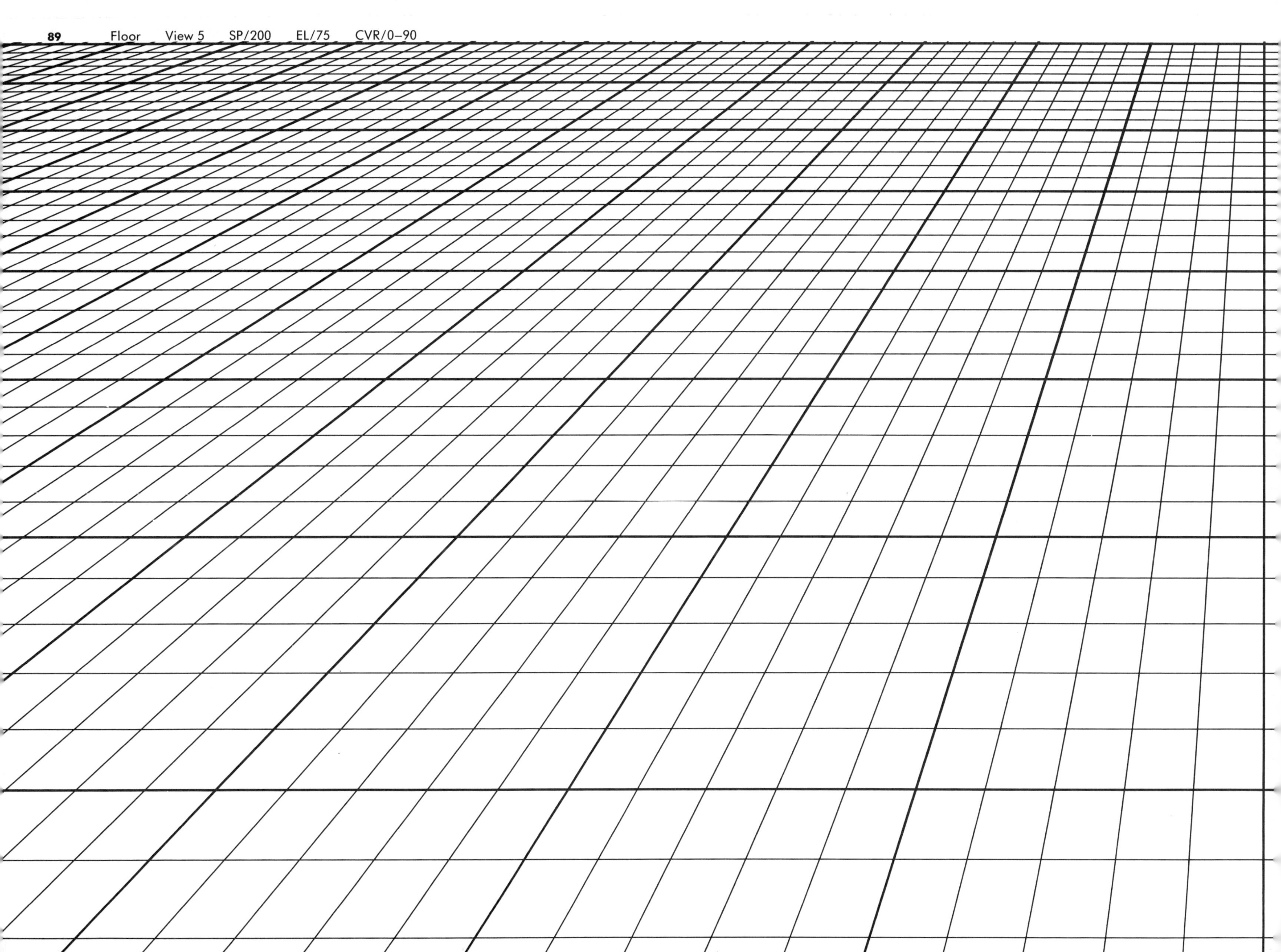

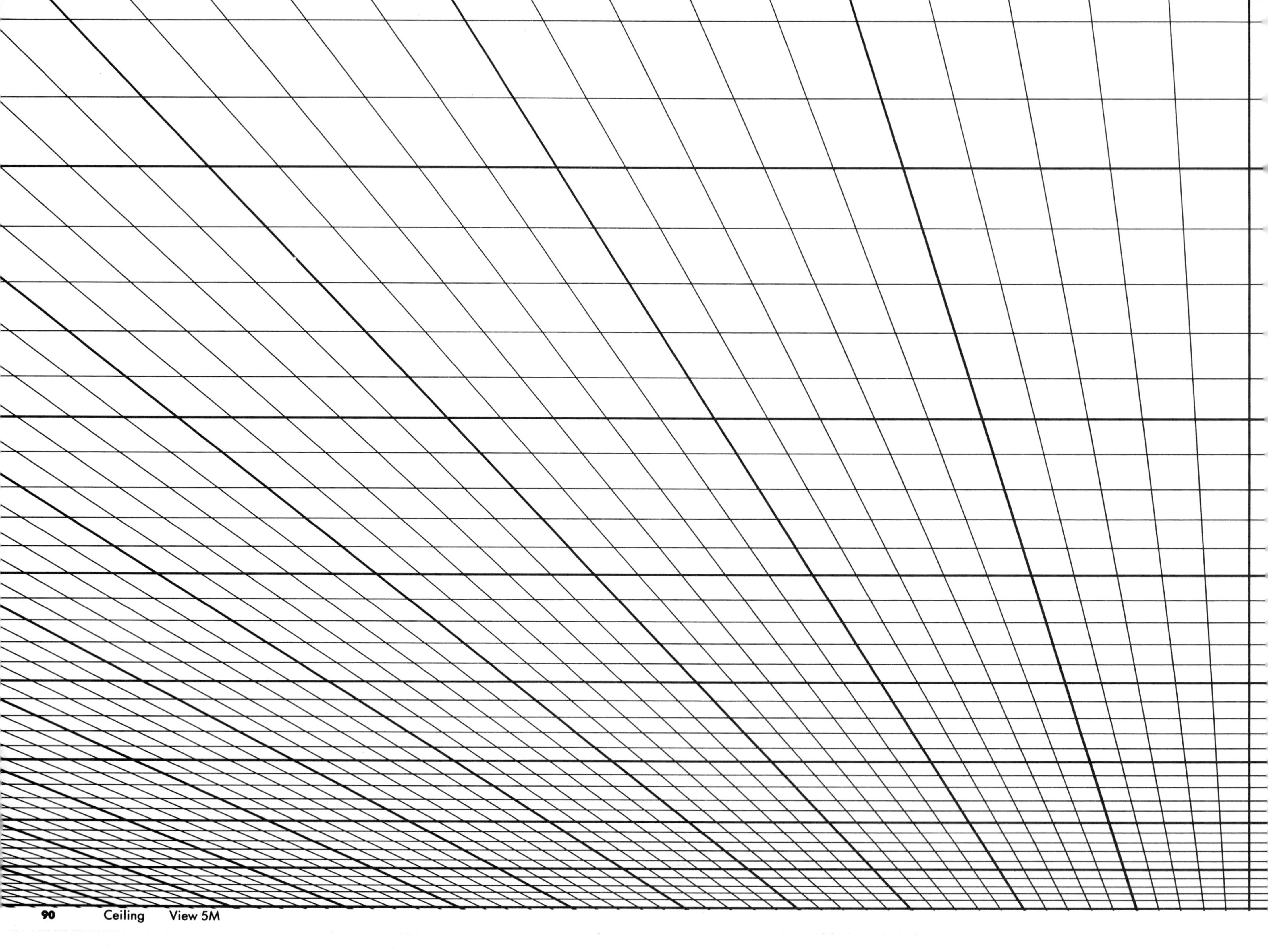

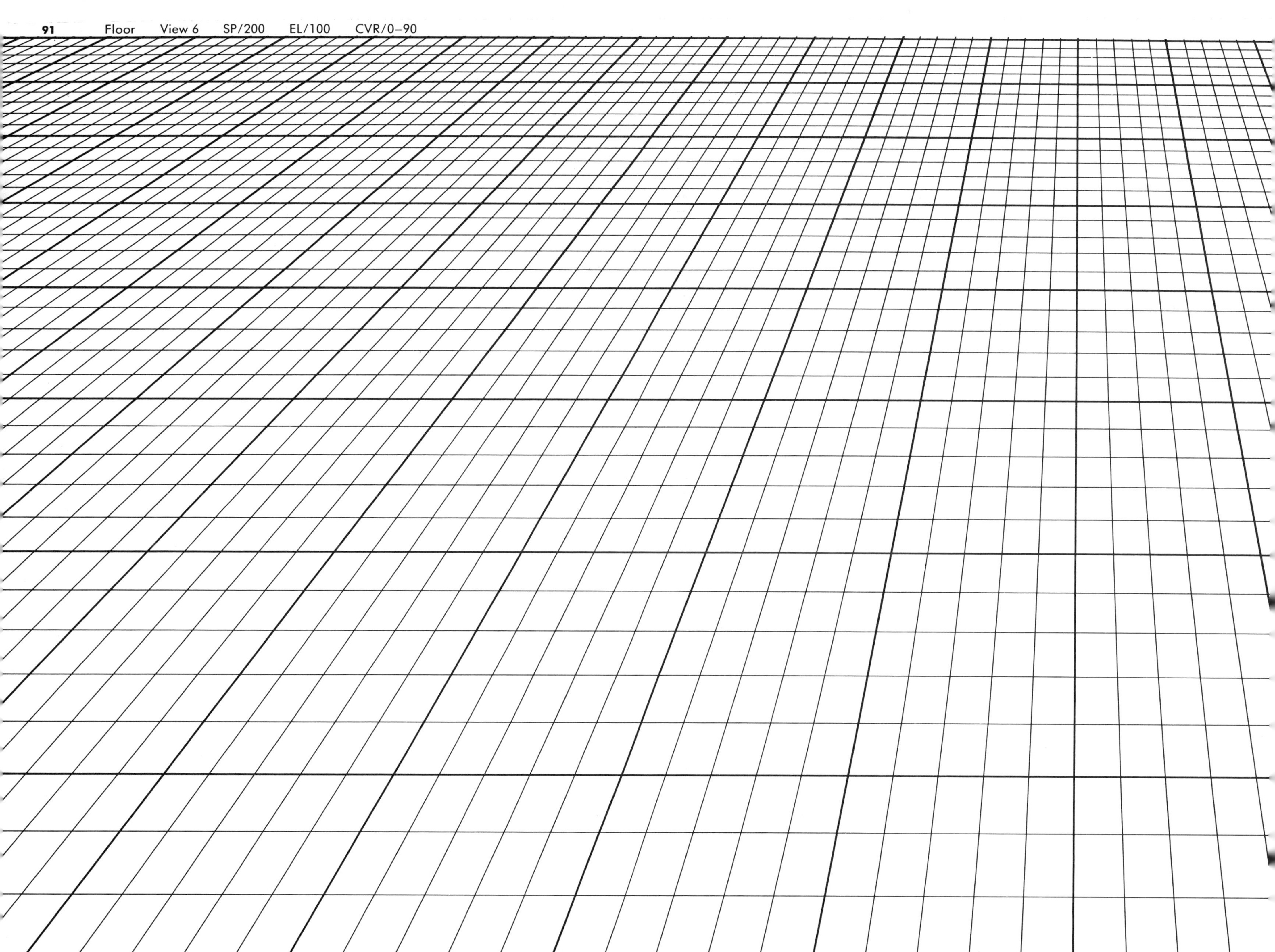

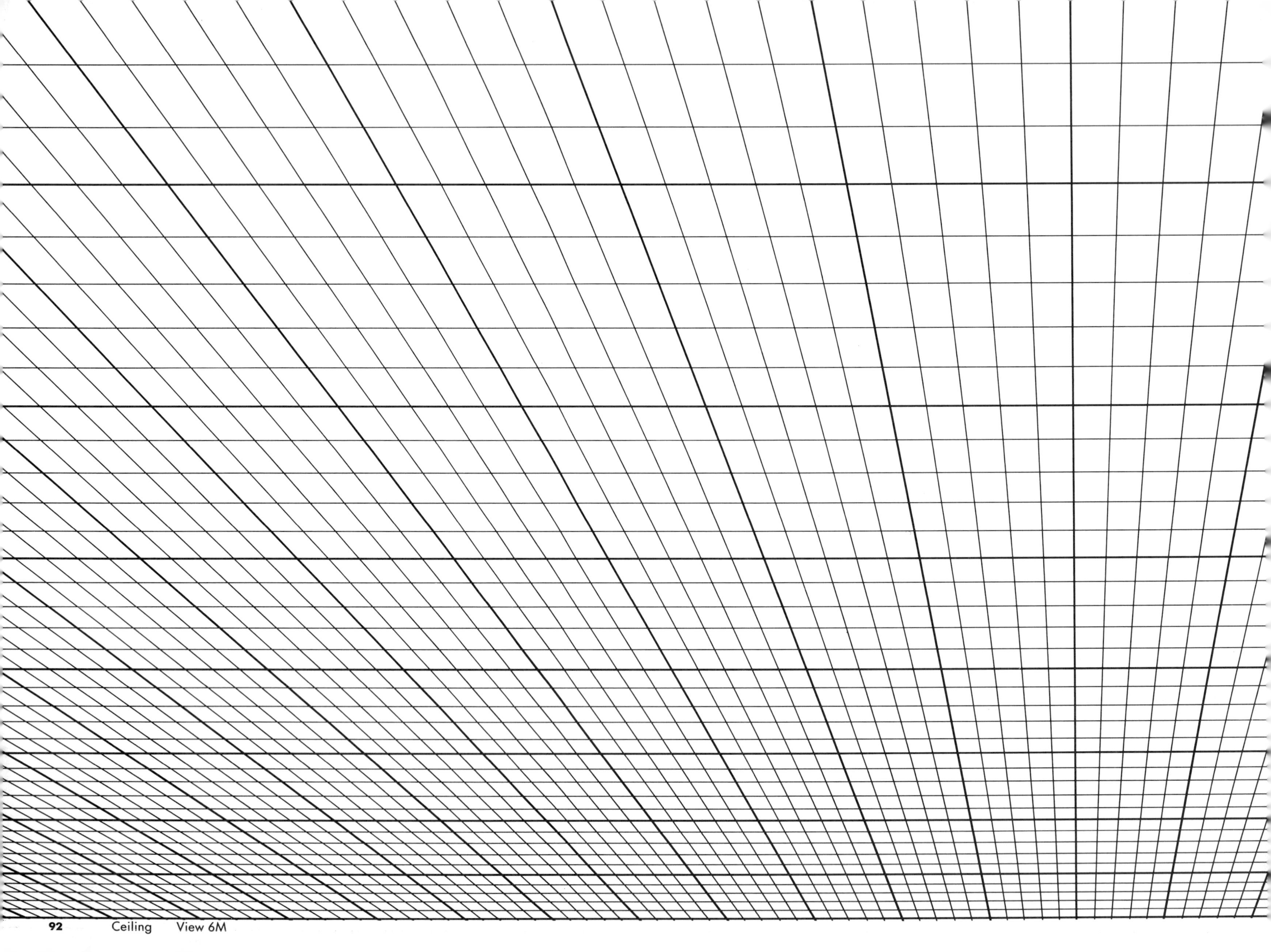

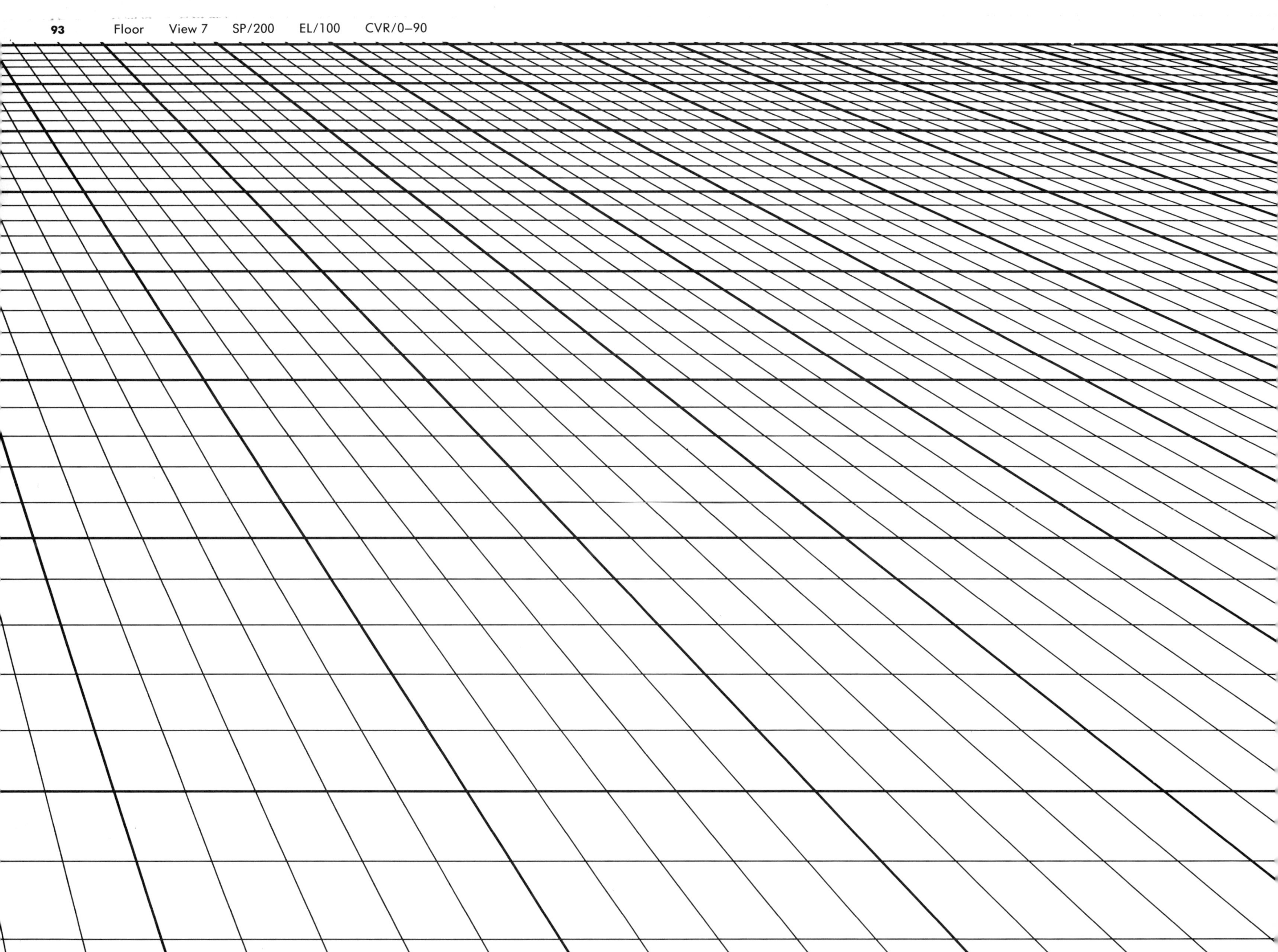

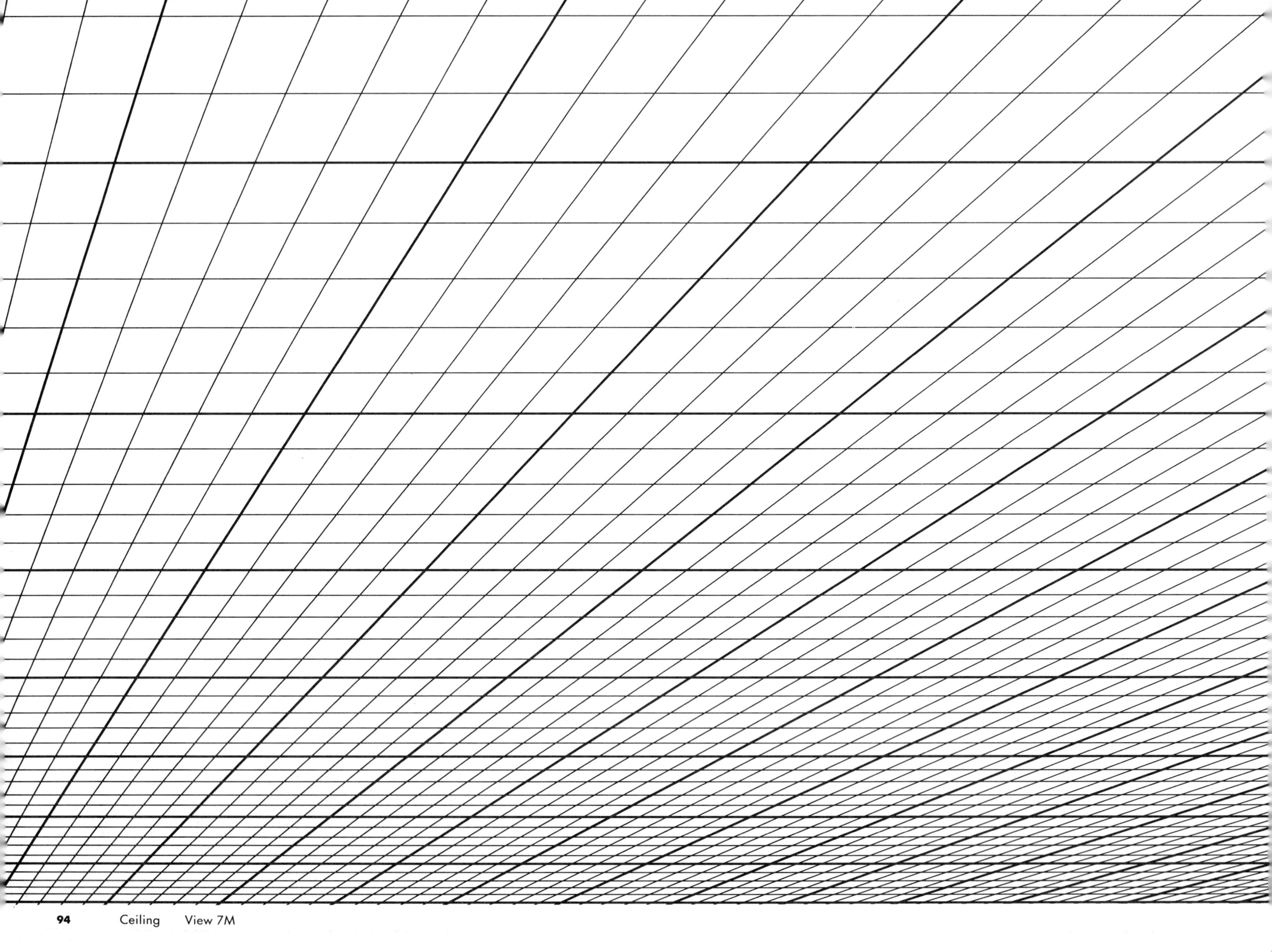

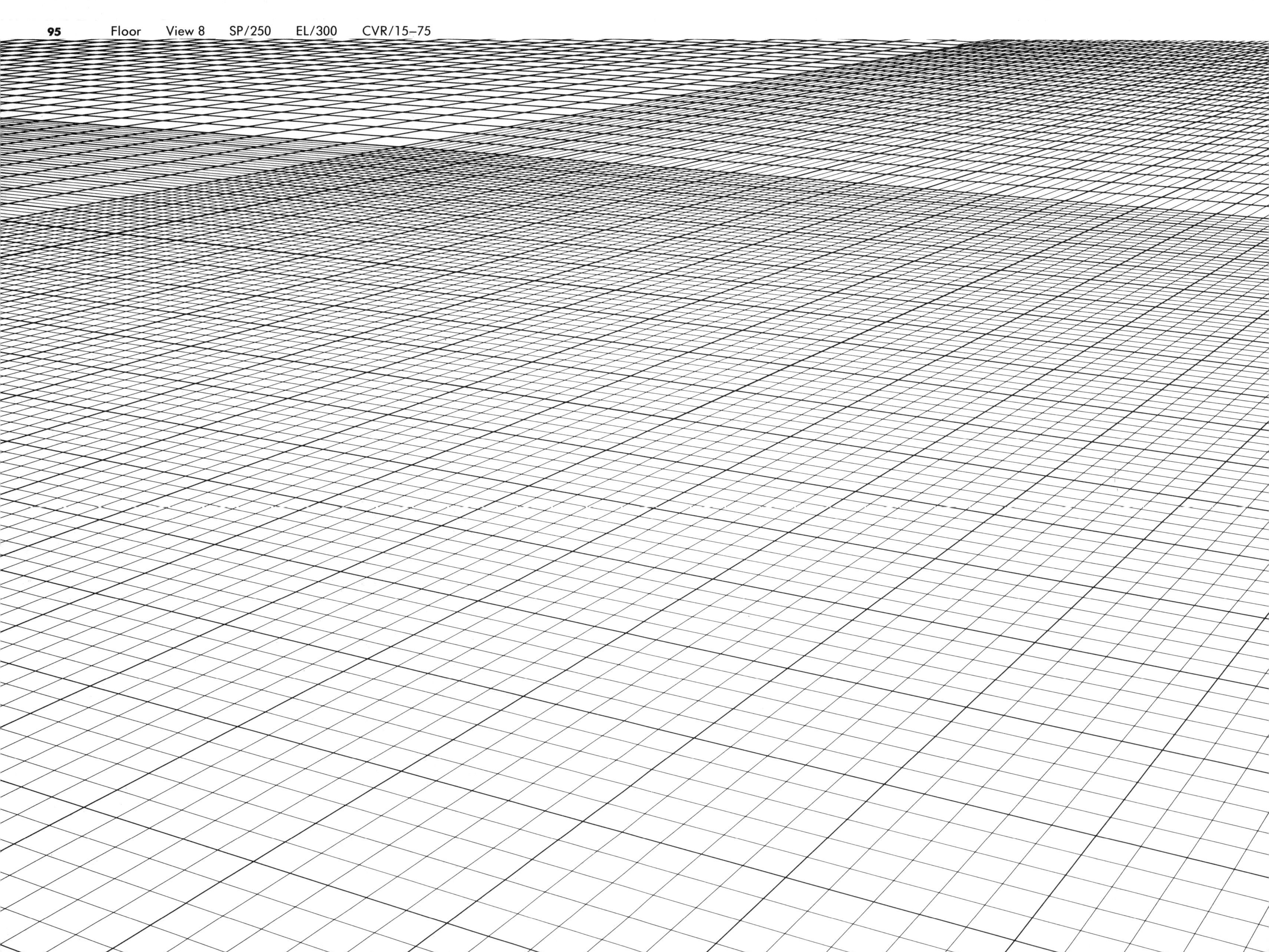

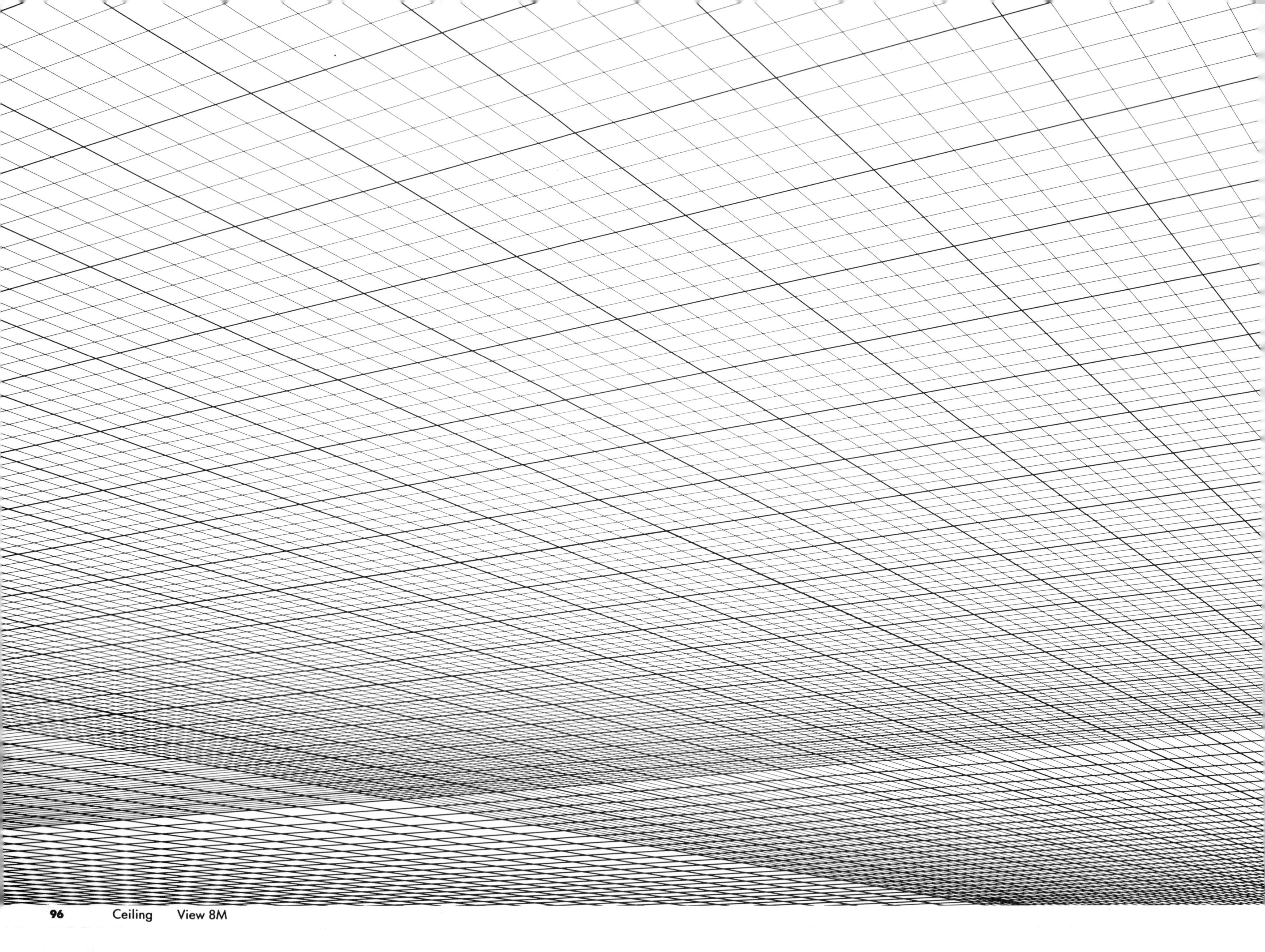

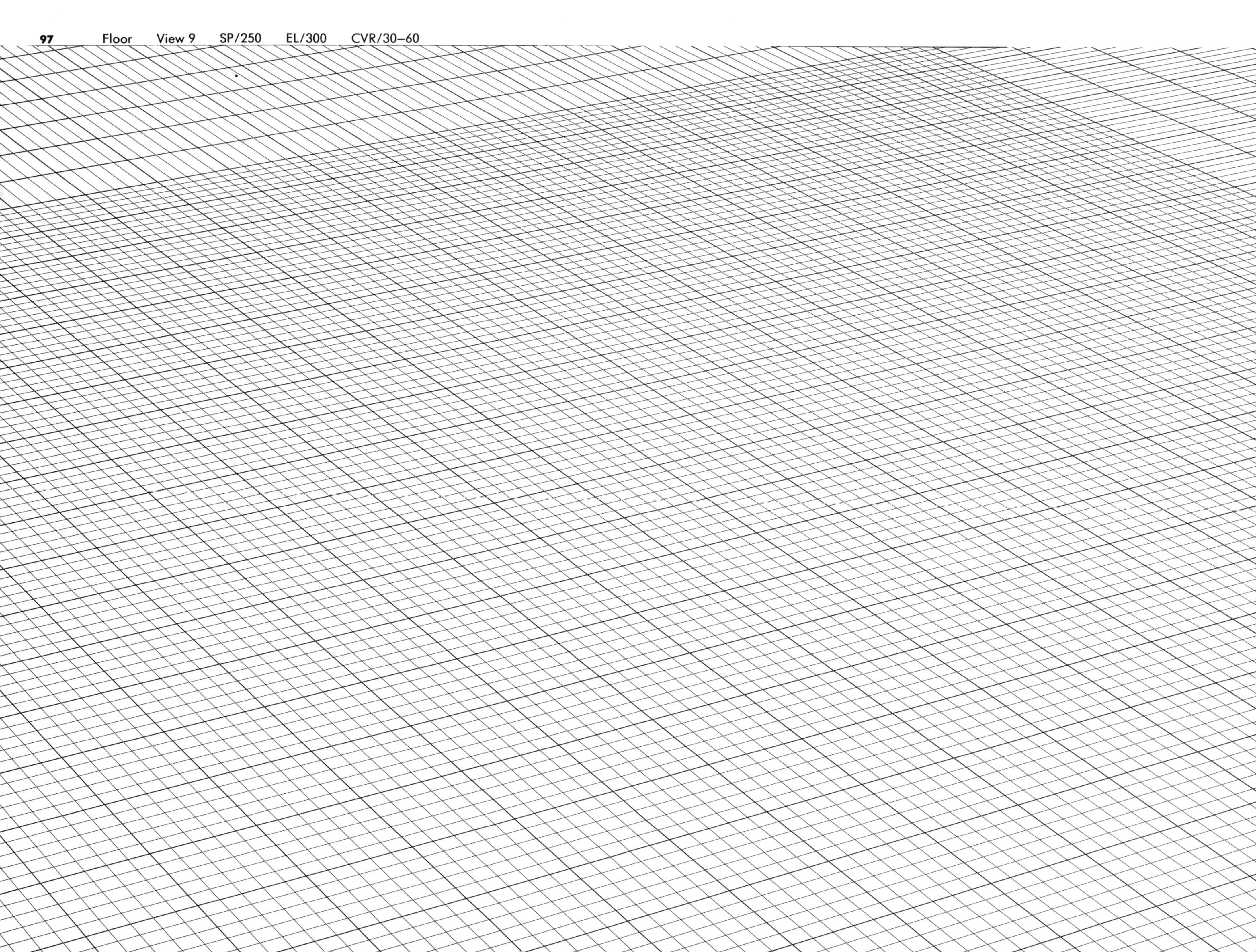

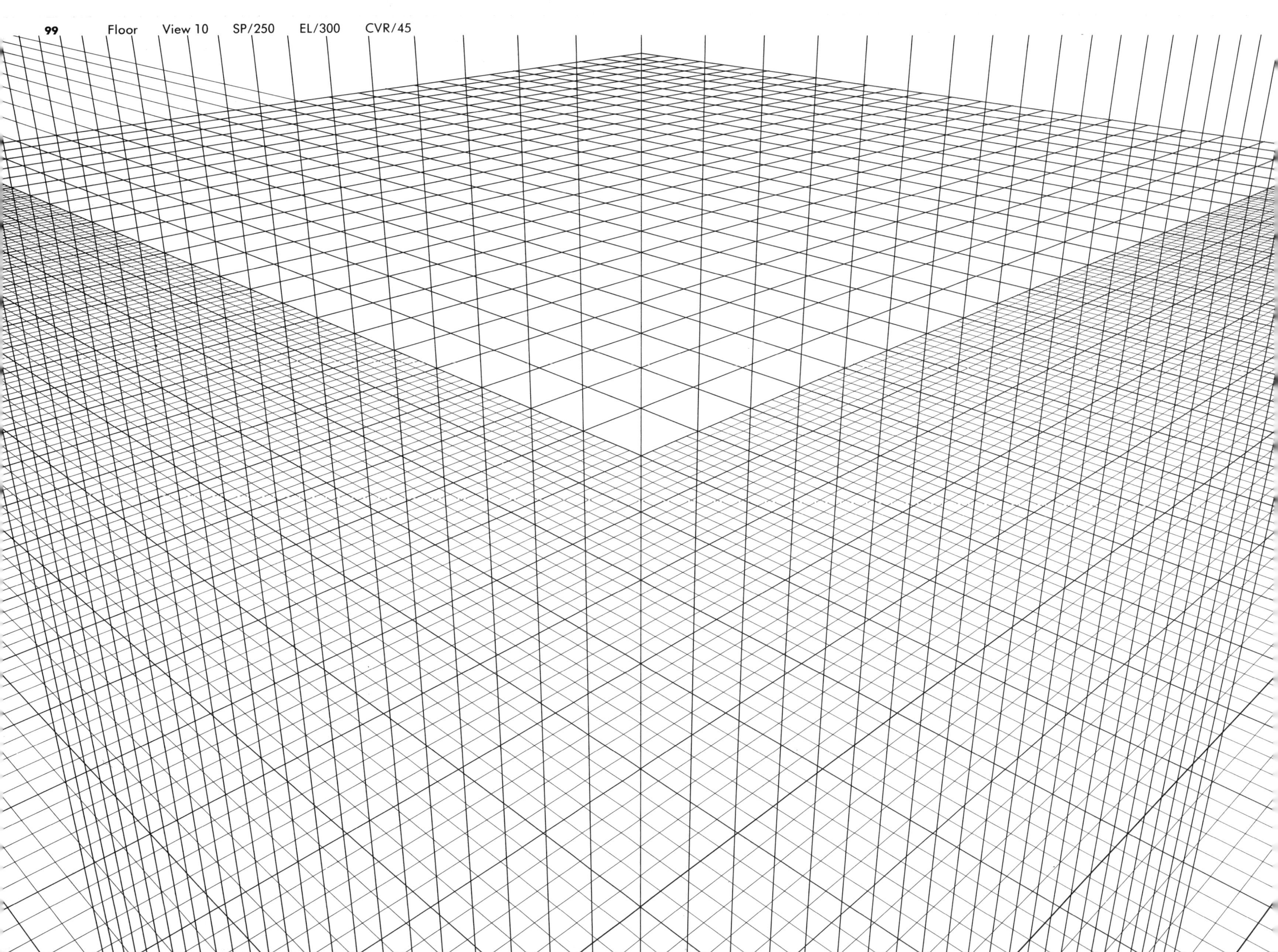

**PART THREE**

# EXTERIOR OBJECT

## Ground and Ceiling Plane

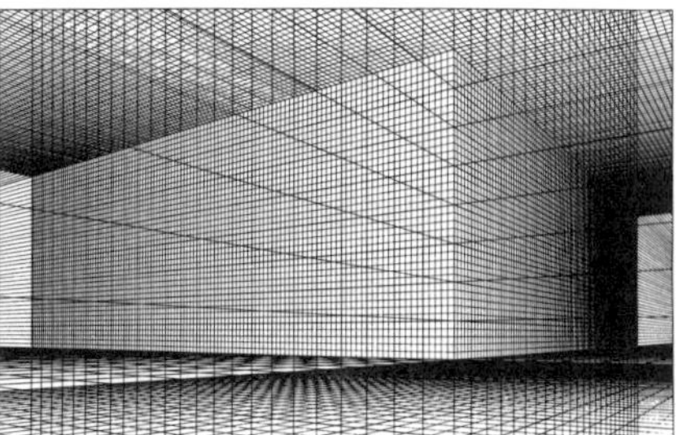

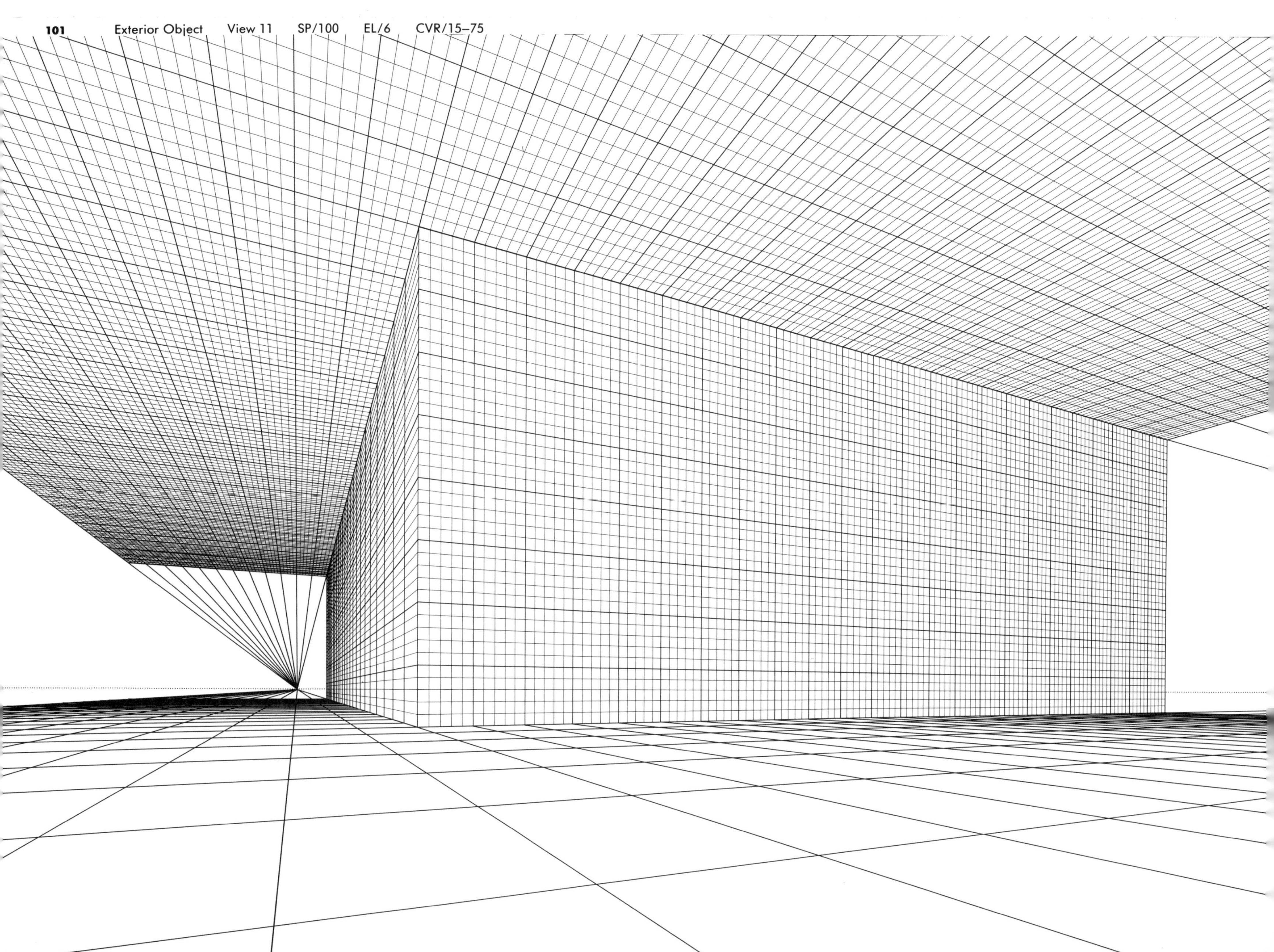

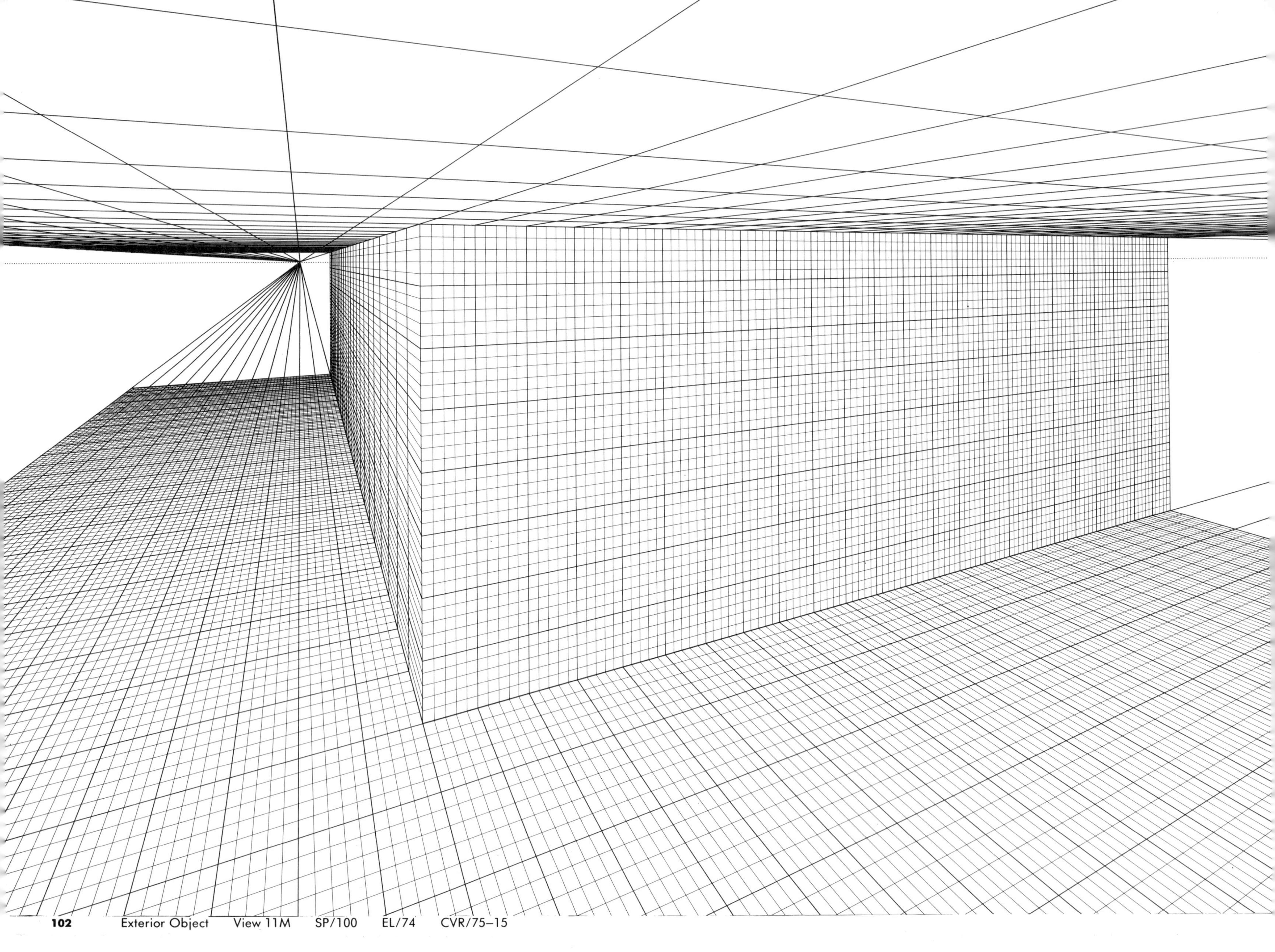

 Exterior Object View 11M SP/100 EL/74 CVR/75–15

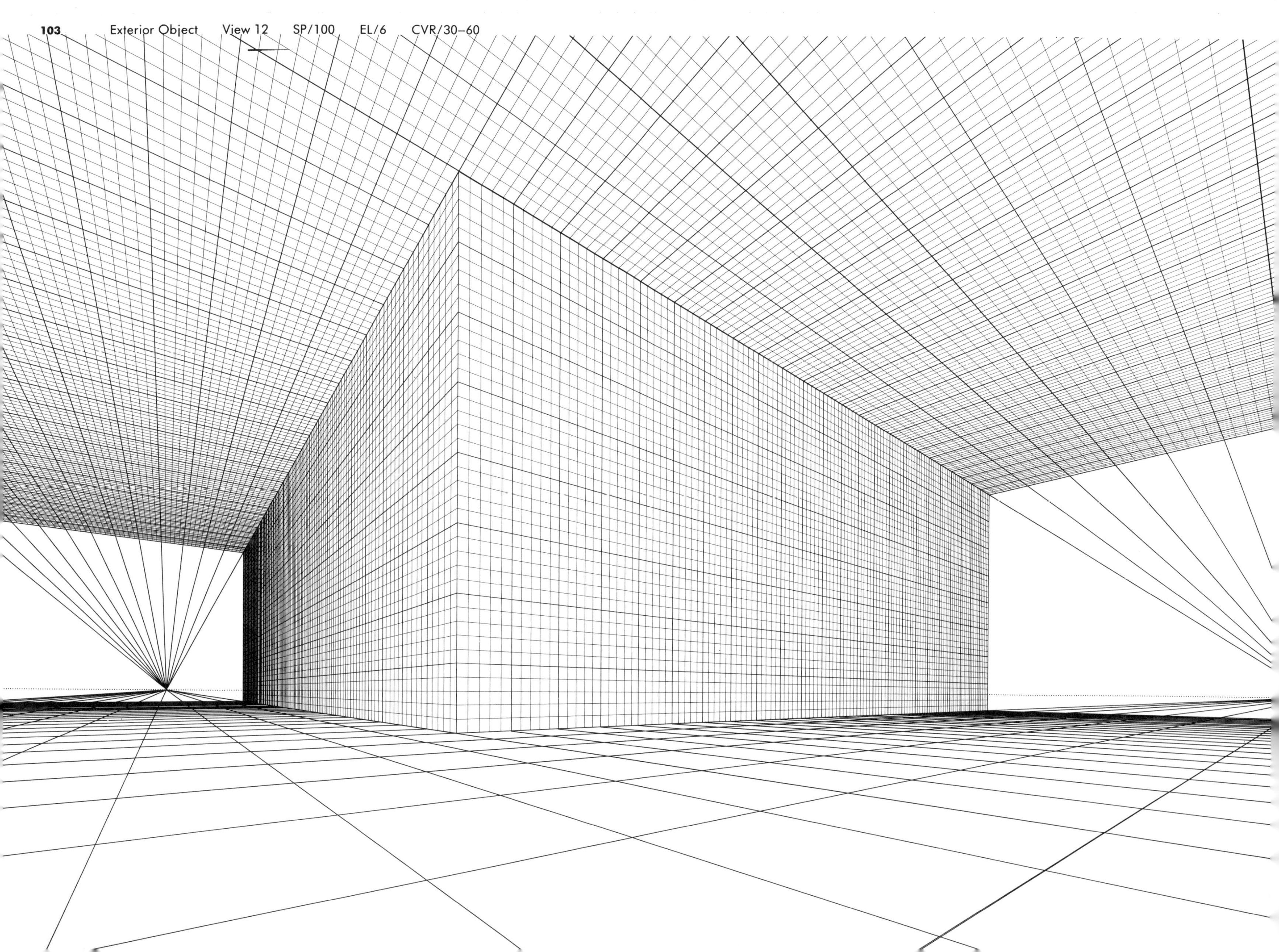

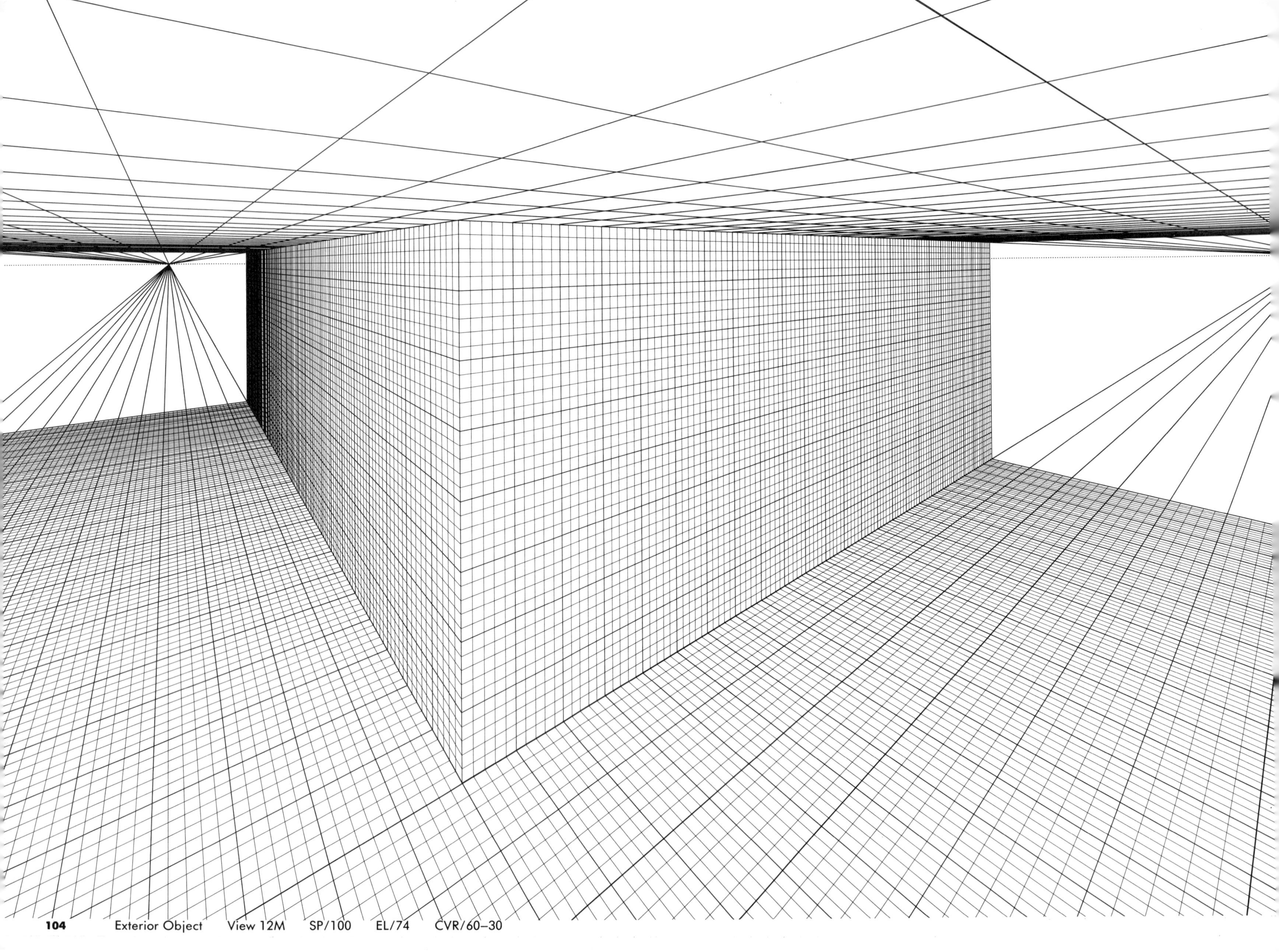

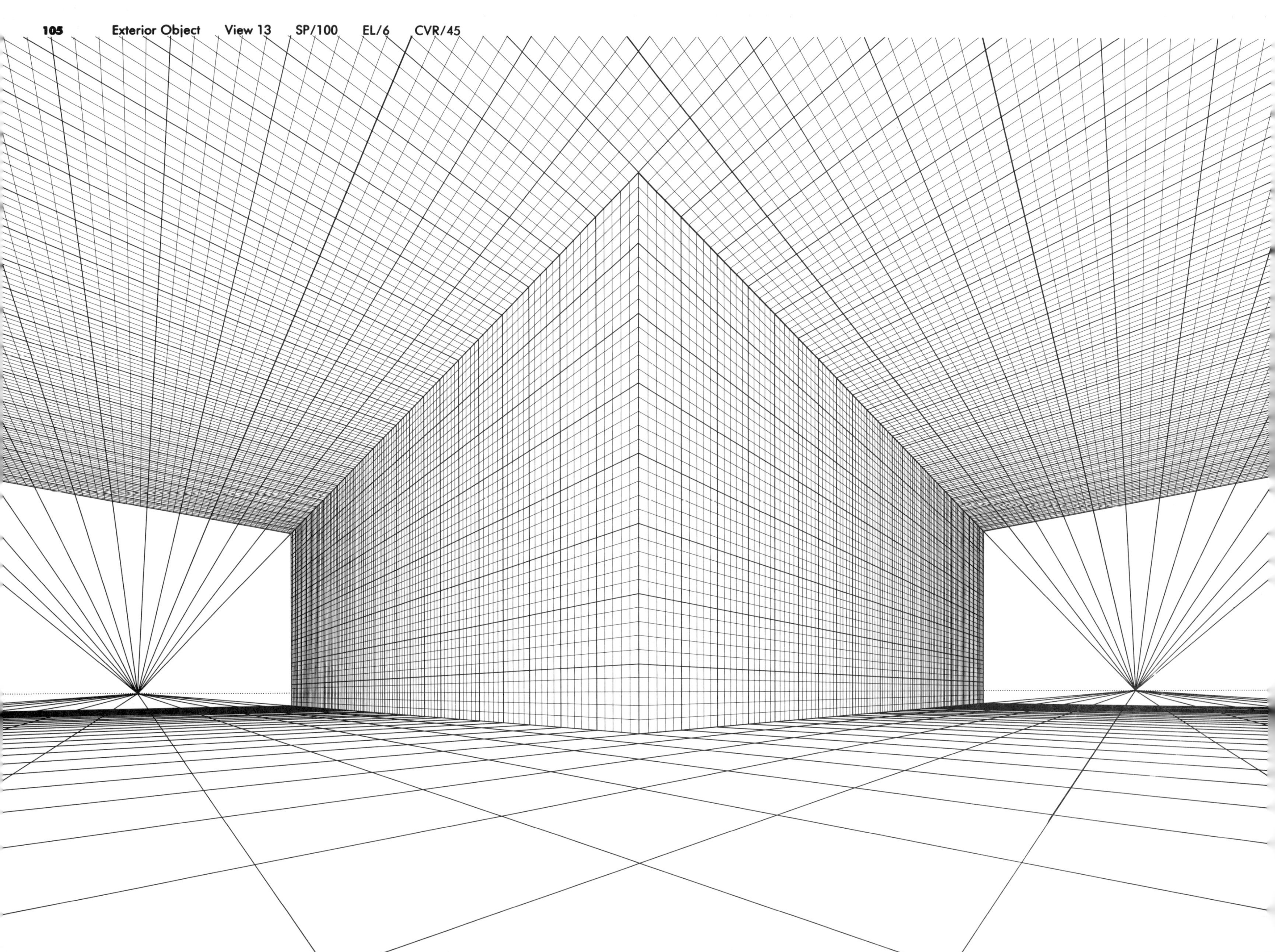

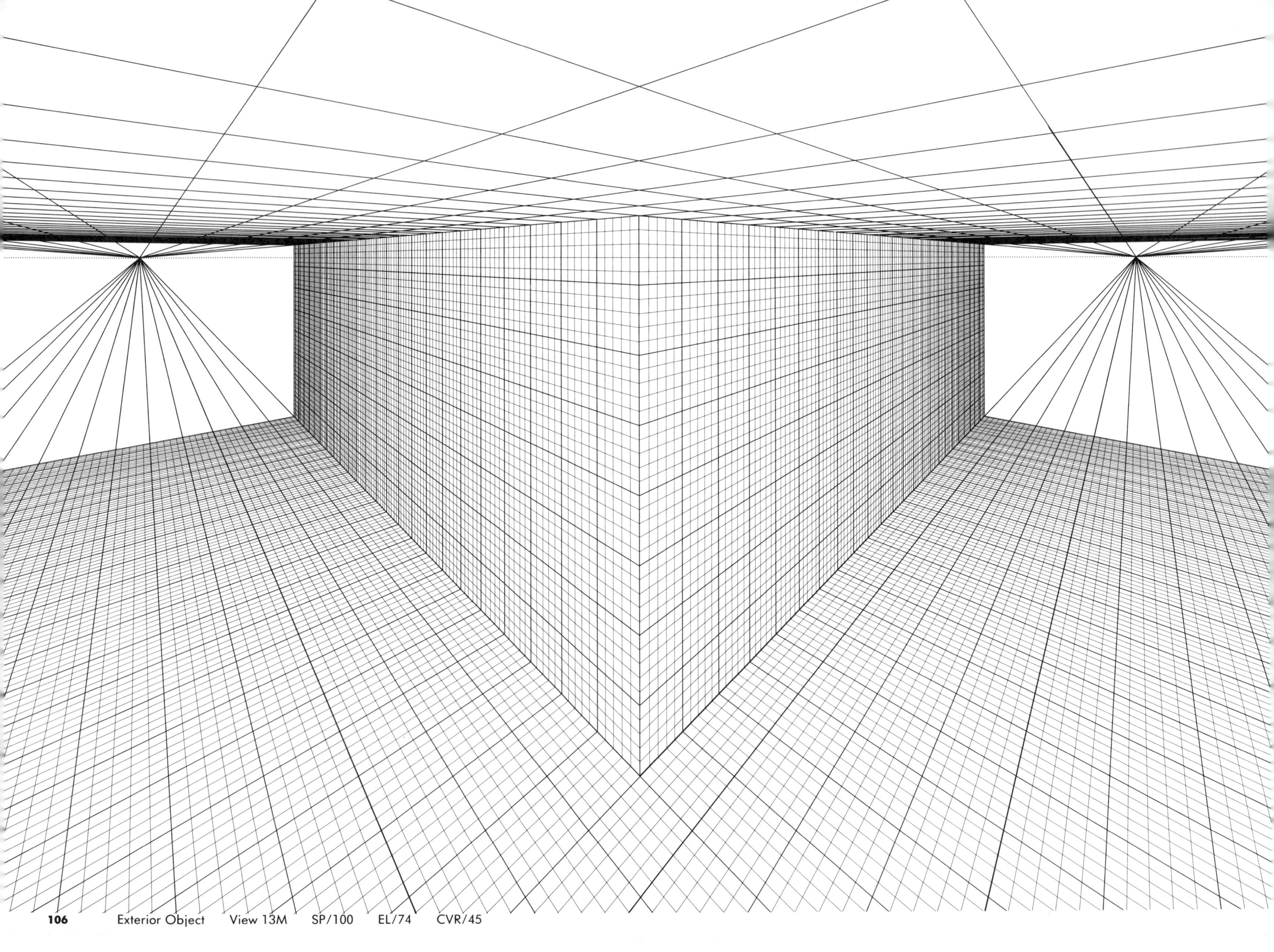

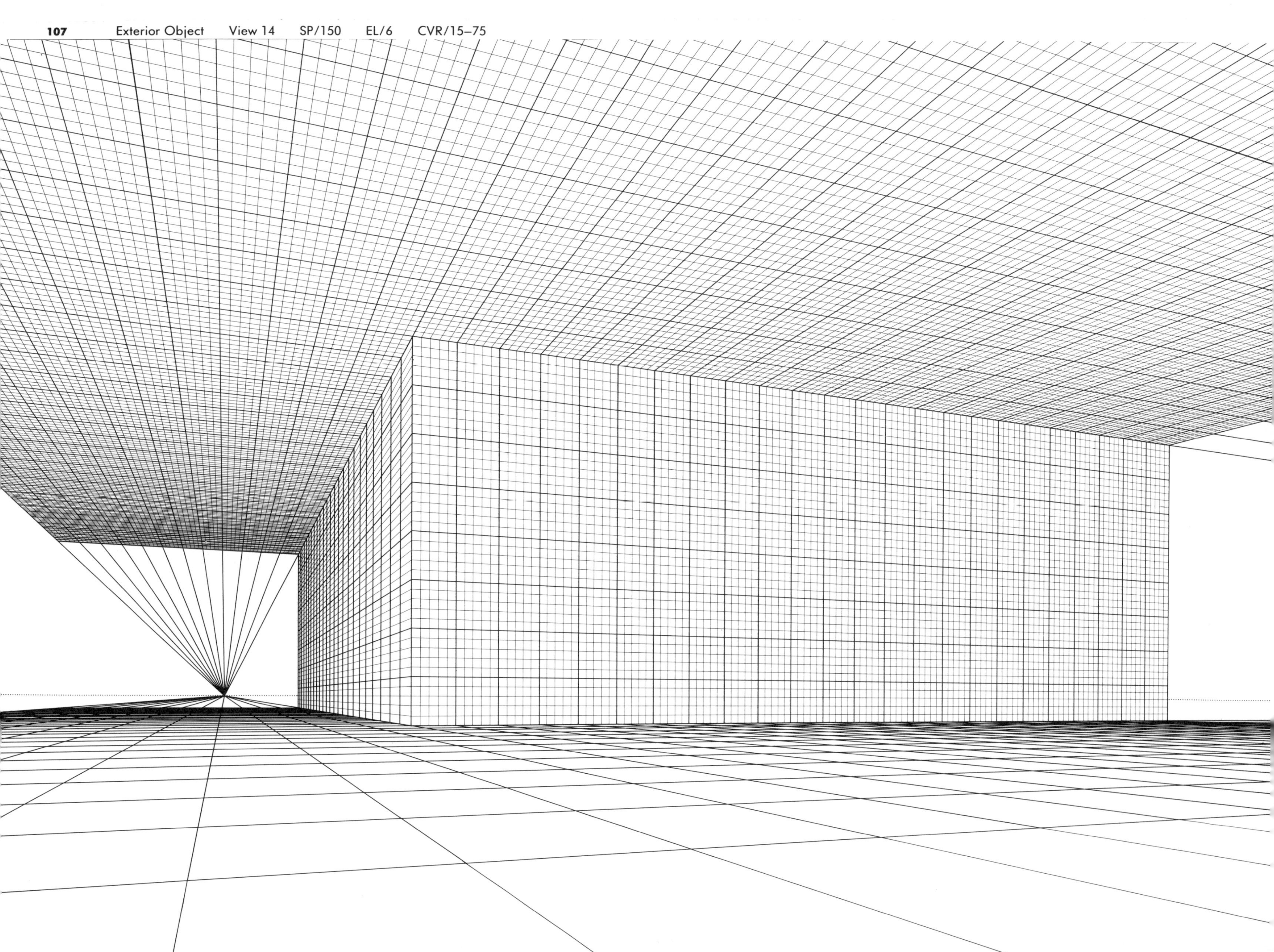

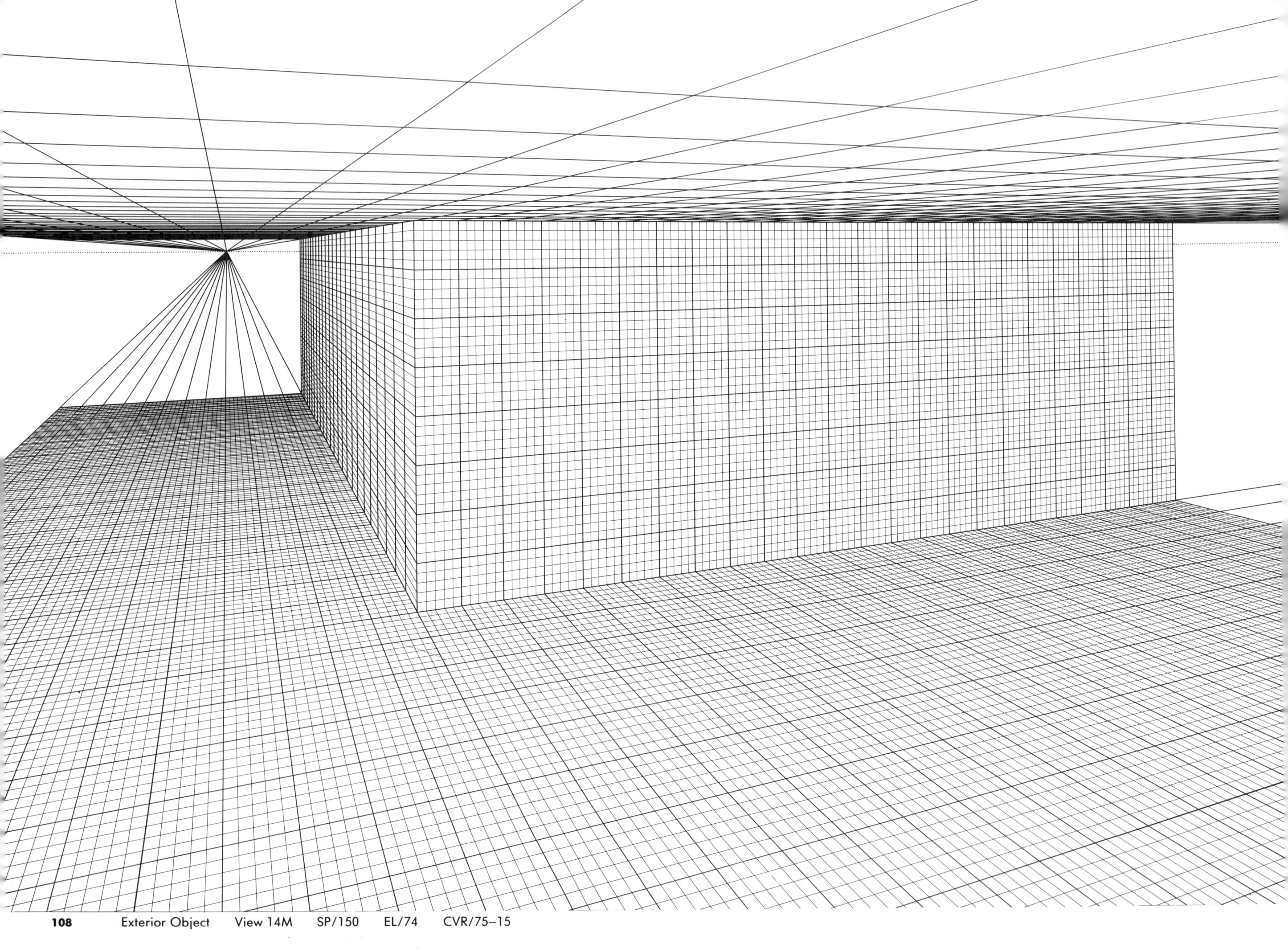



Exterior Object View 14M SP/150 EL/74 CVR/75–15

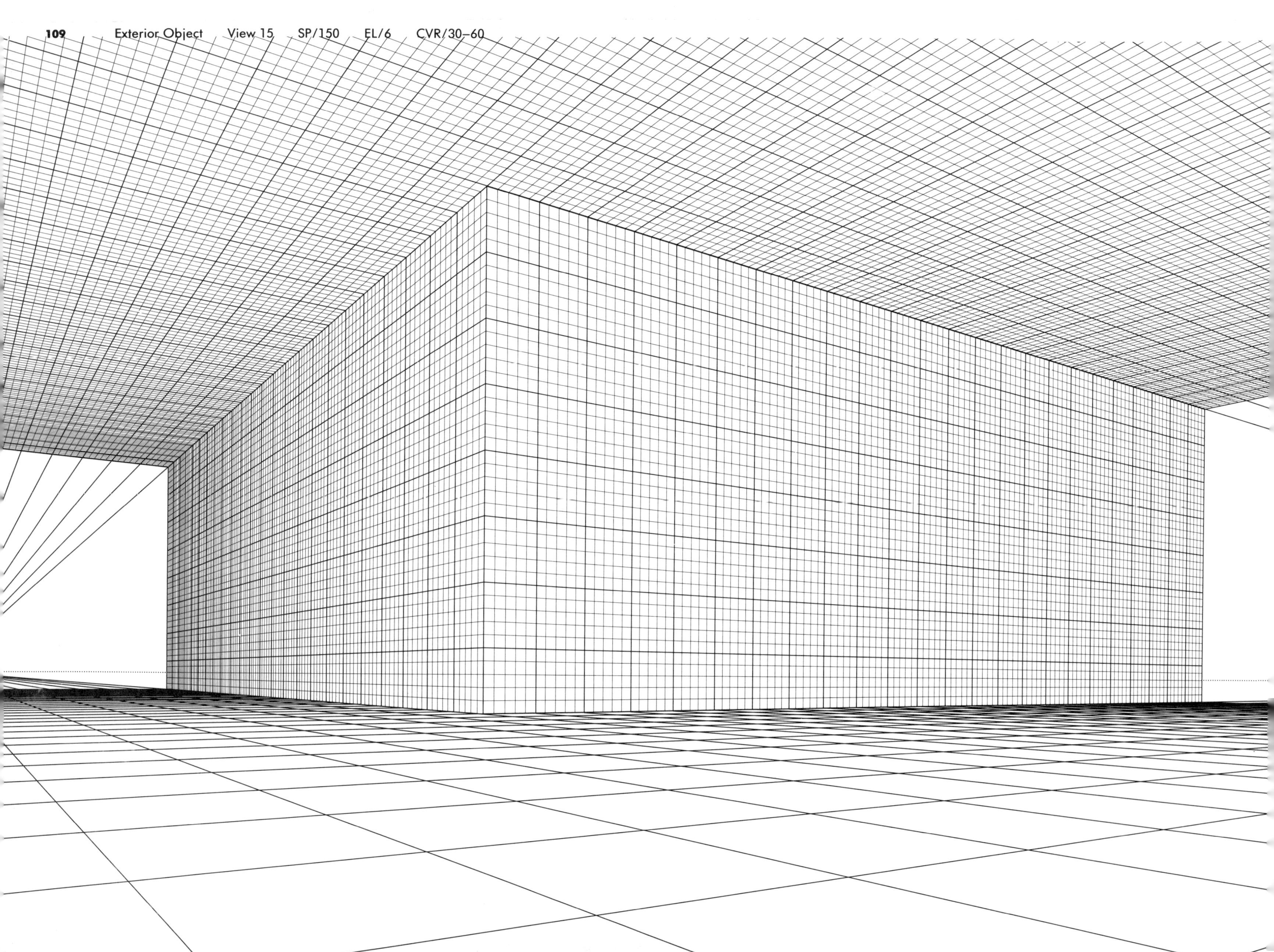

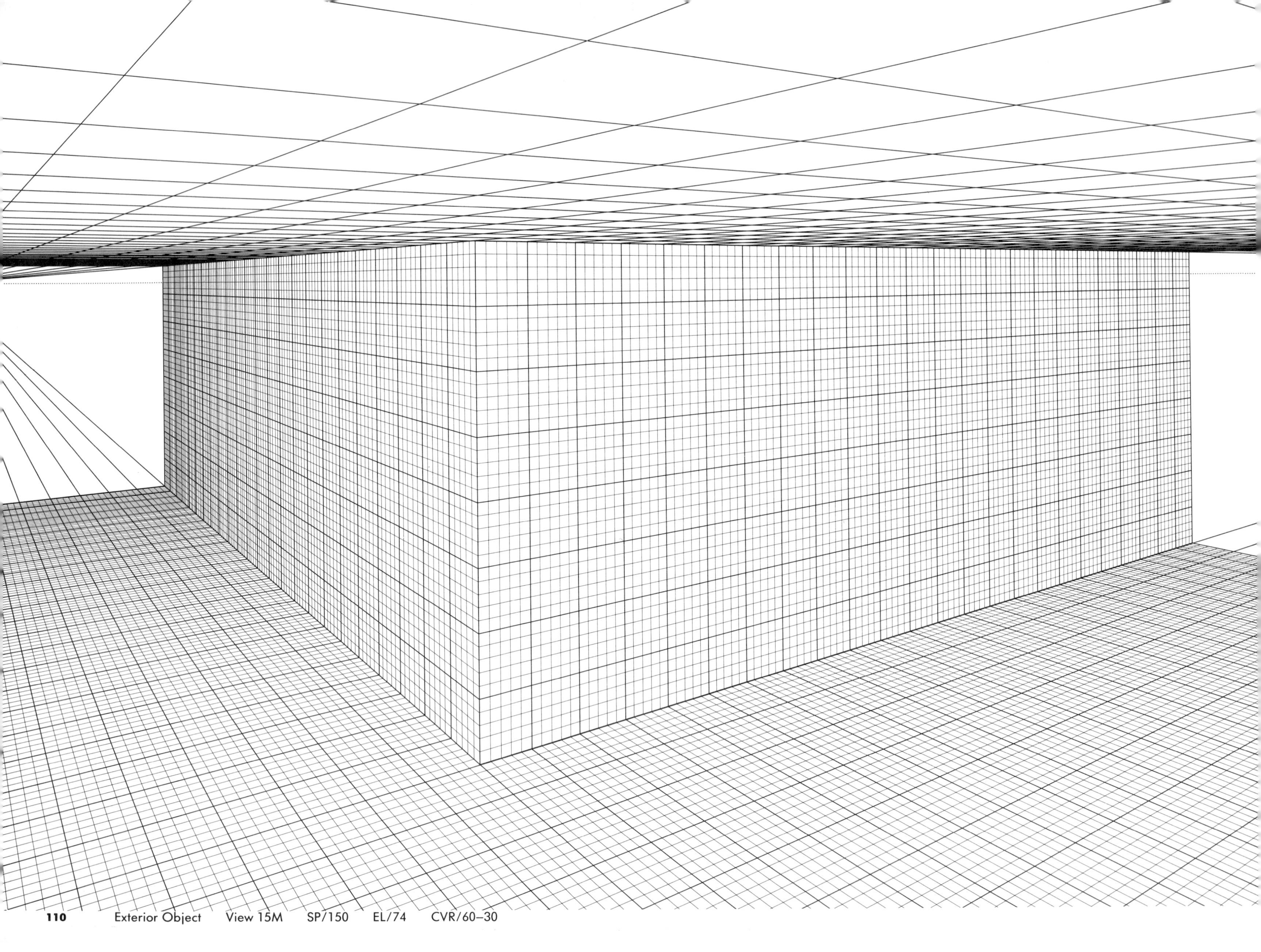

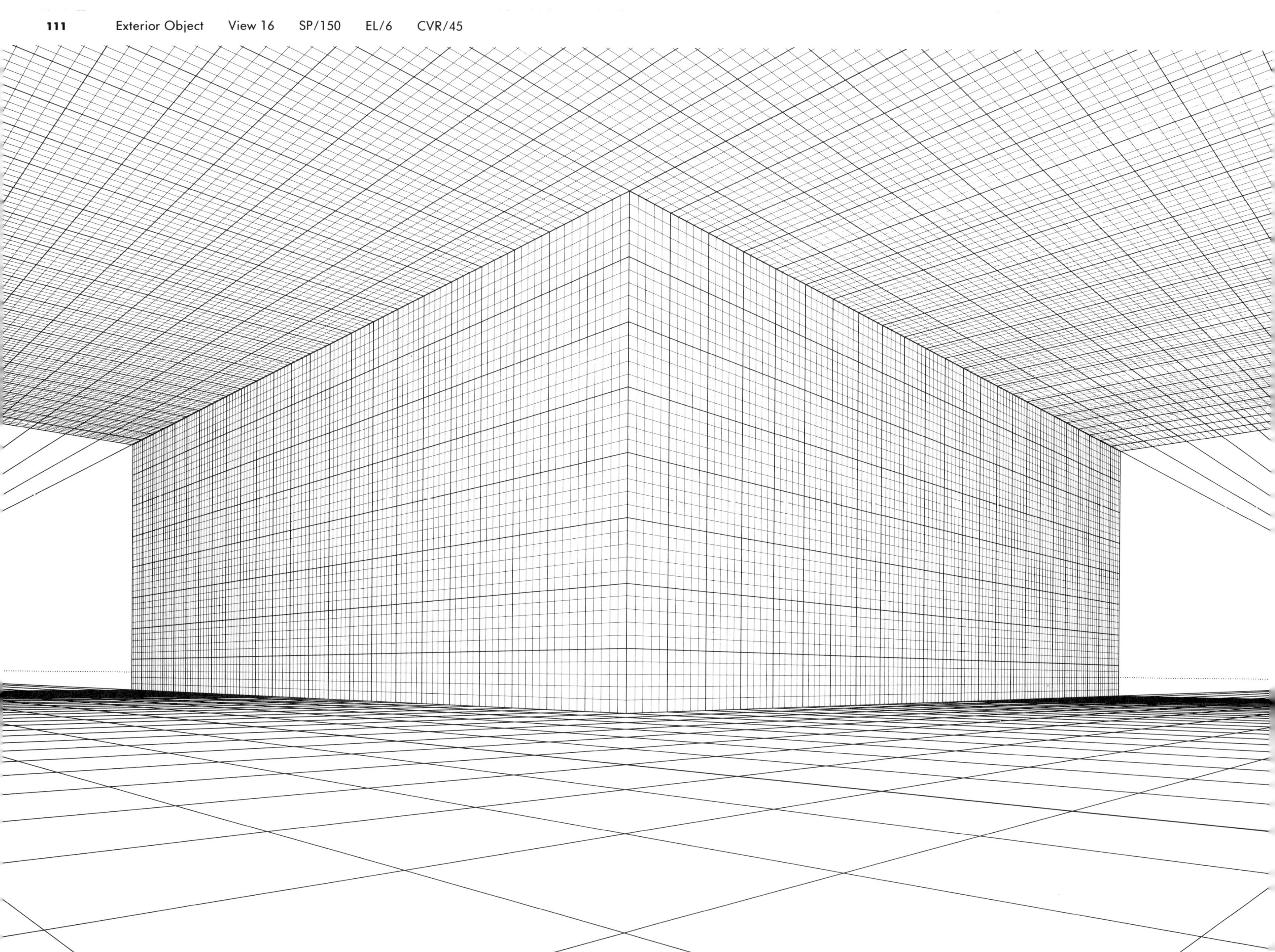

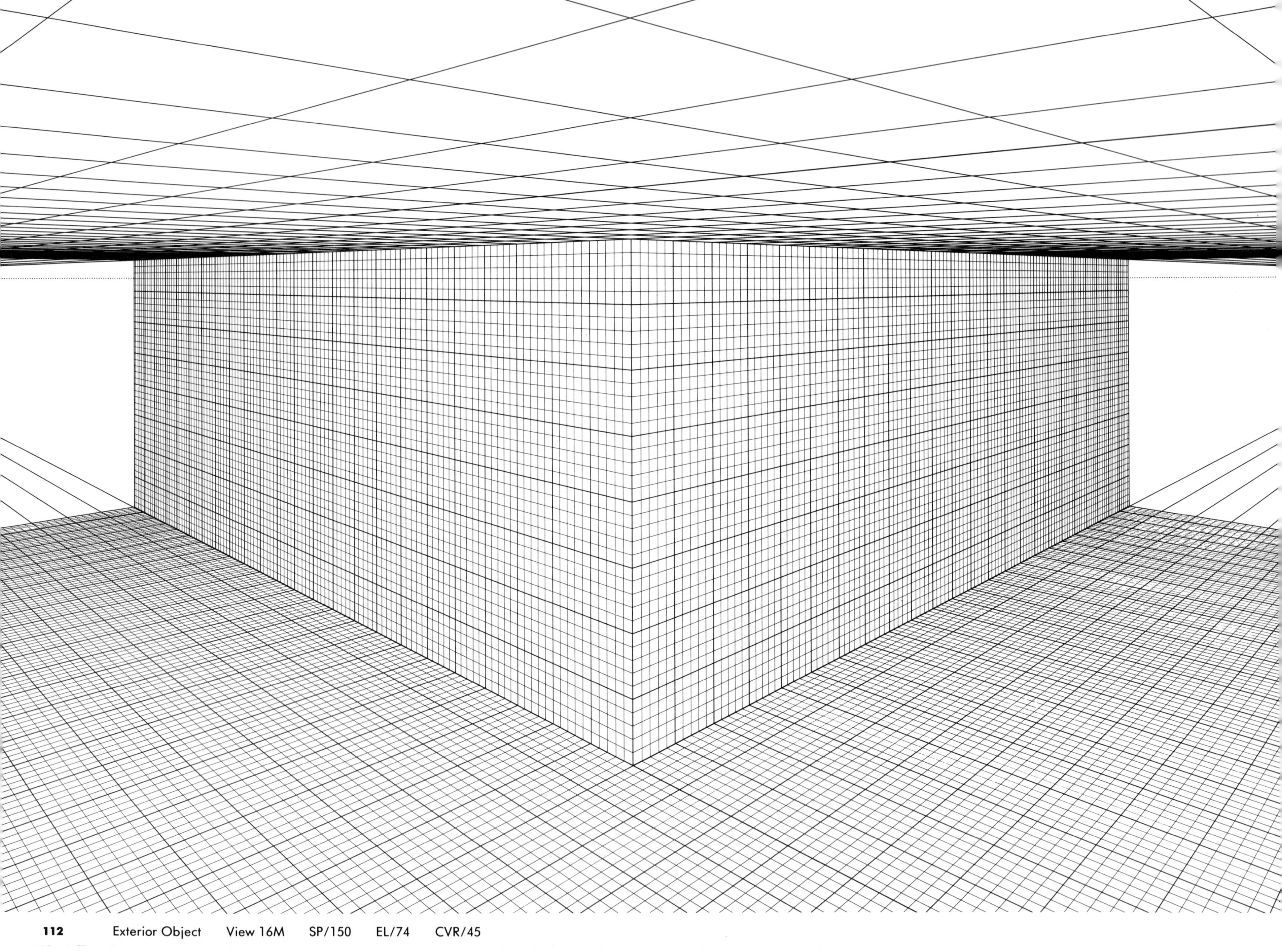

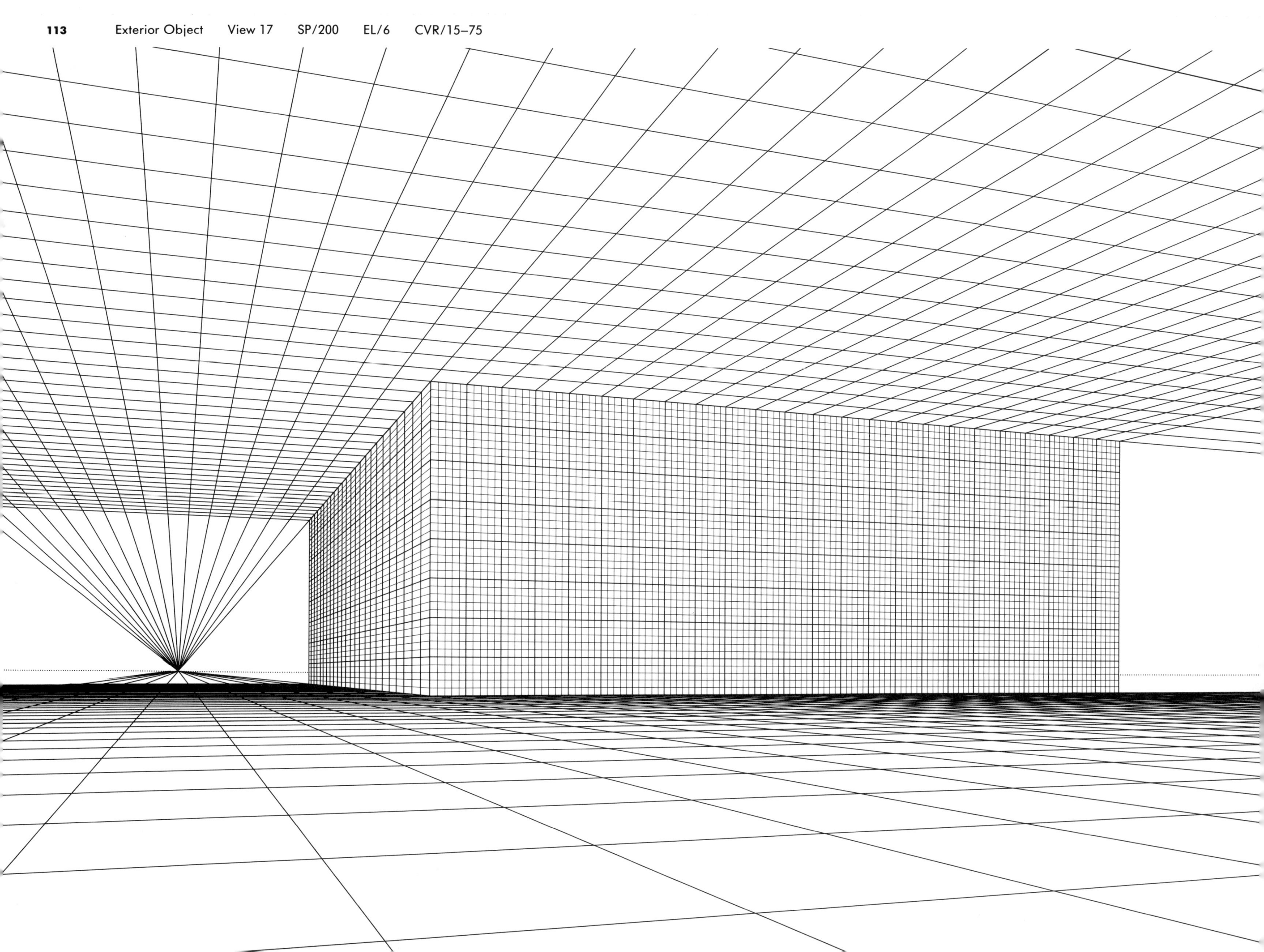

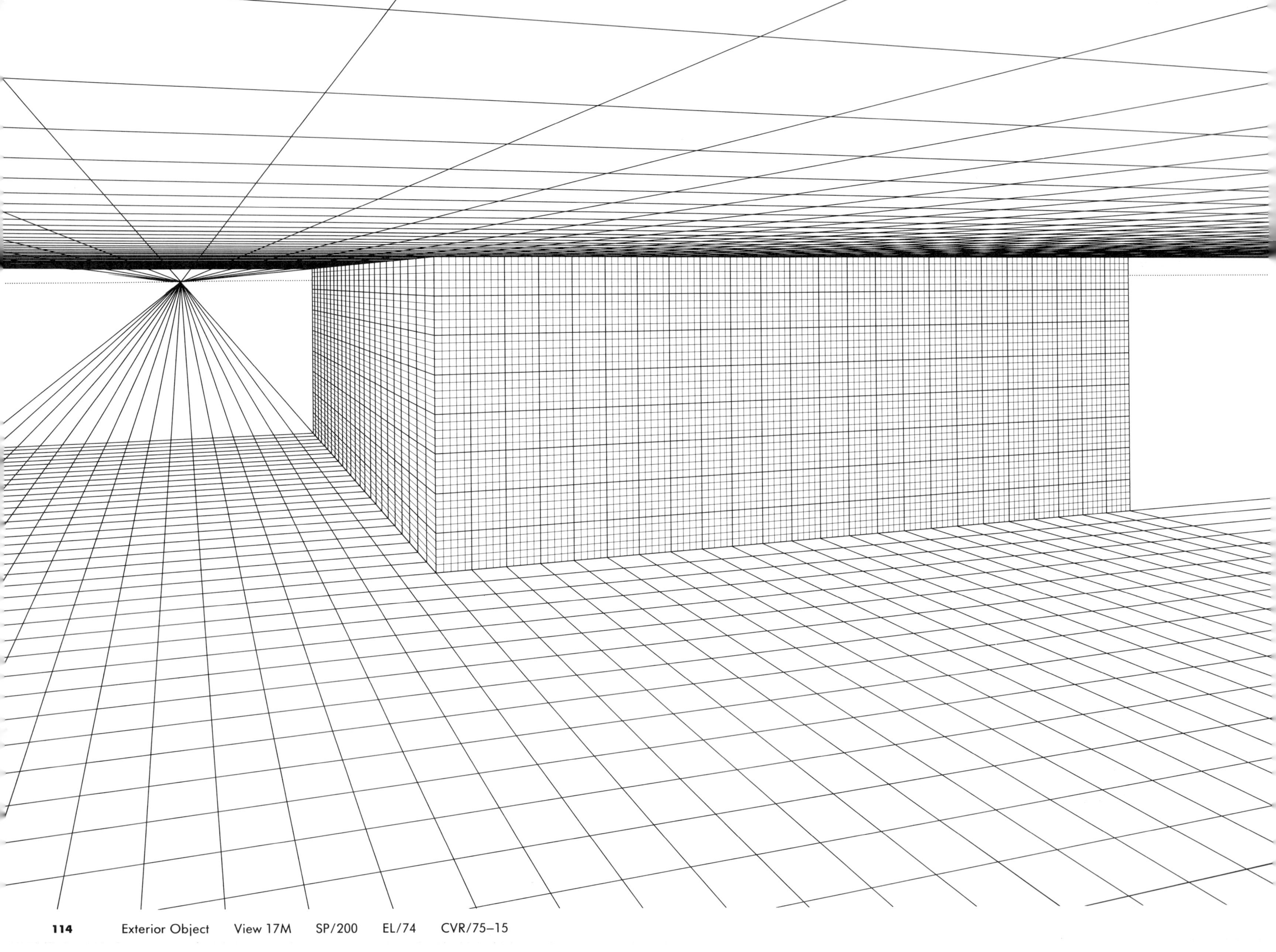

 Exterior Object View 17M SP/200 EL/74 CVR/75–15

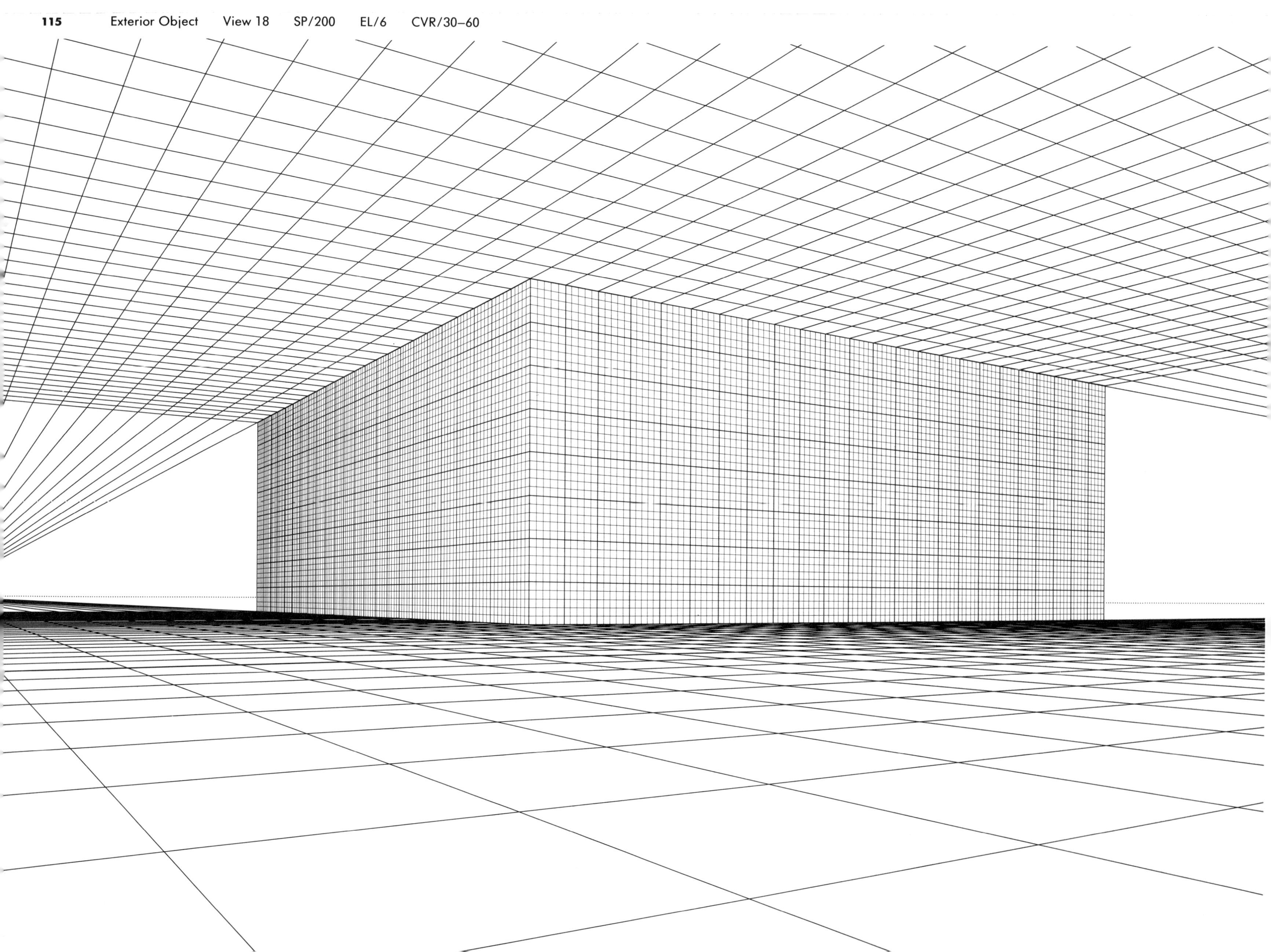

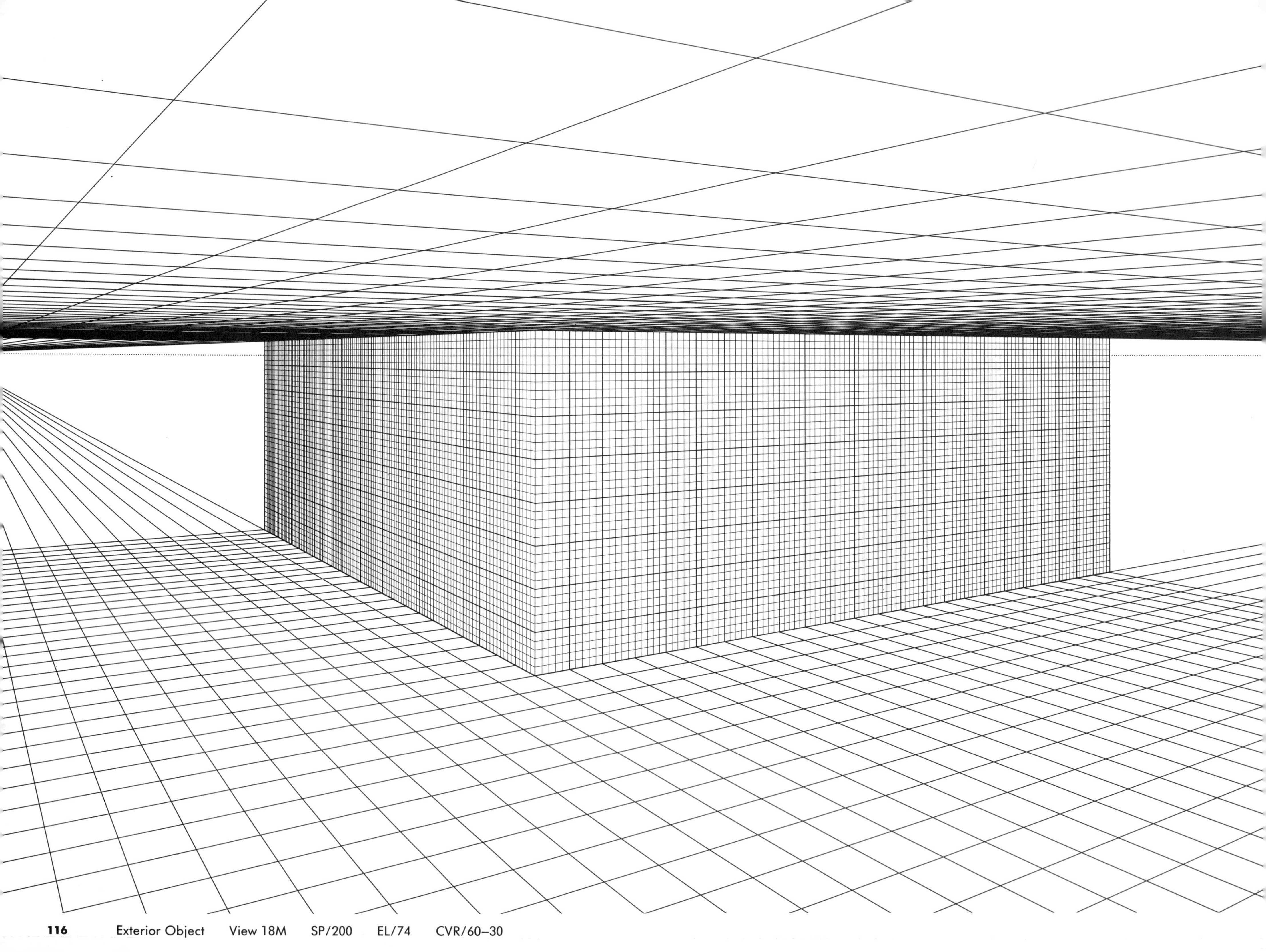

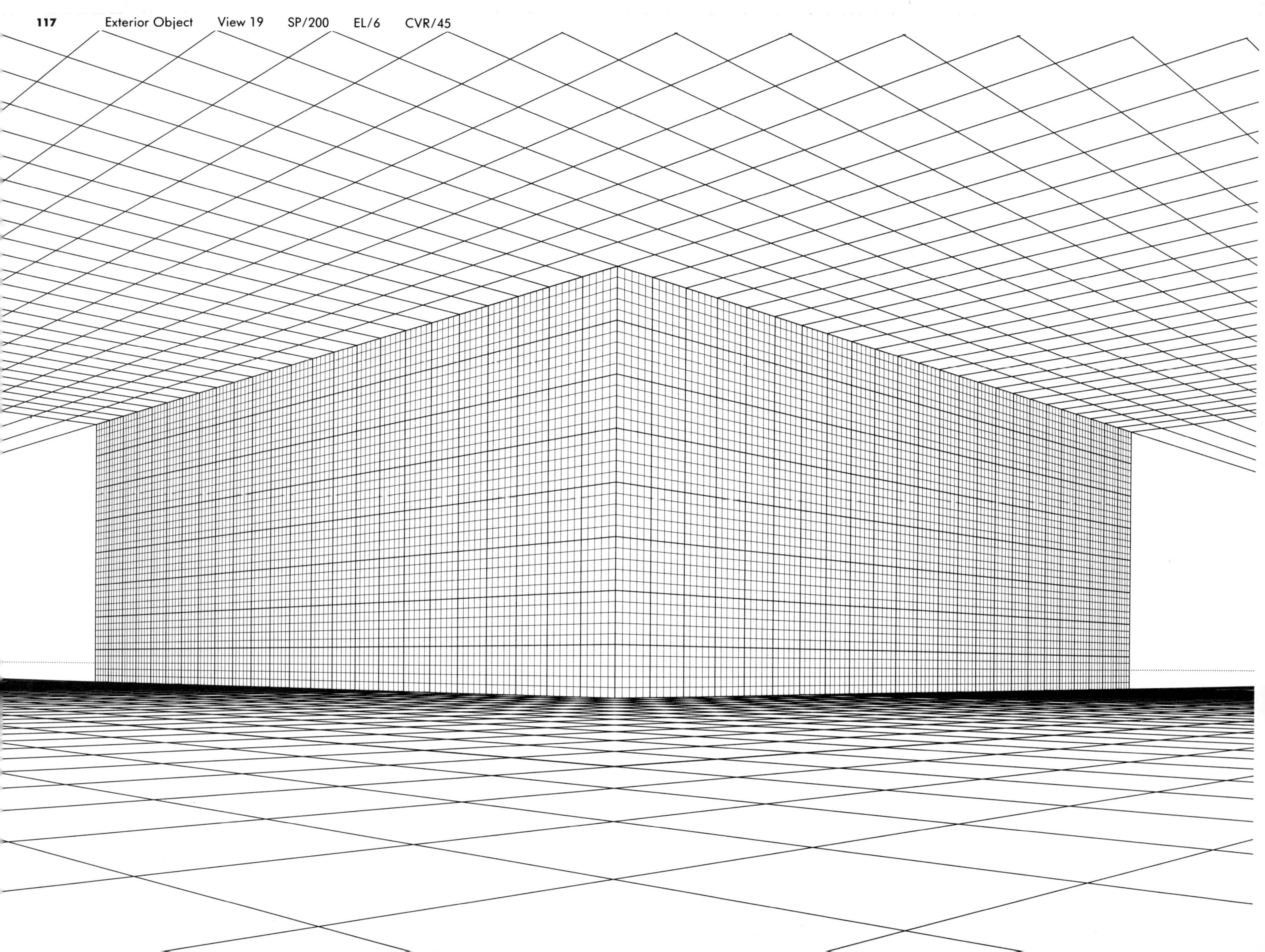

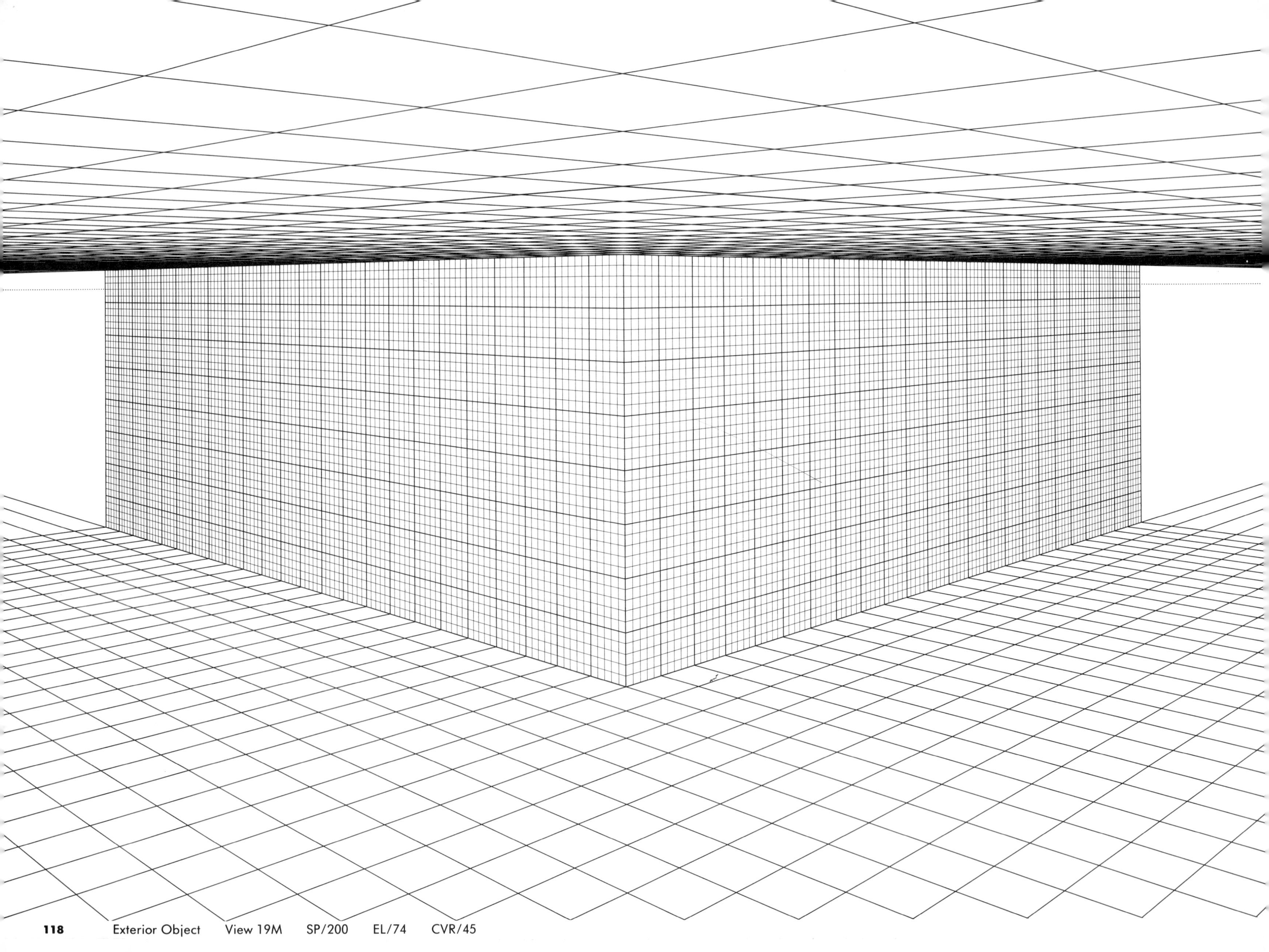

 Exterior Object View 19M SP/200 EL/74 CVR/45

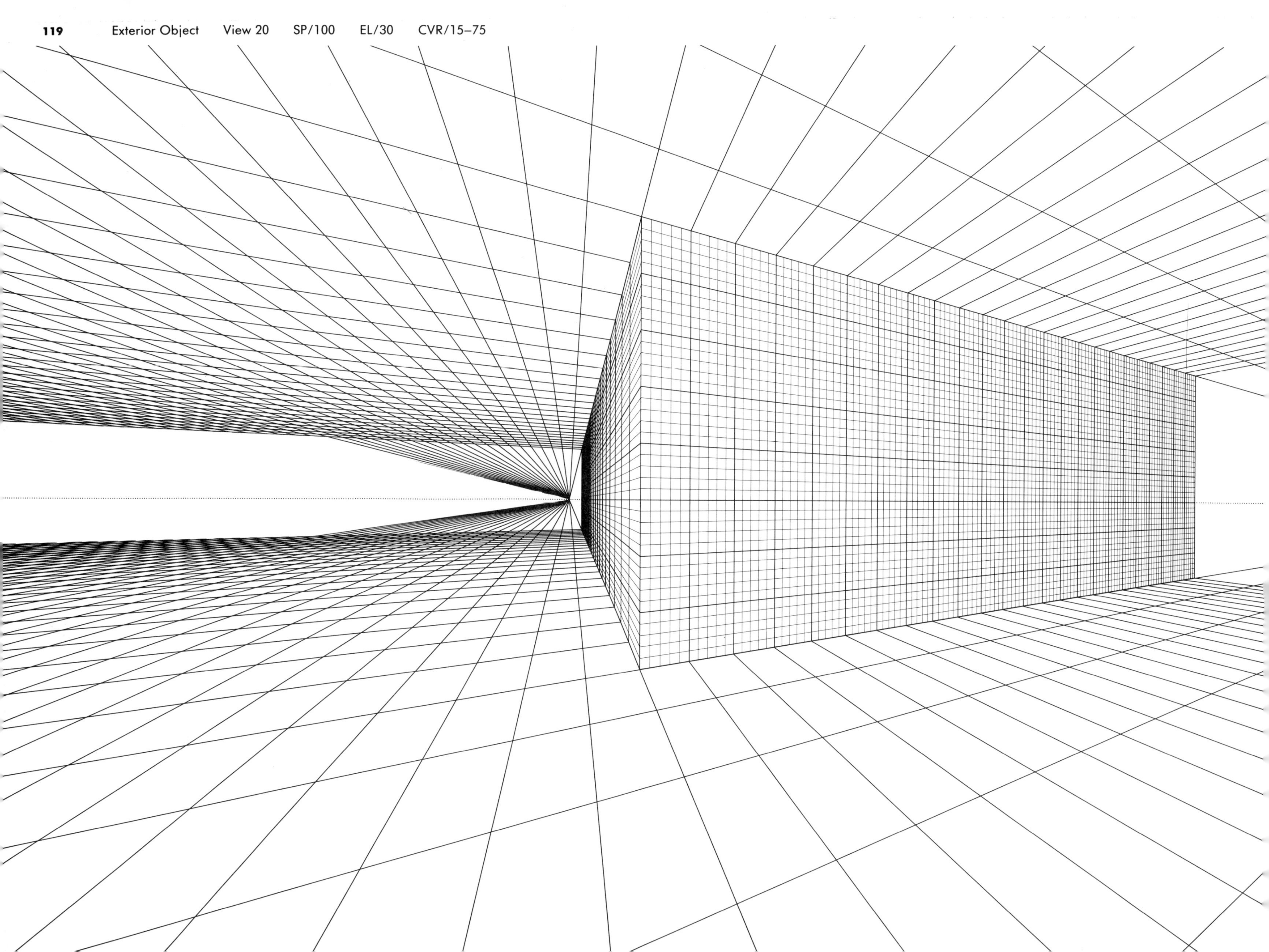

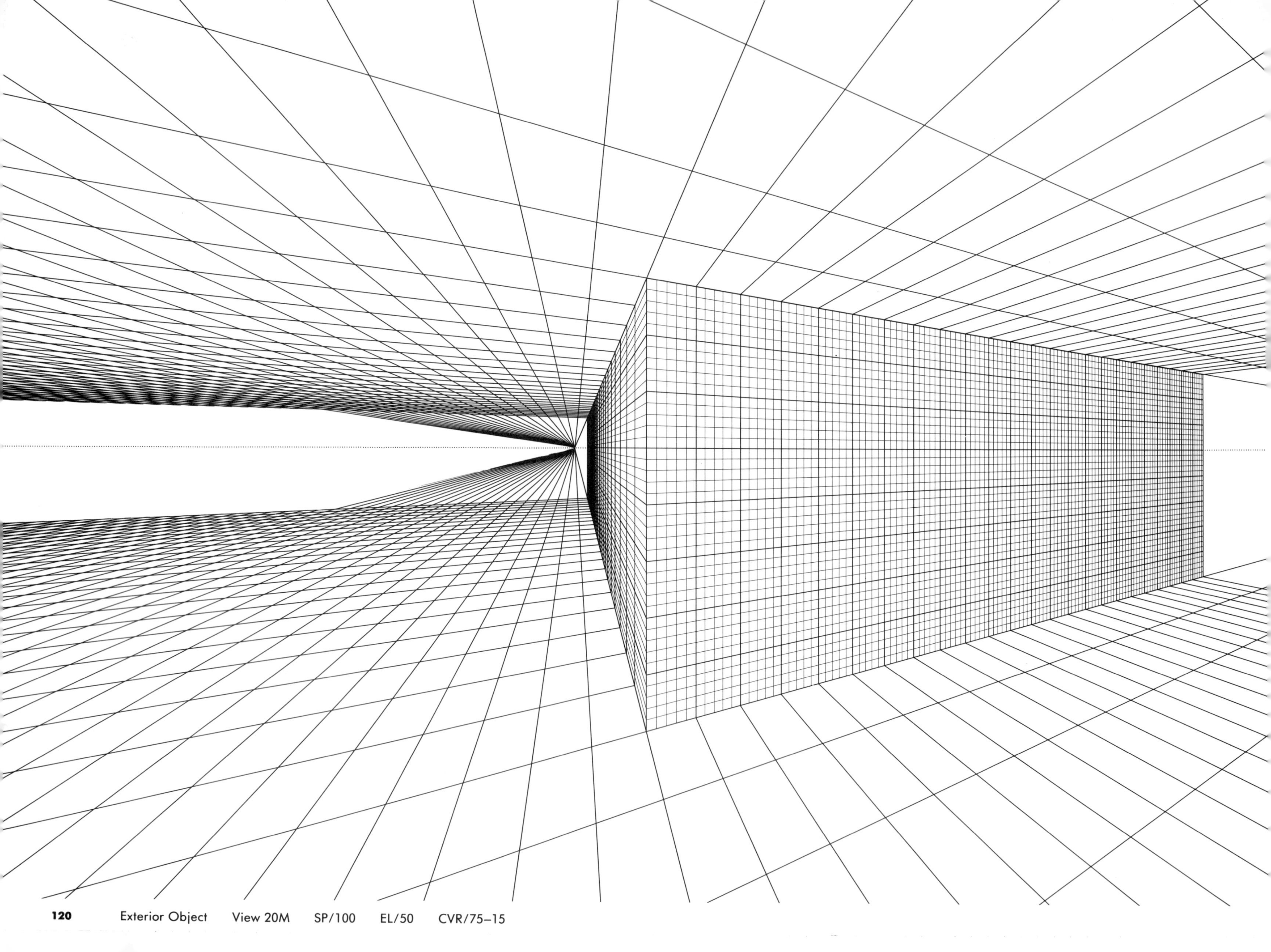



Exterior Object View 20M SP/100 EL/50 CVR/75–15

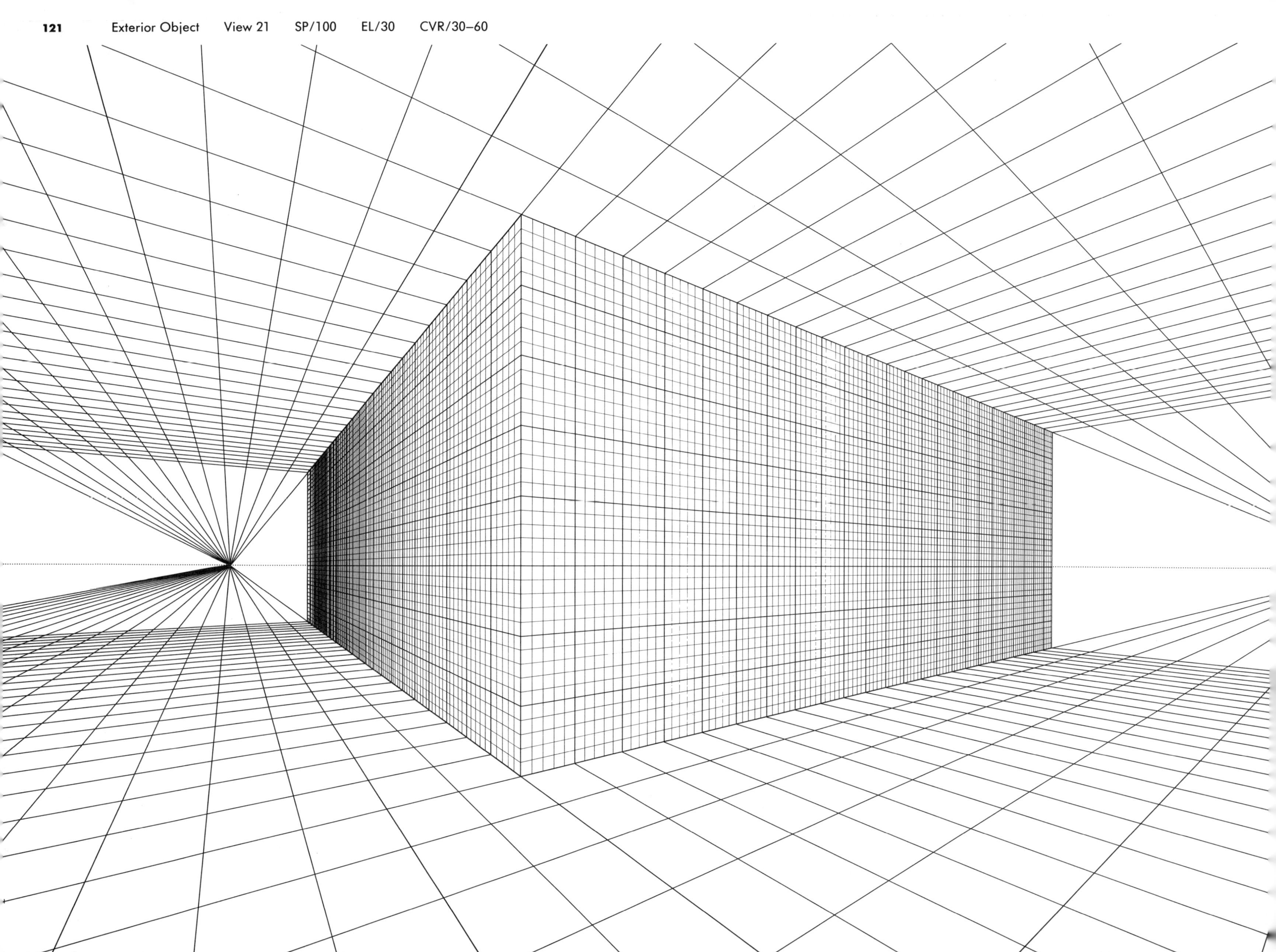

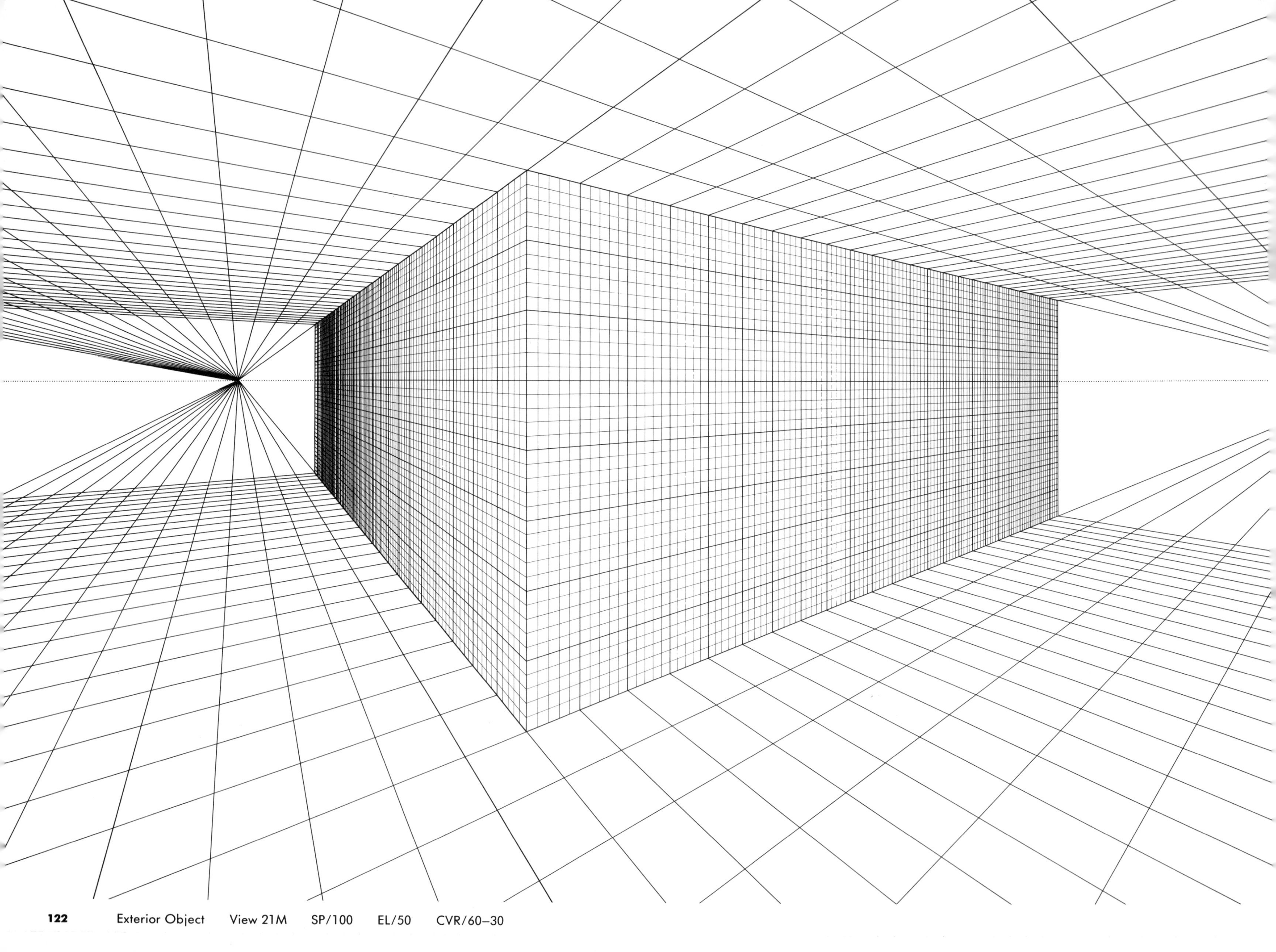

 Exterior Object View 21M SP/100 EL/50 CVR/60–30

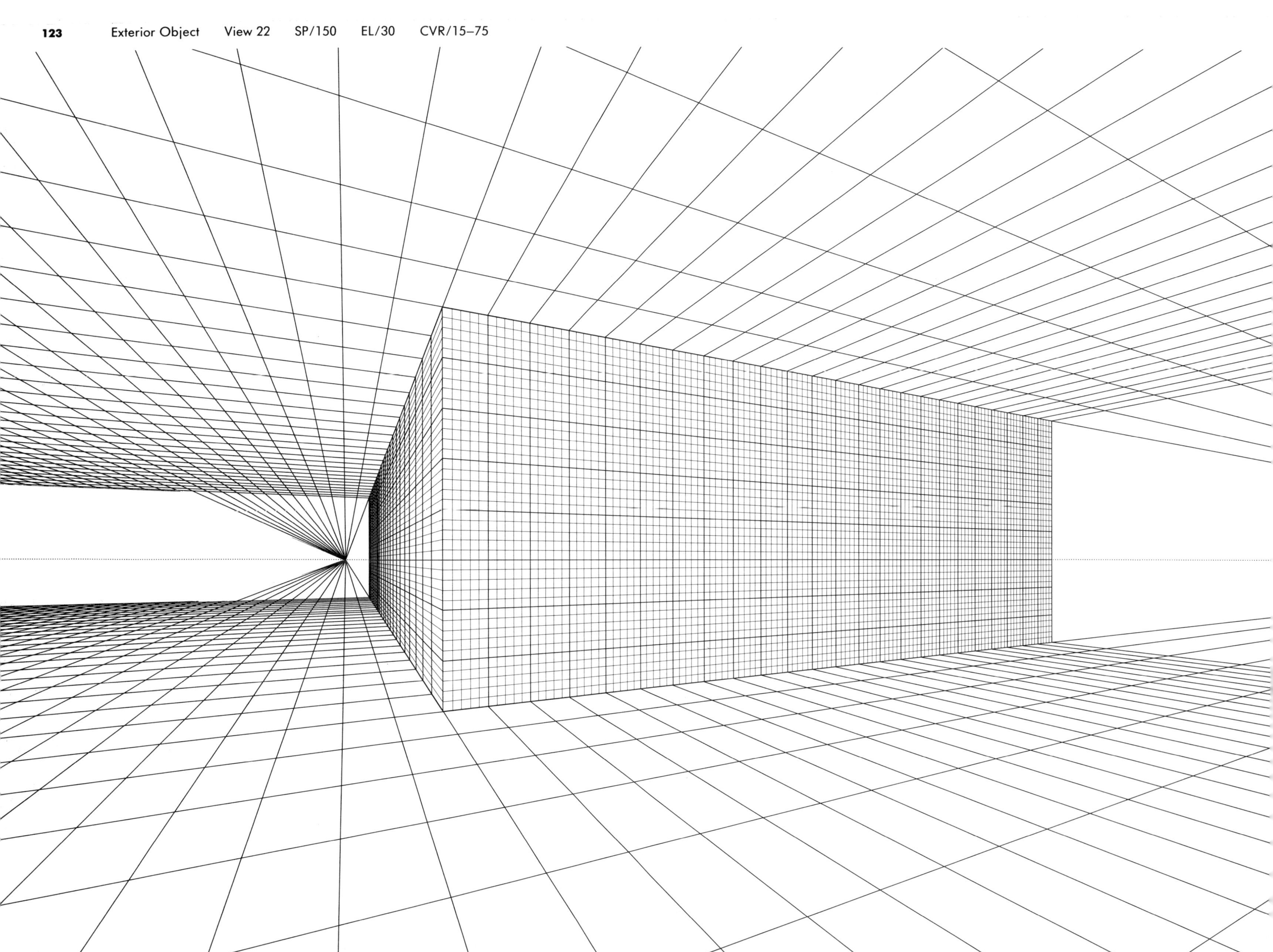

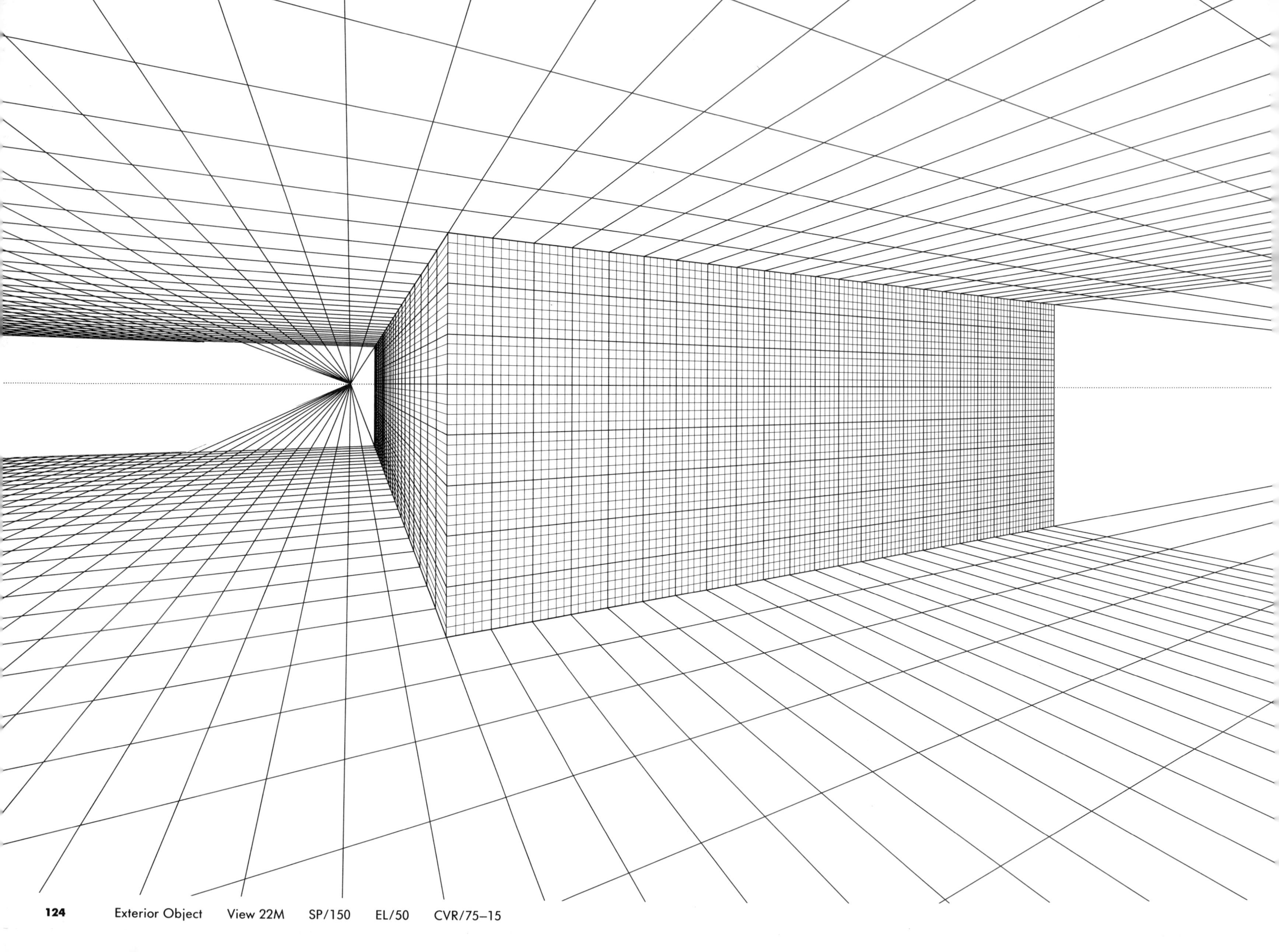

 Exterior Object View 22M SP/150 EL/50 CVR/75–15

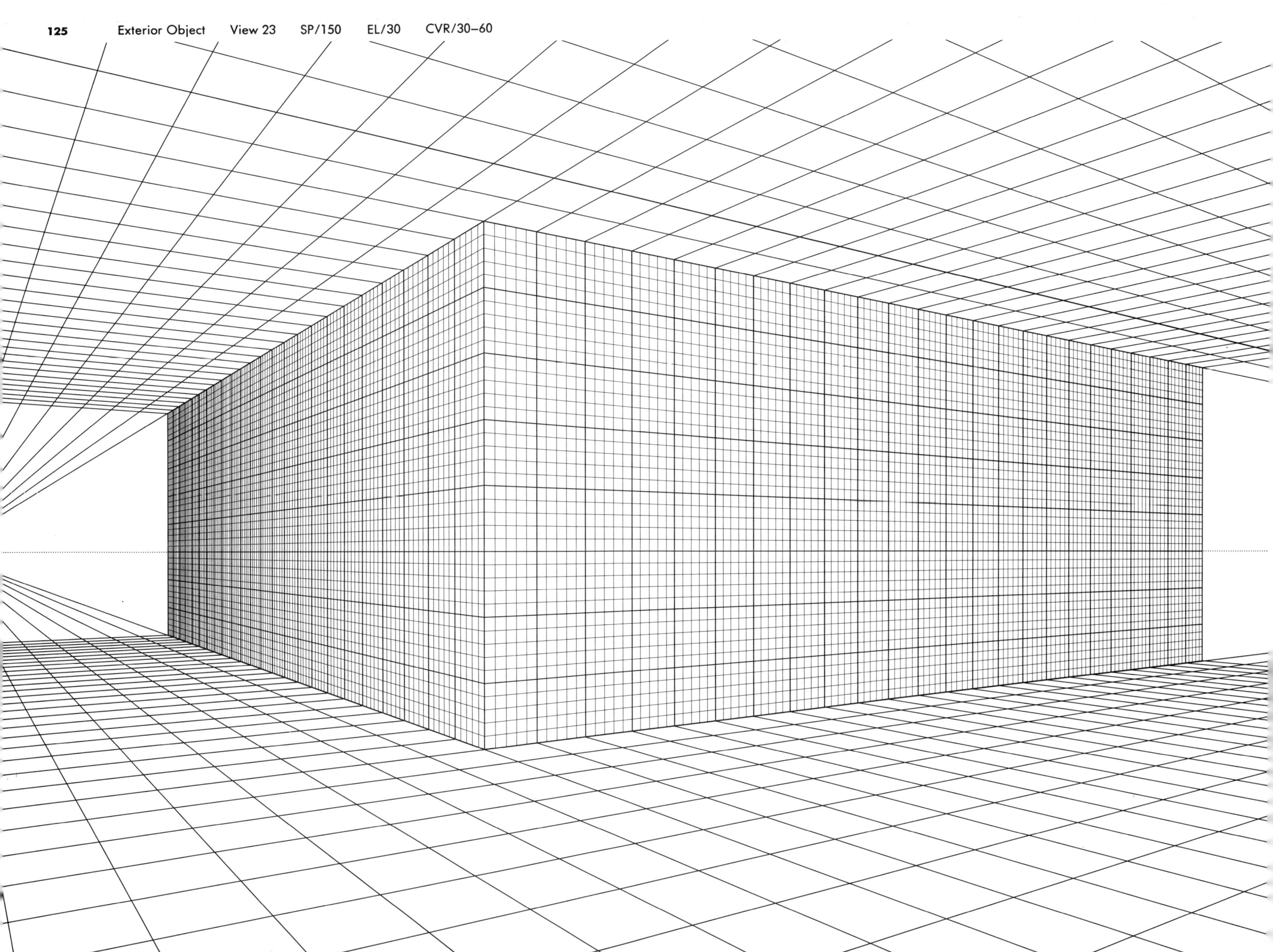

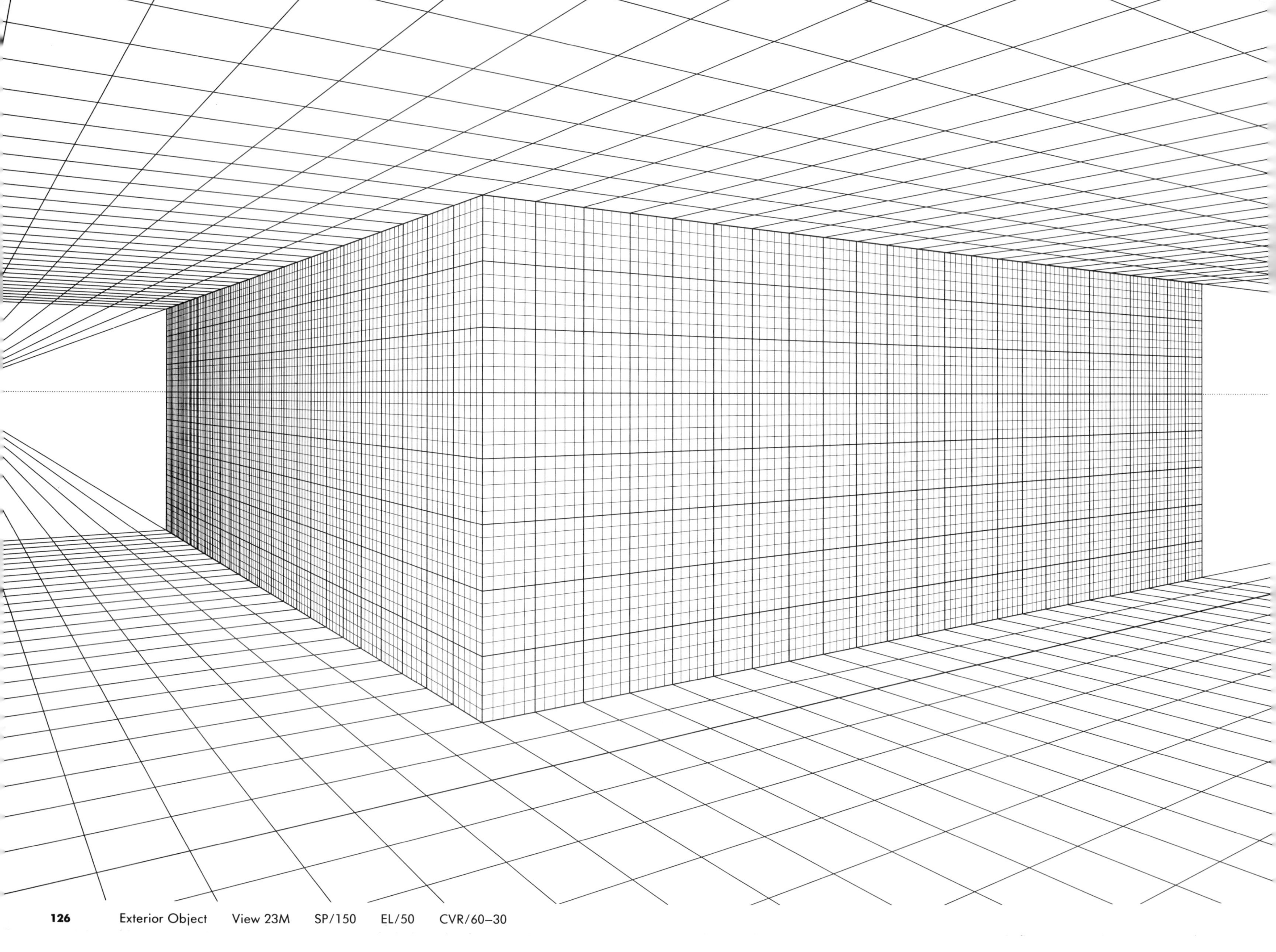



Exterior Object View 23M SP/150 EL/50 CVR/60–30

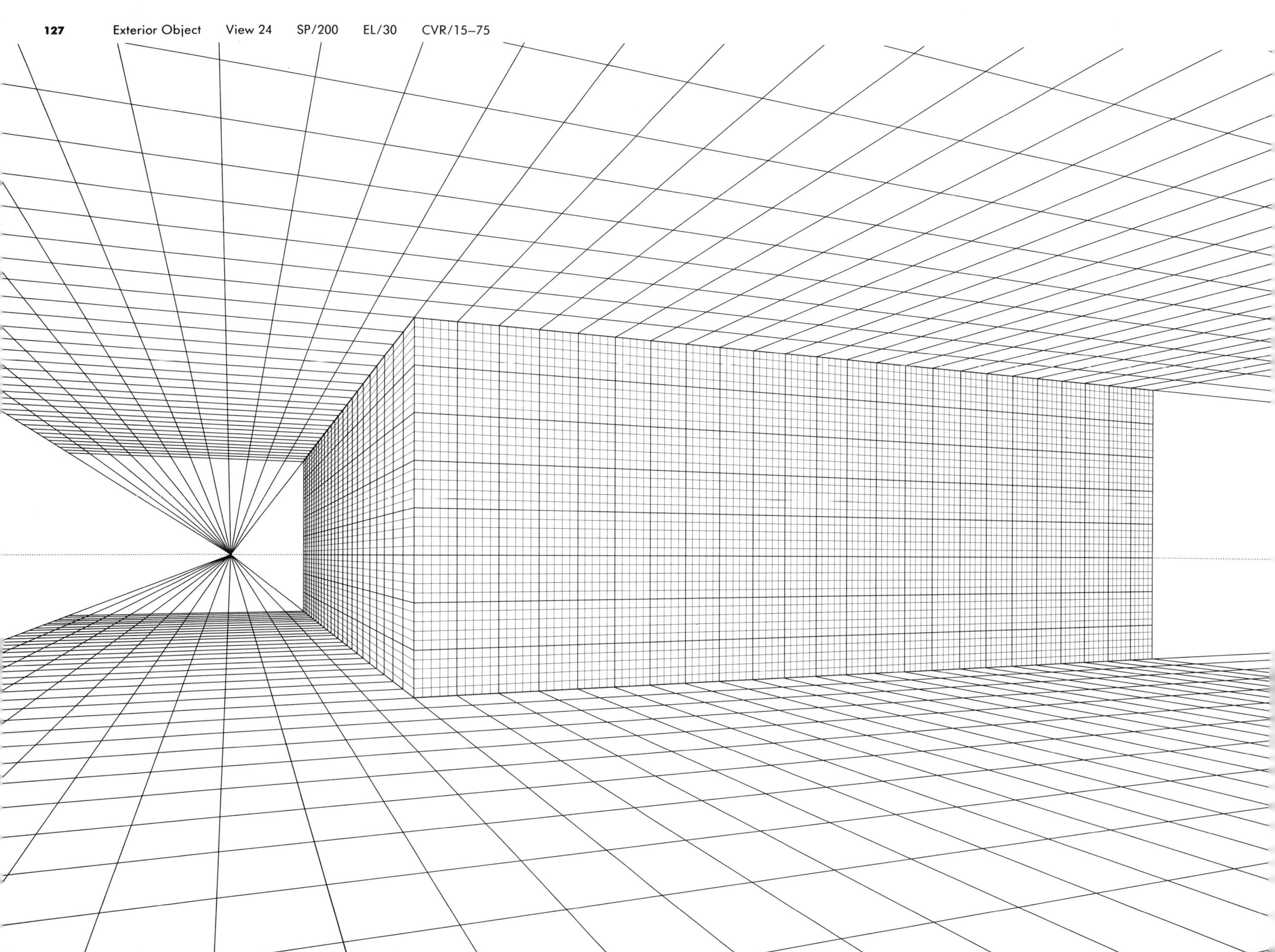

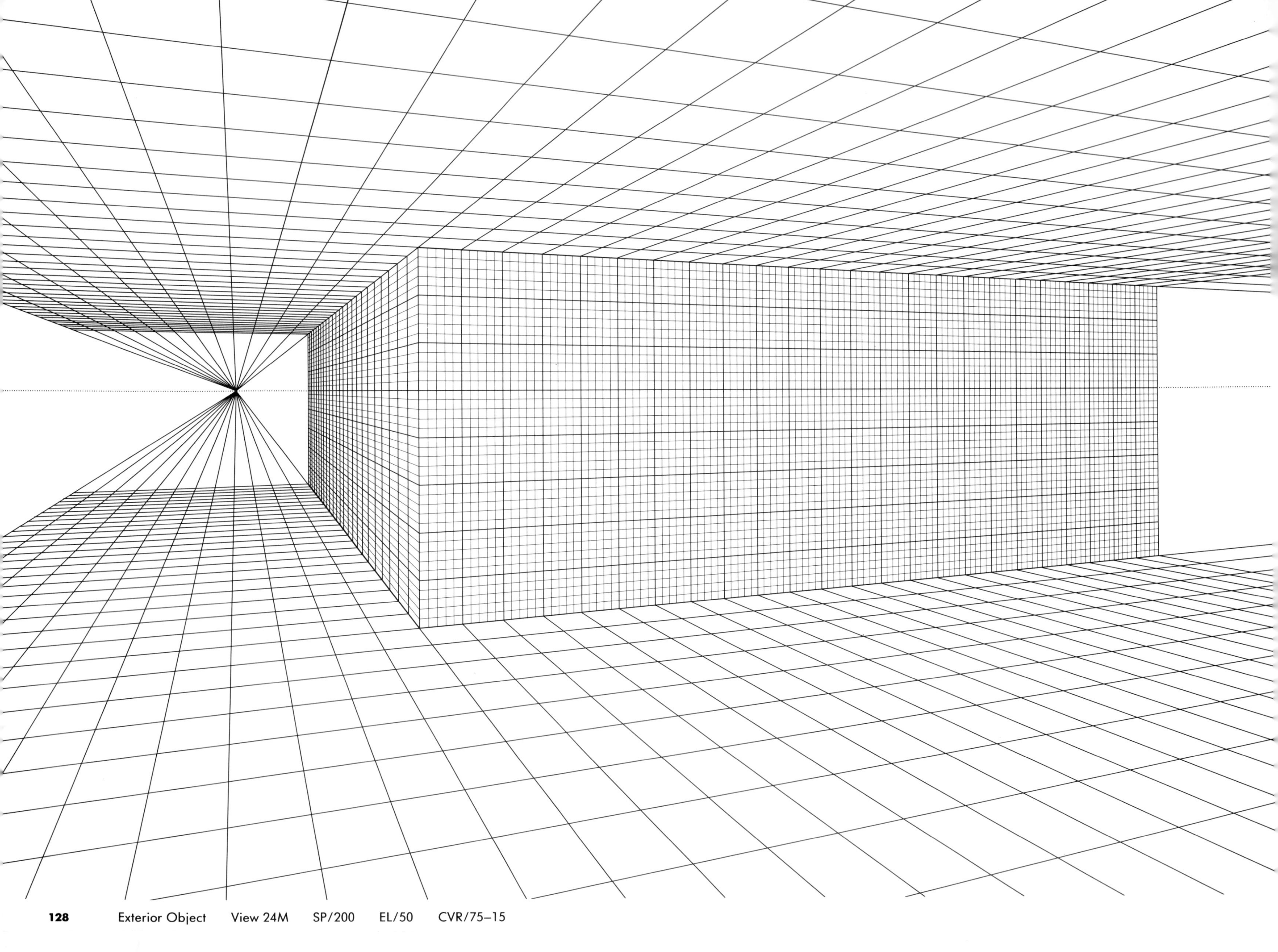

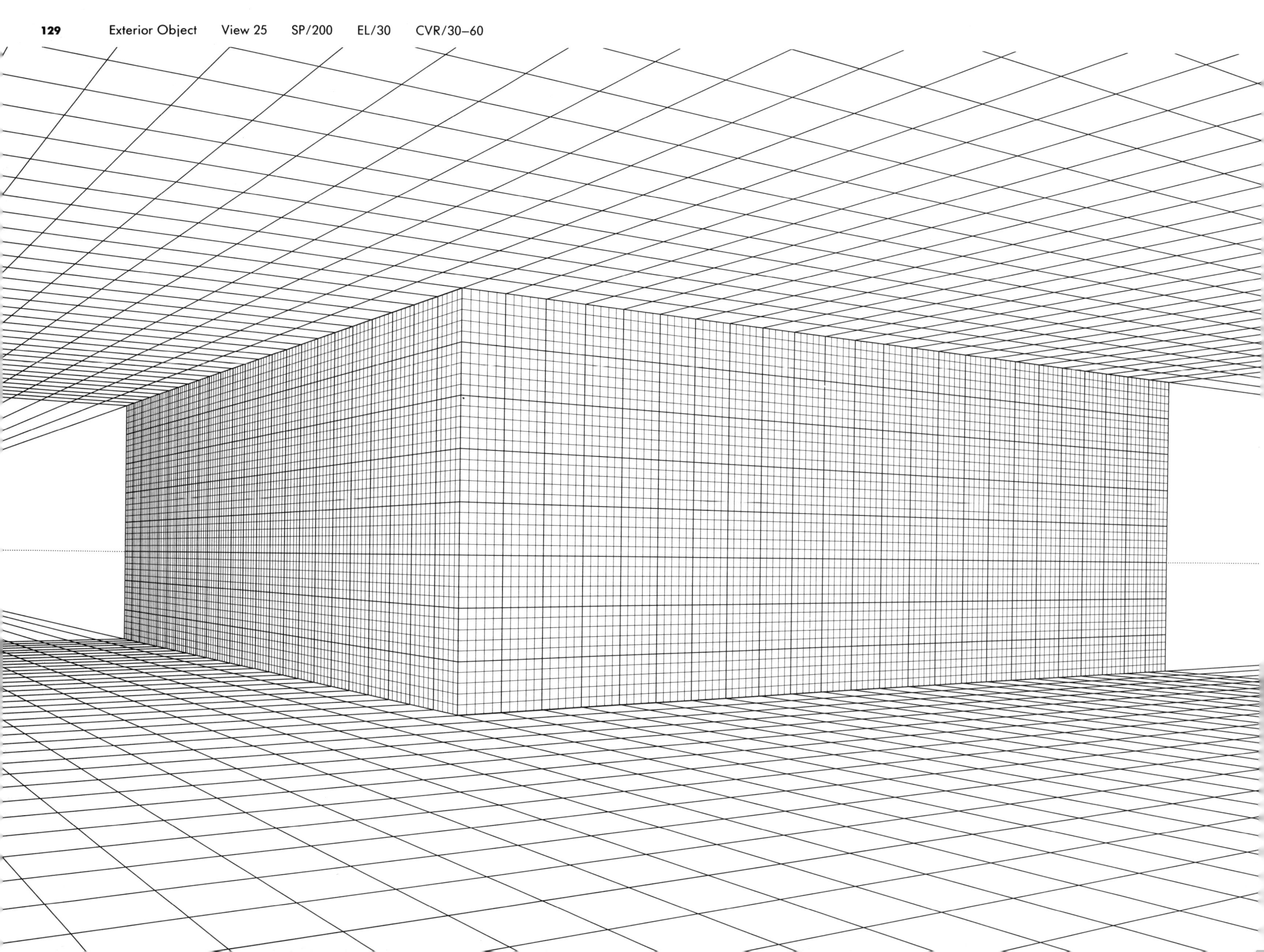

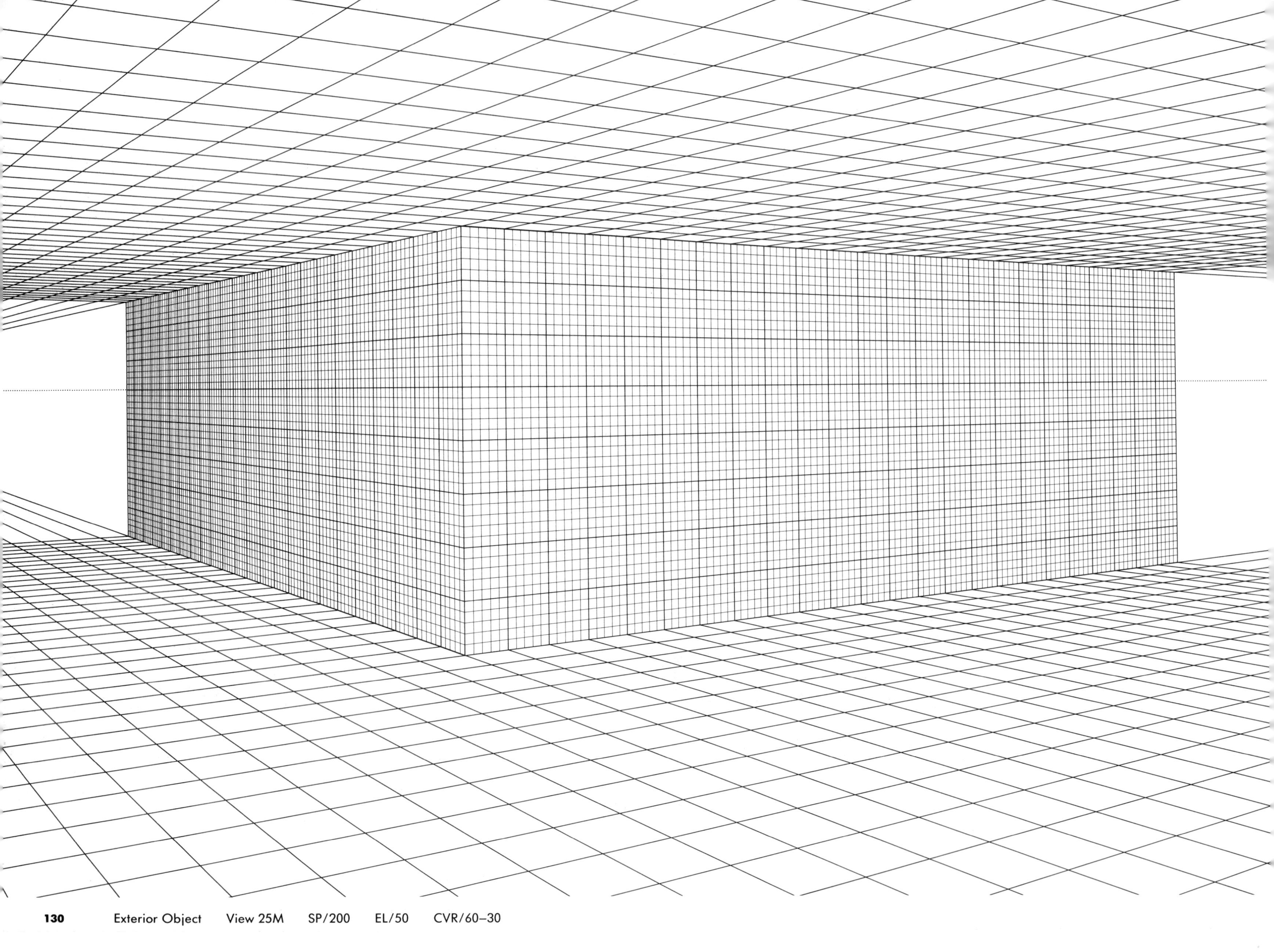



Exterior Object View 25M SP/200 EL/50 CVR/60–30

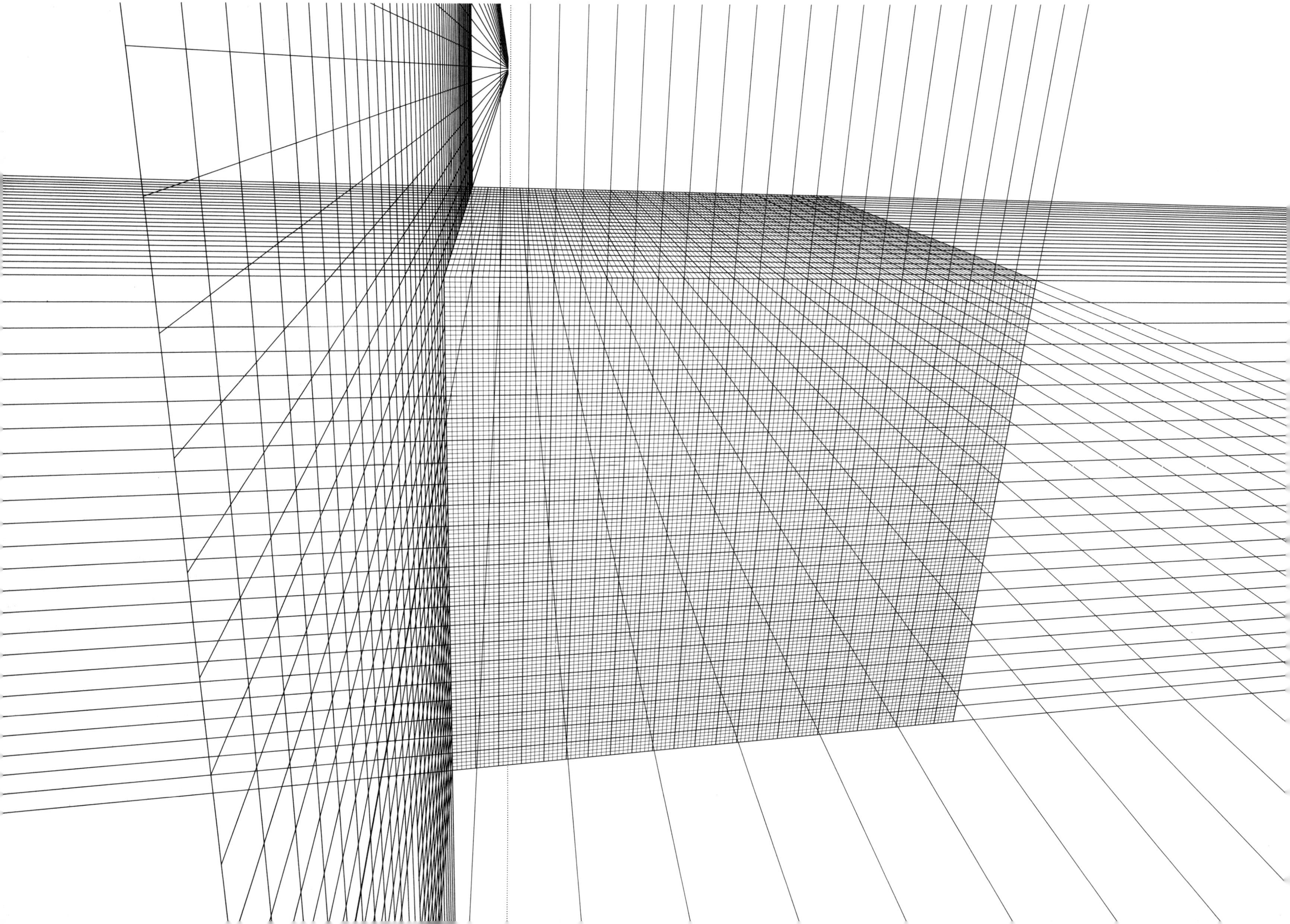

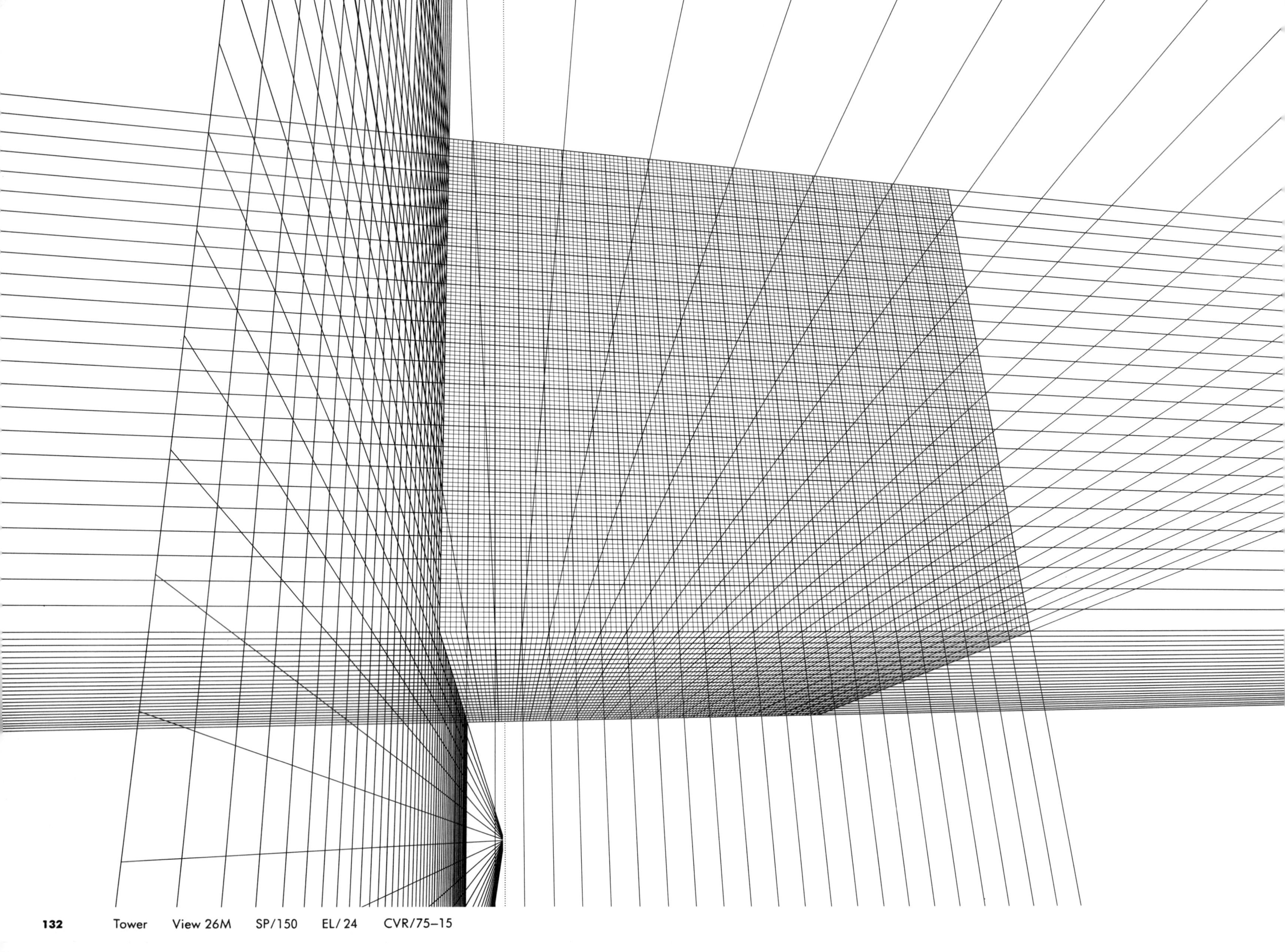



Tower View 26M SP/150 EL/ 24 CVR/75–15

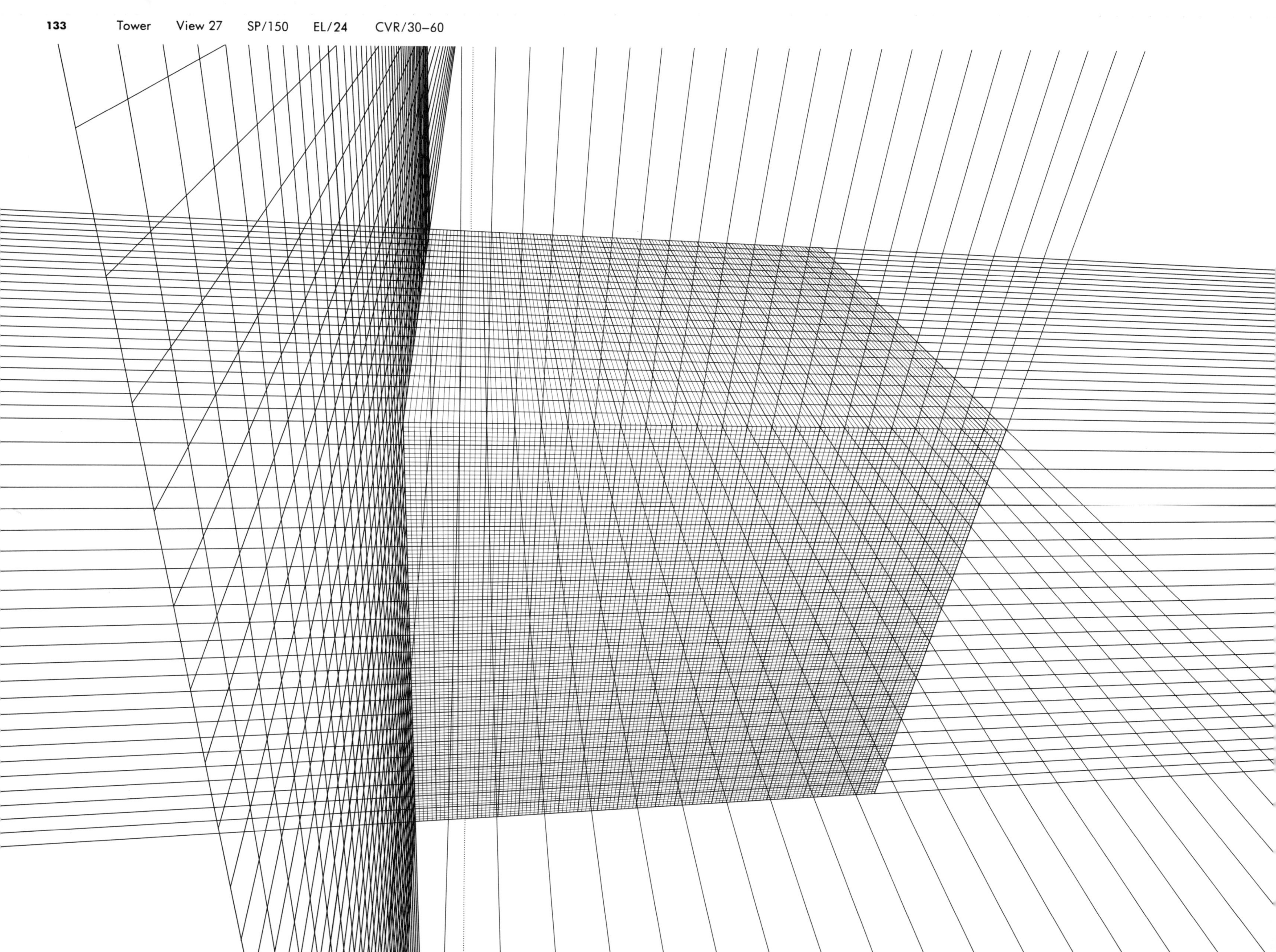

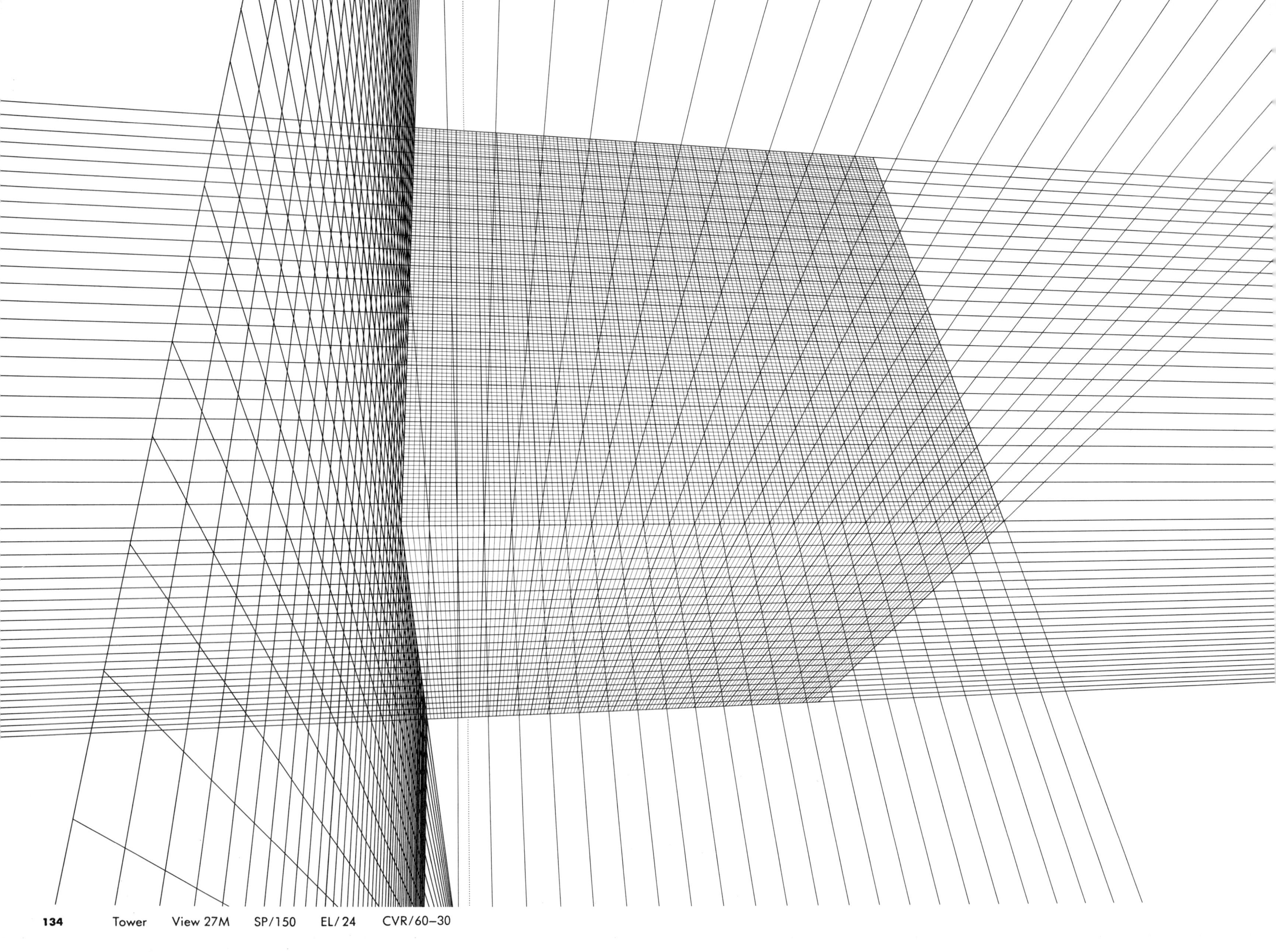

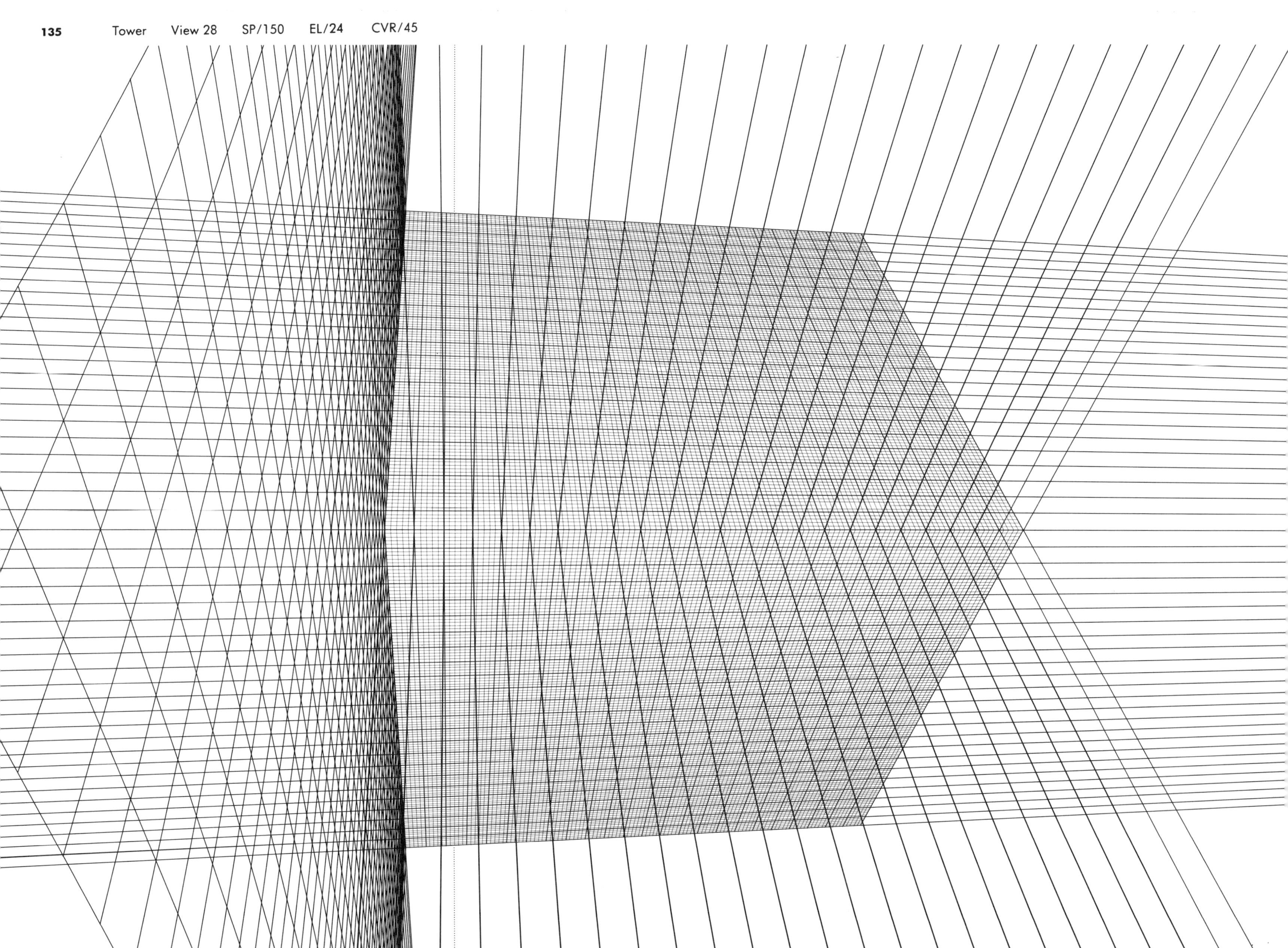

**PART FOUR**

# MULTIPLE GRIDS

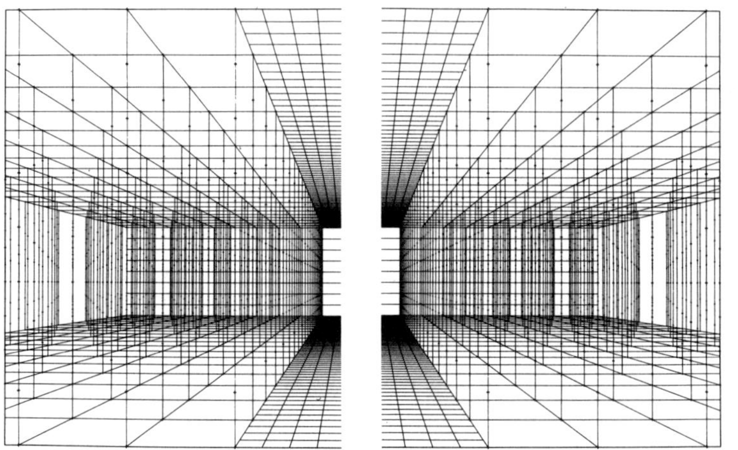

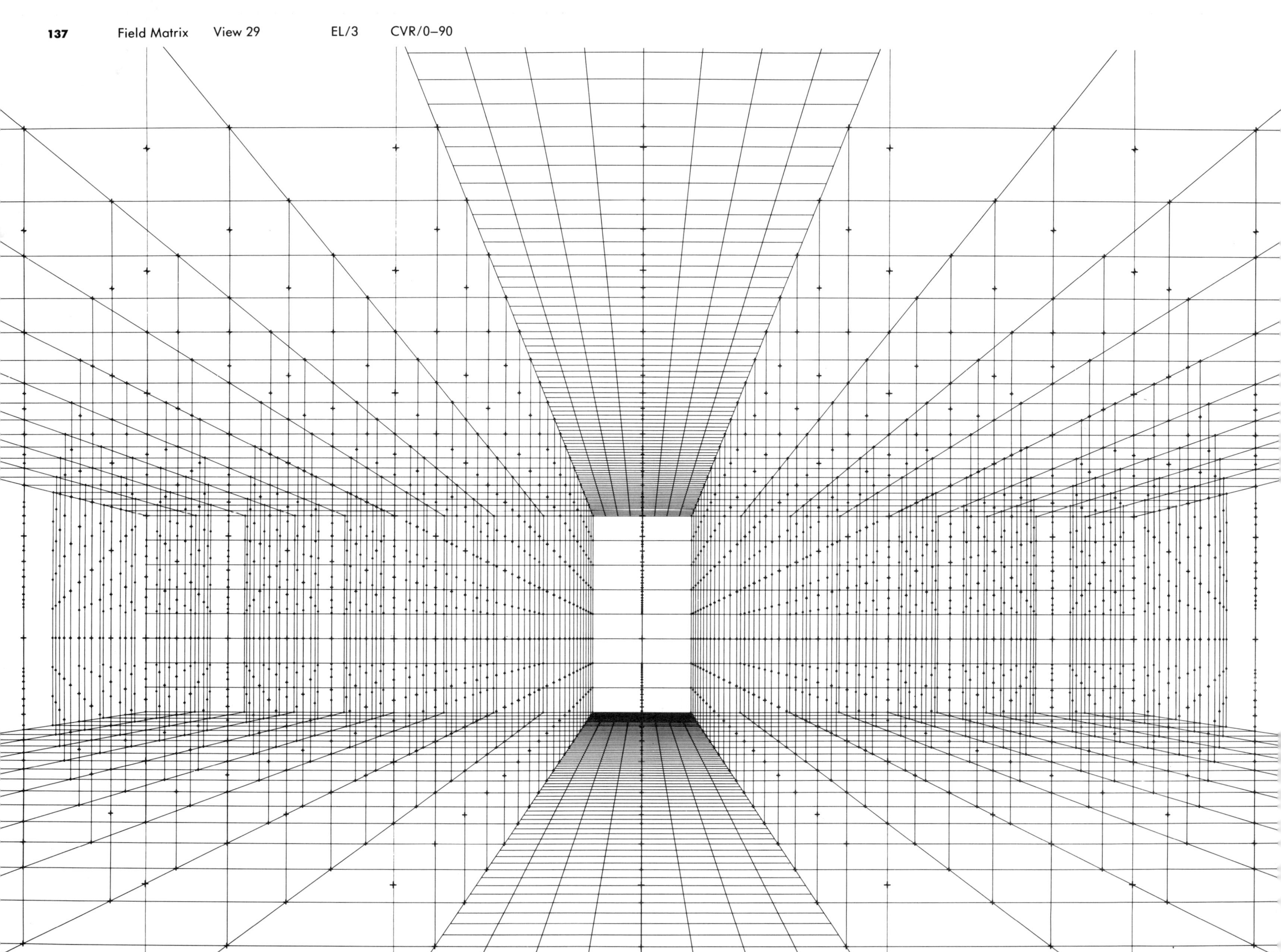

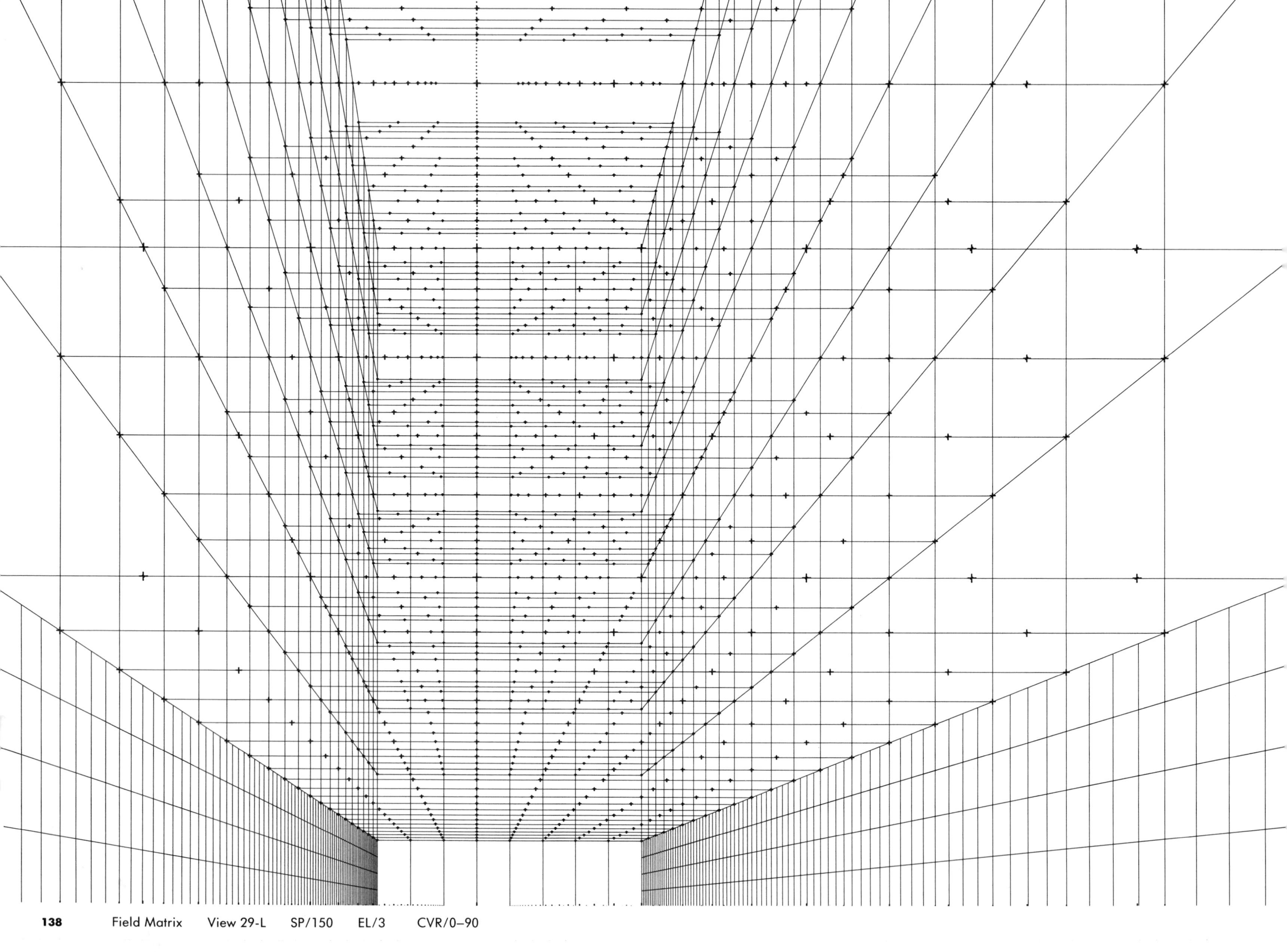

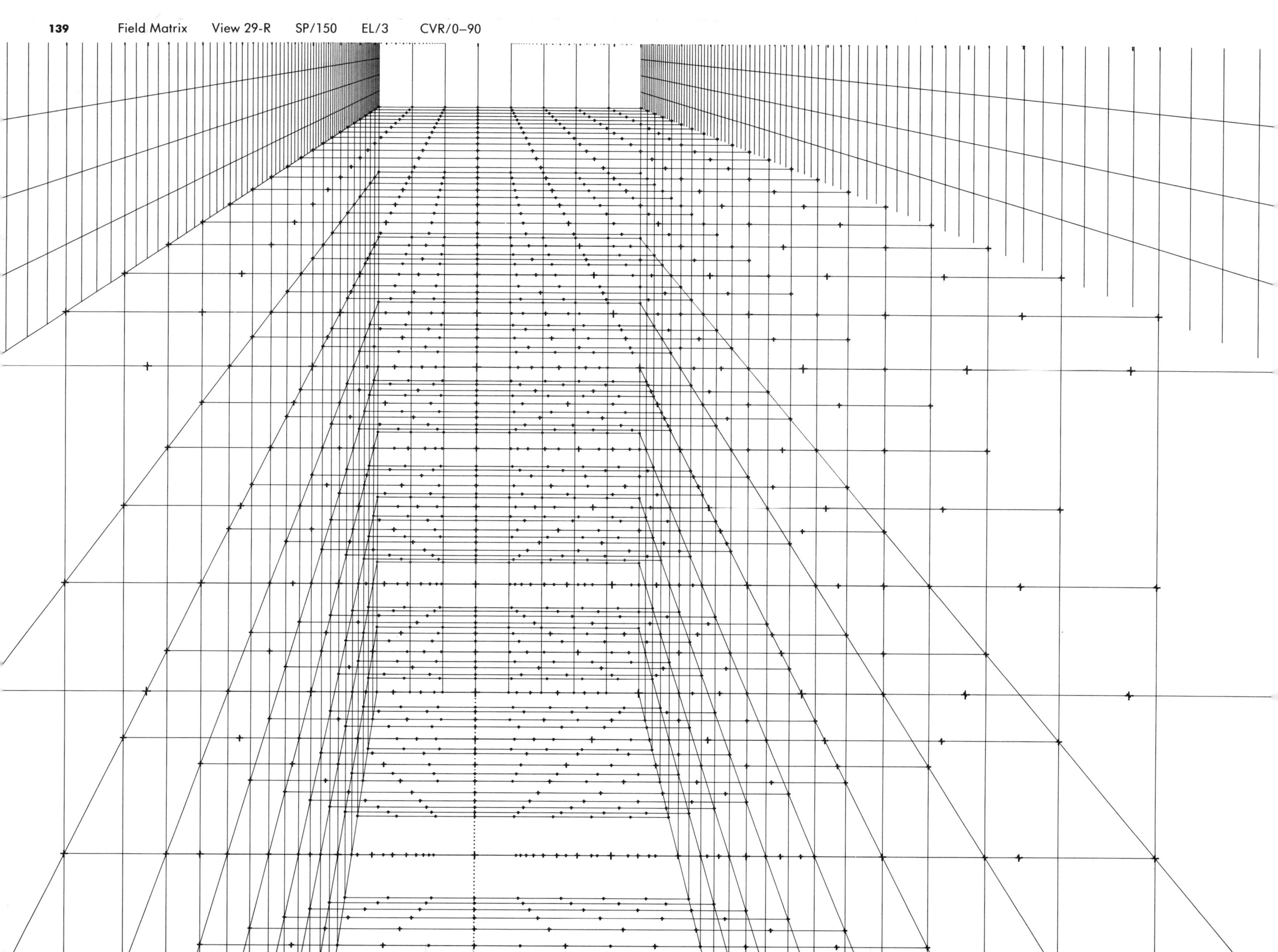

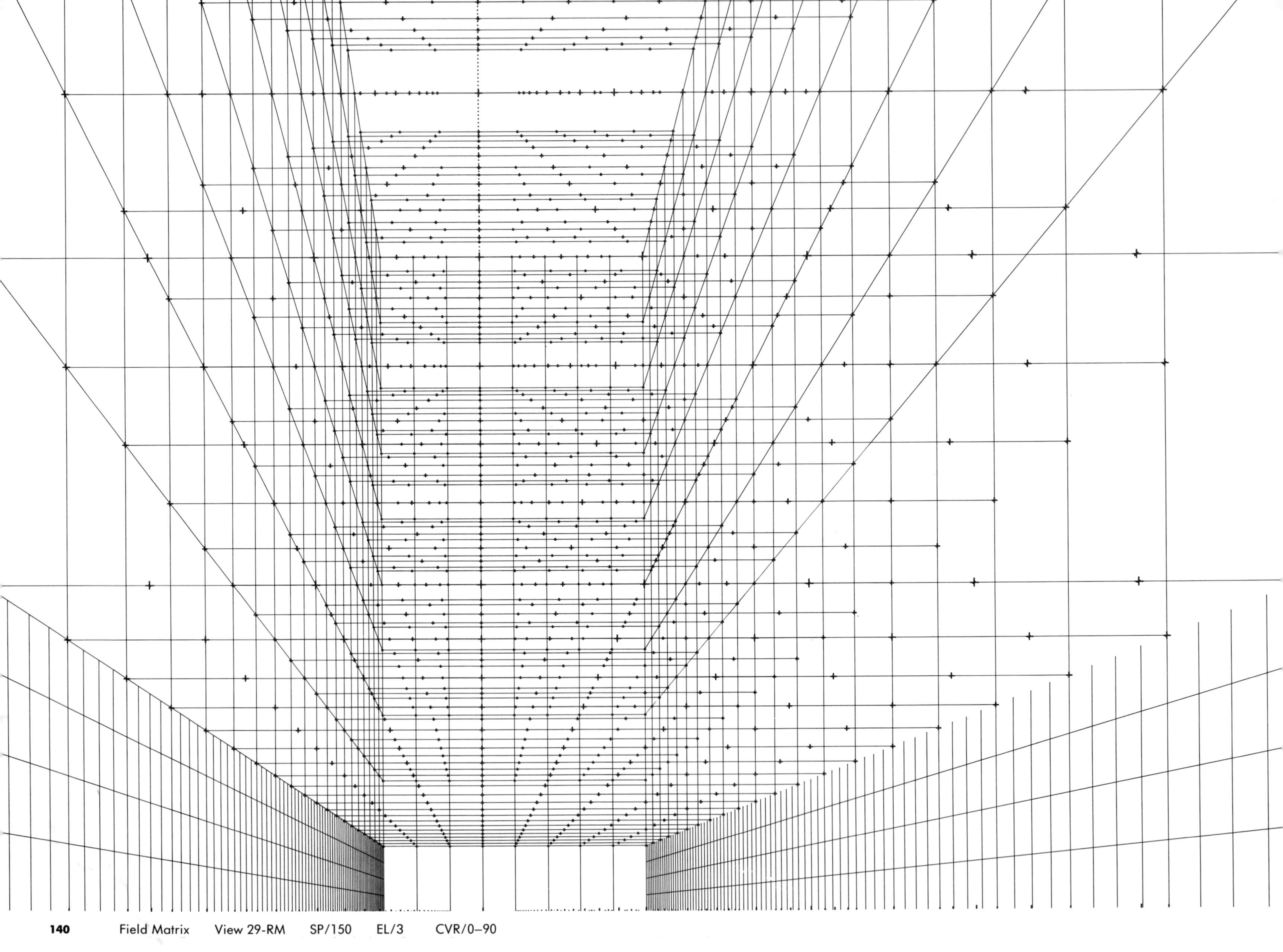

 Field Matrix View 29-RM SP/150 EL/3 CVR/0–90

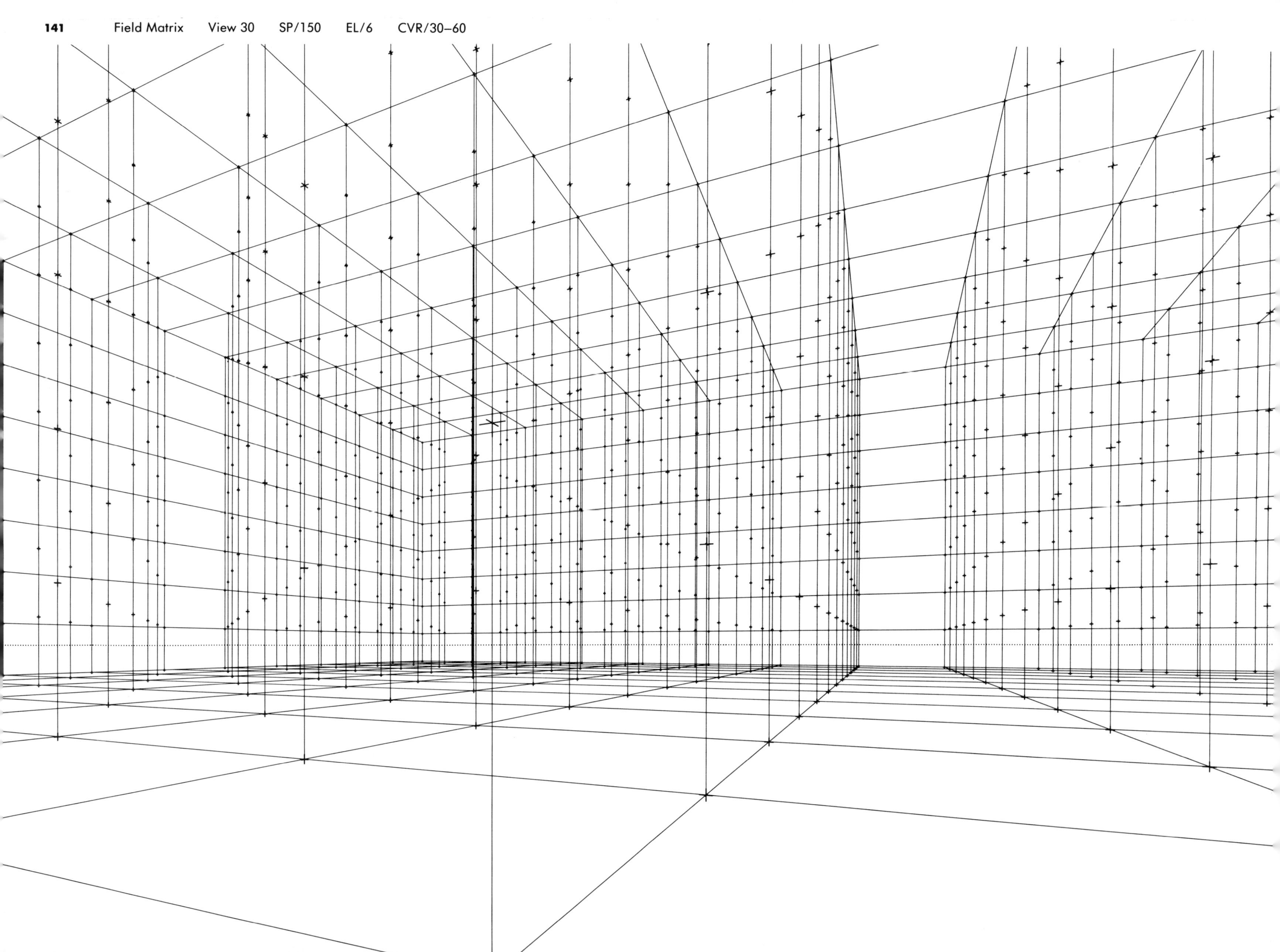

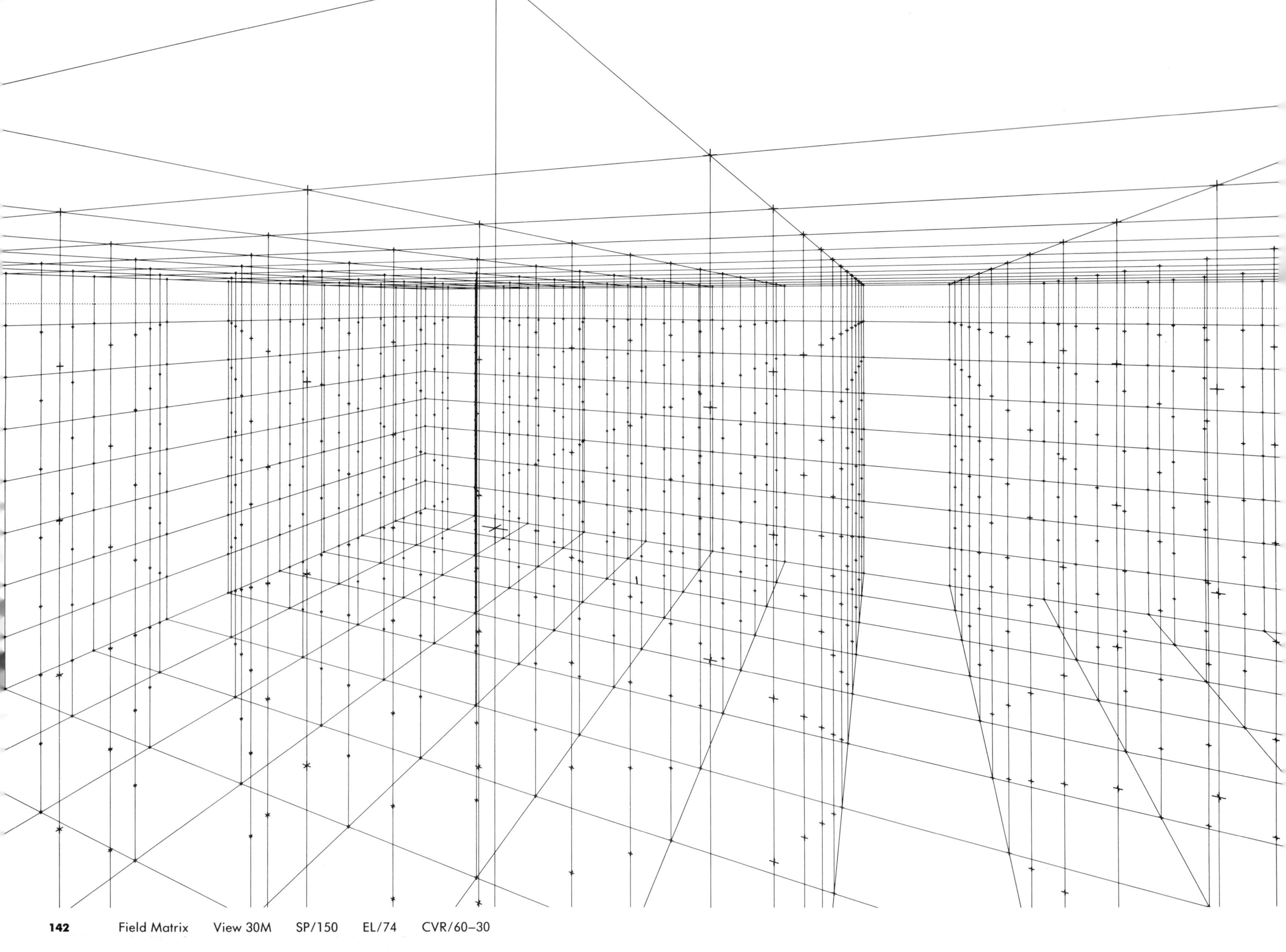

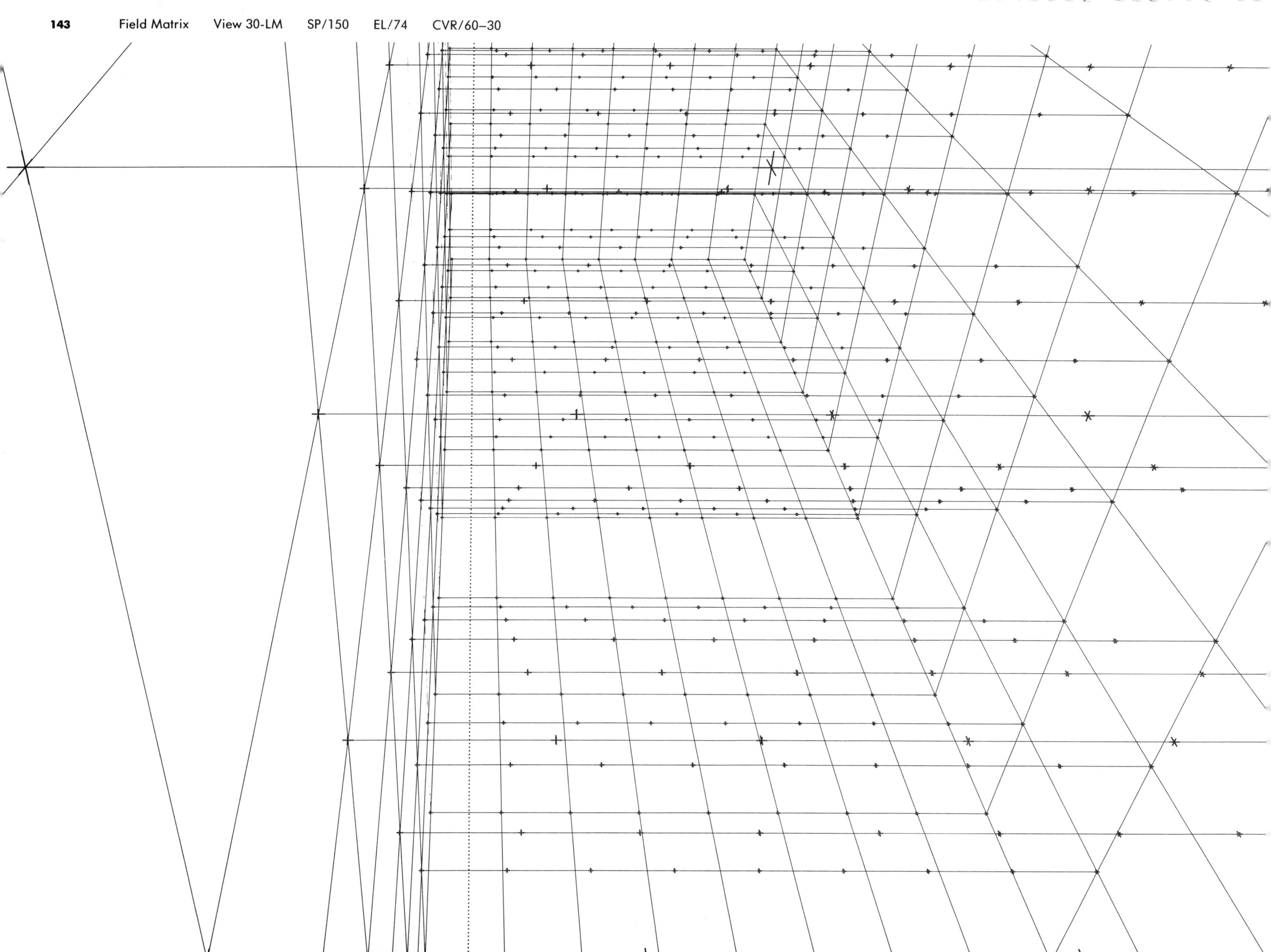

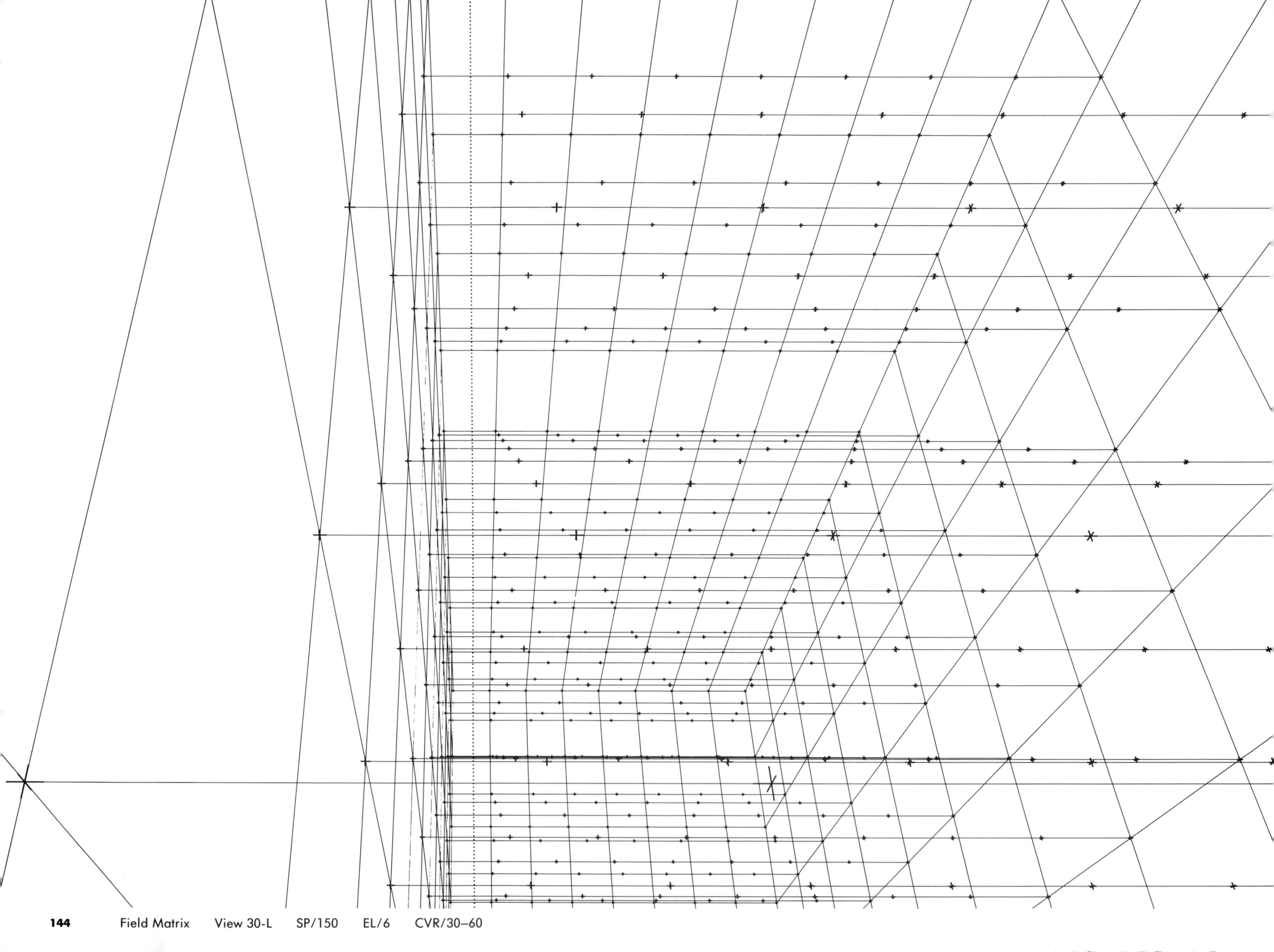

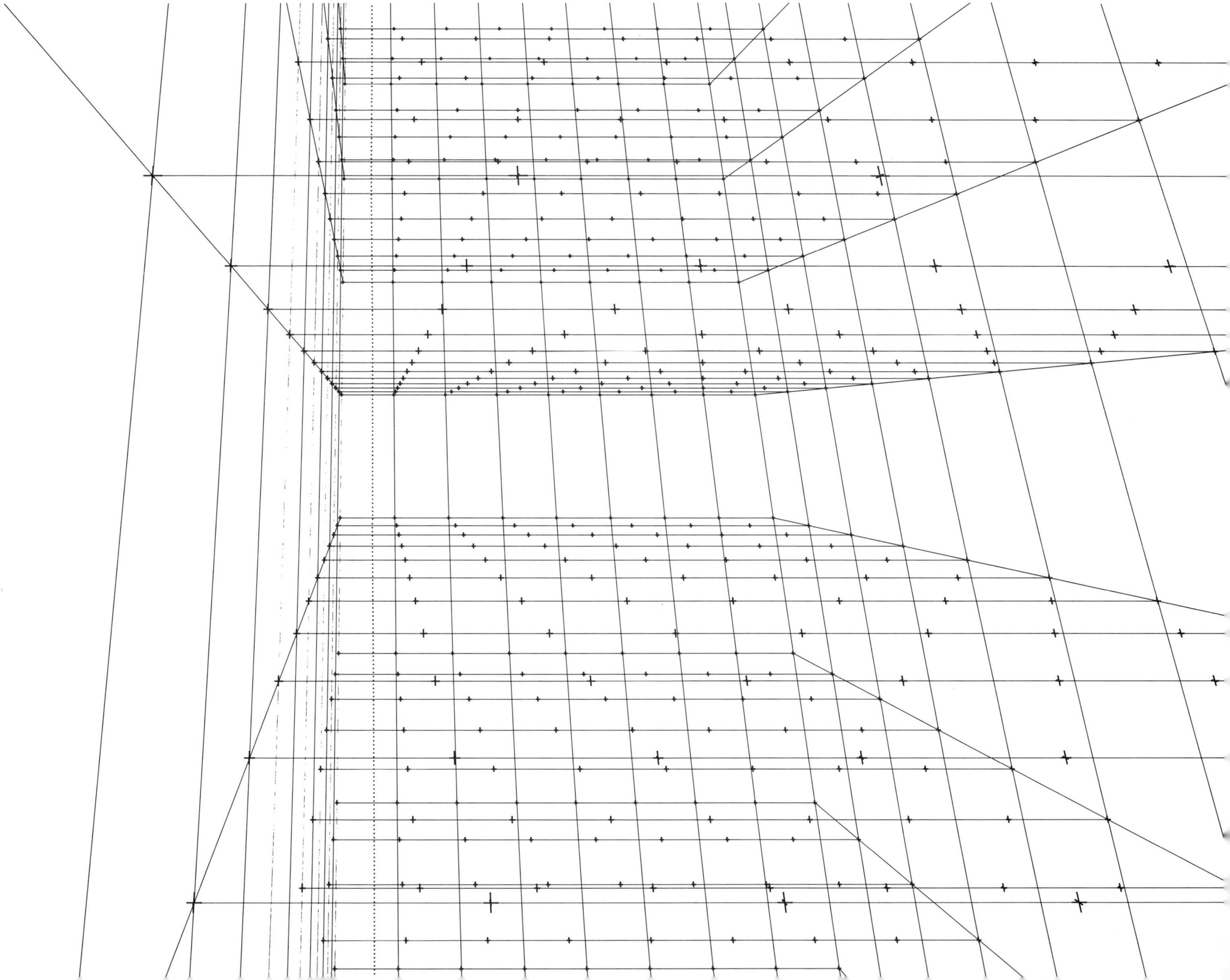

Field Matrix View 30-RM SP/150 EL/74 CVR/60–30

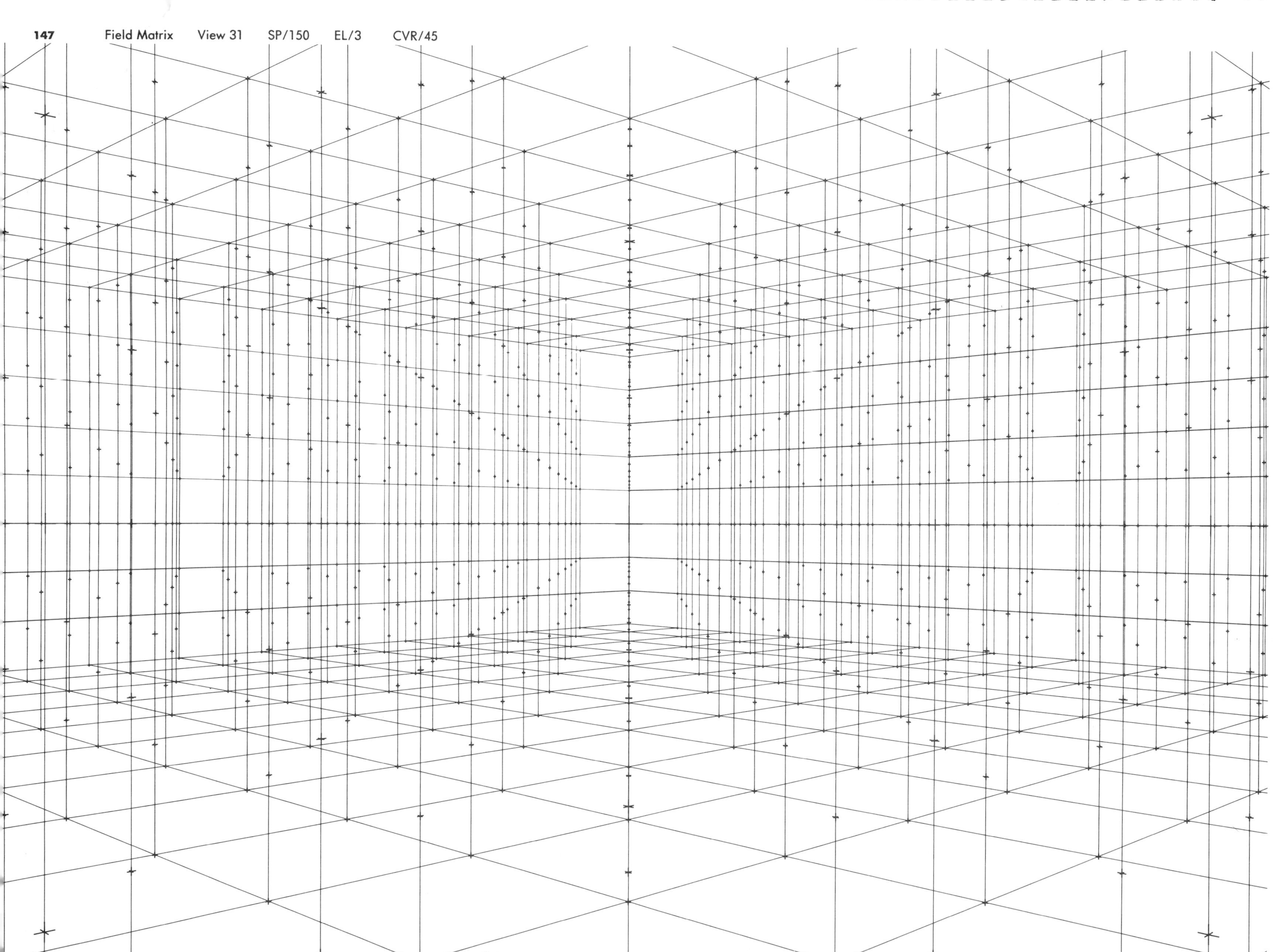

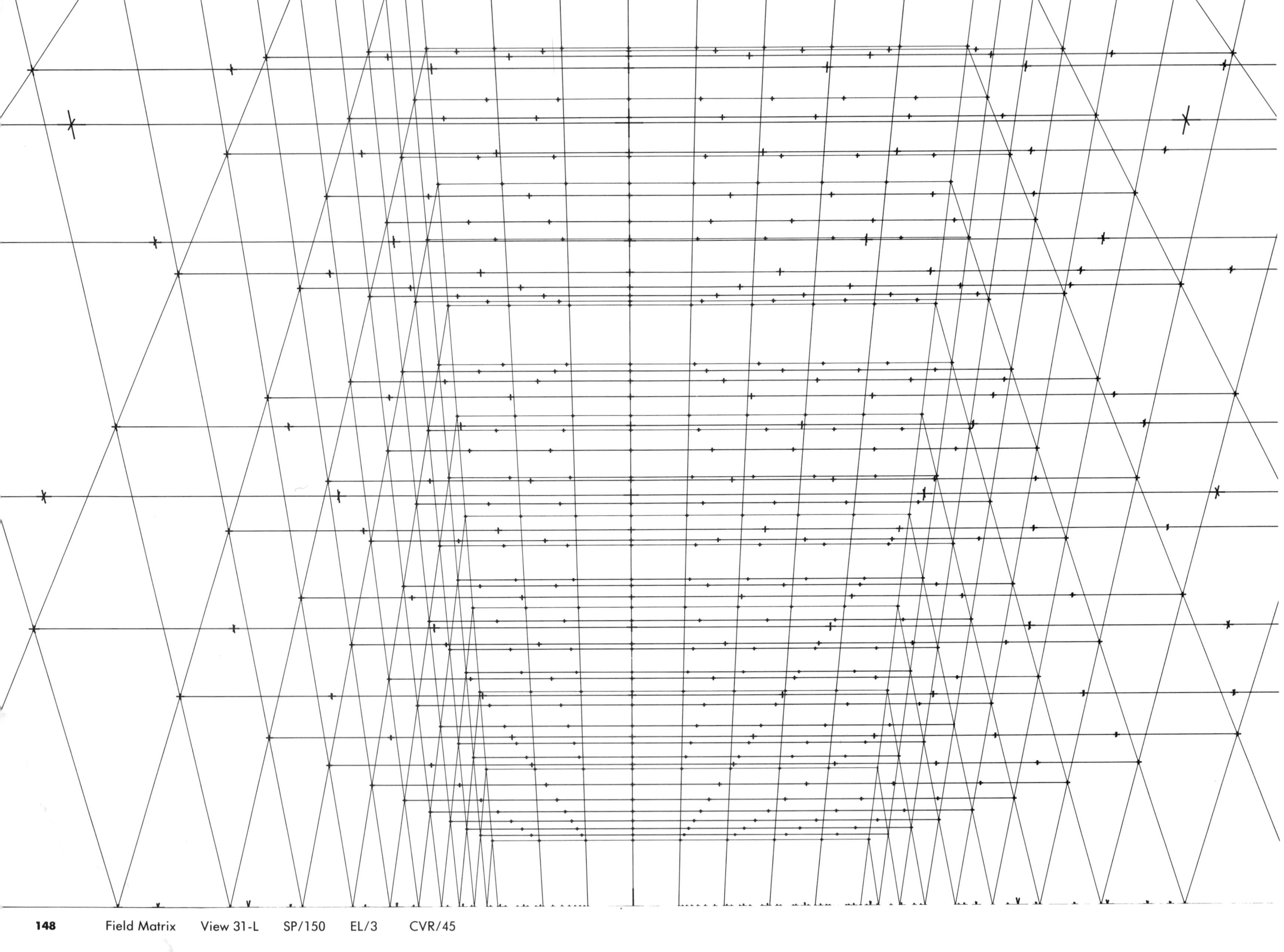

 Field Matrix View 31-L SP/150 EL/3 CVR/45

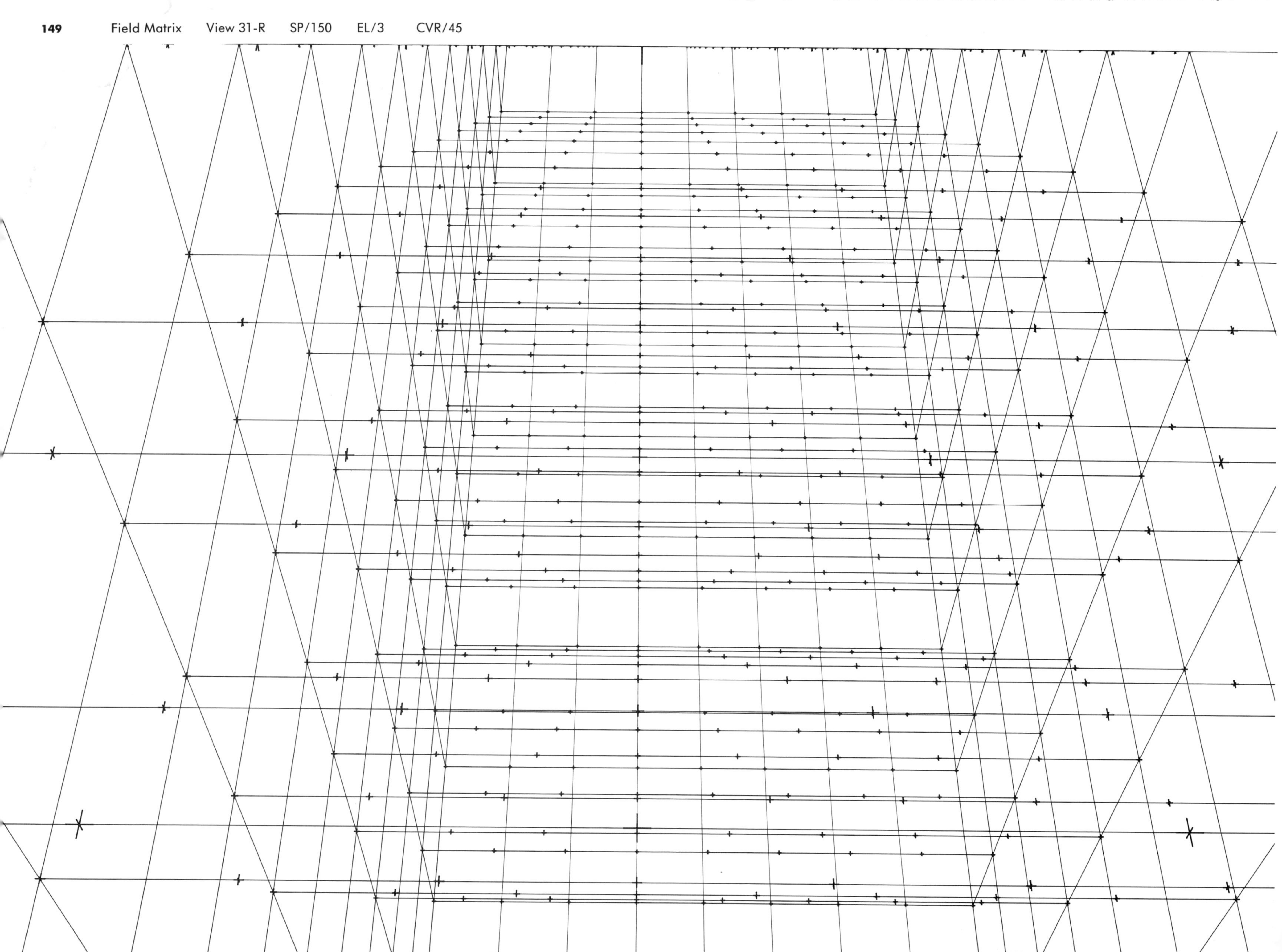

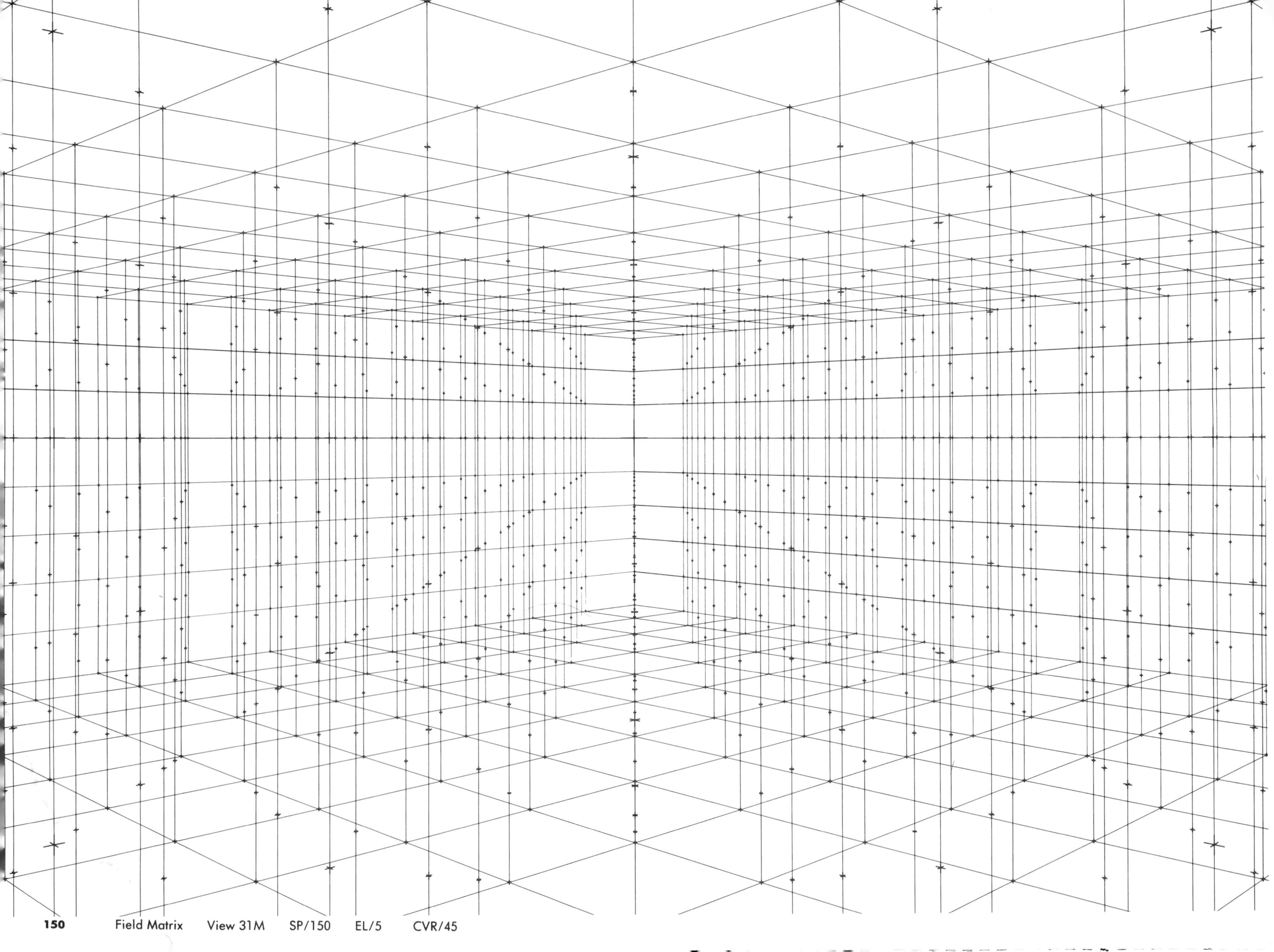

 Field Matrix View 31M SP/150 EL/5 CVR/45

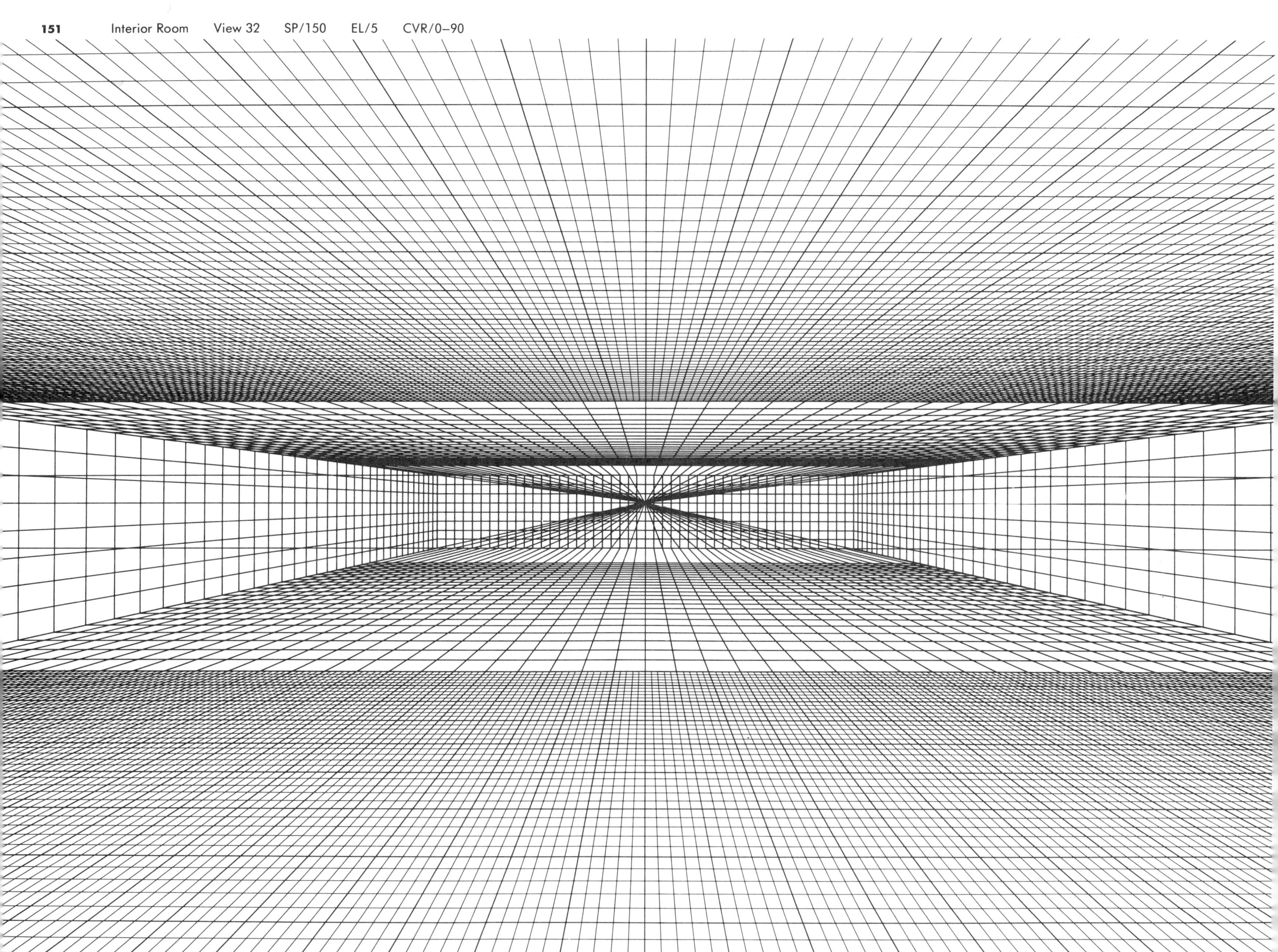

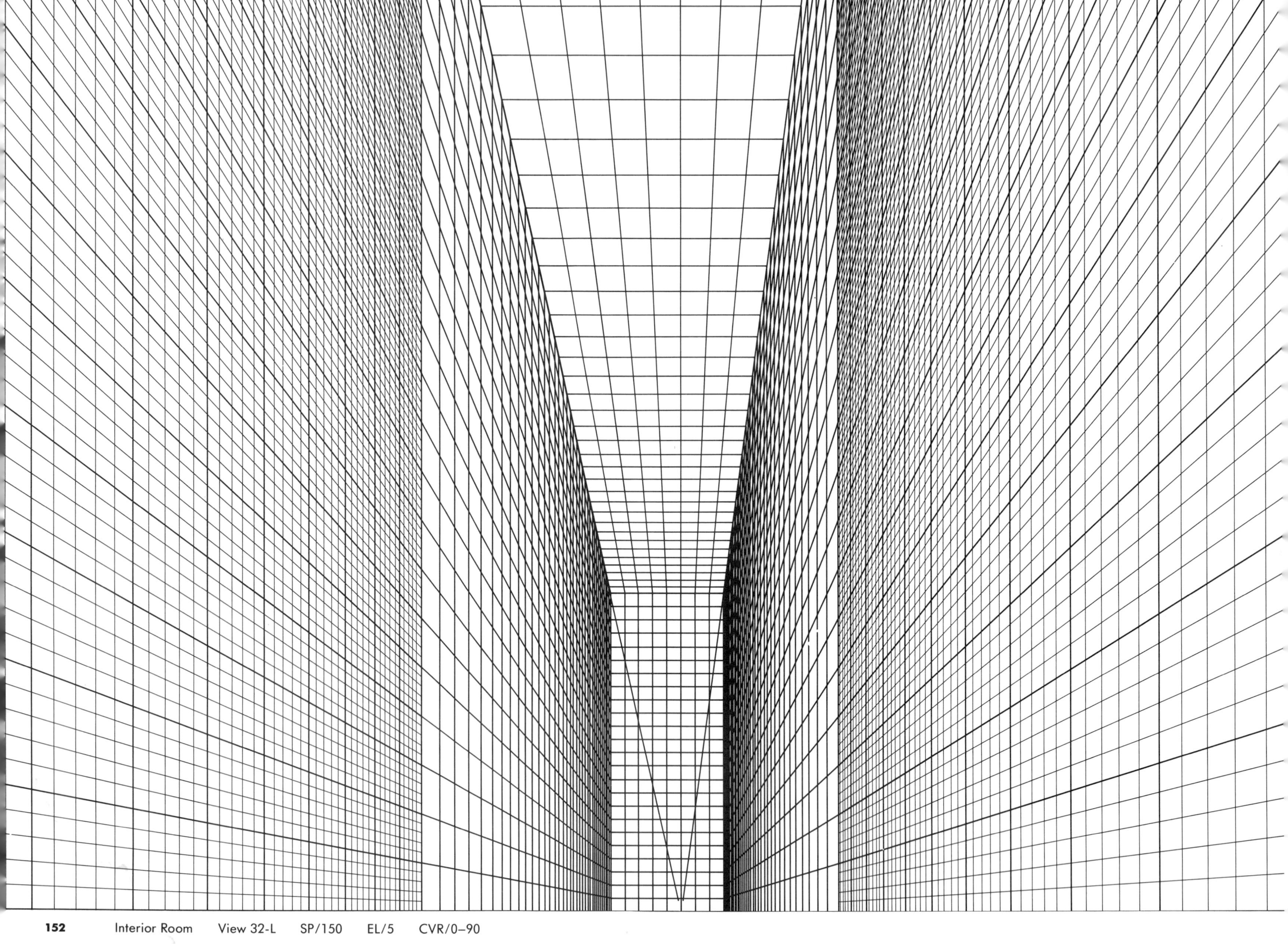



Interior Room View 32-L SP/150 EL/5 CVR/0–90

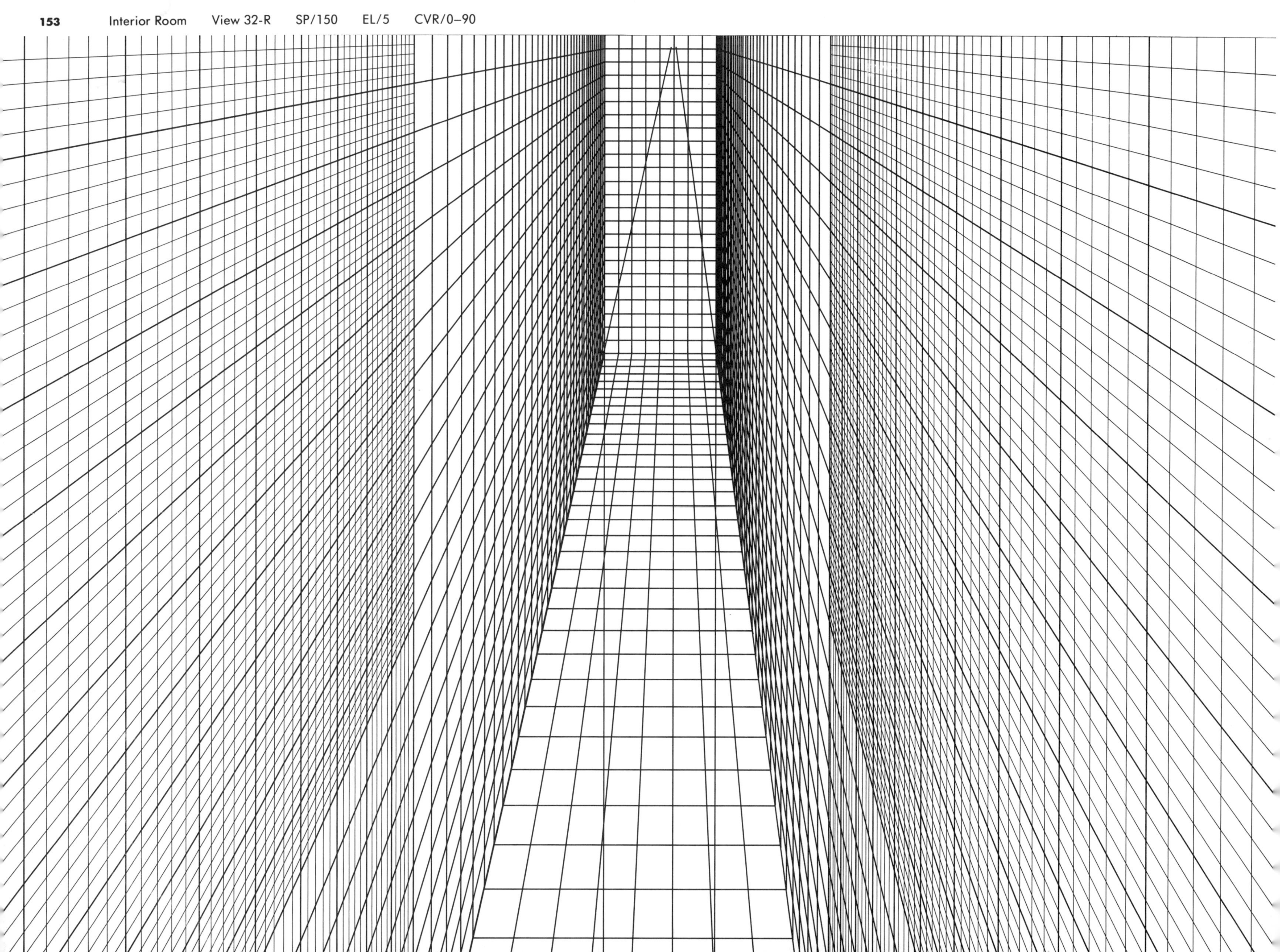

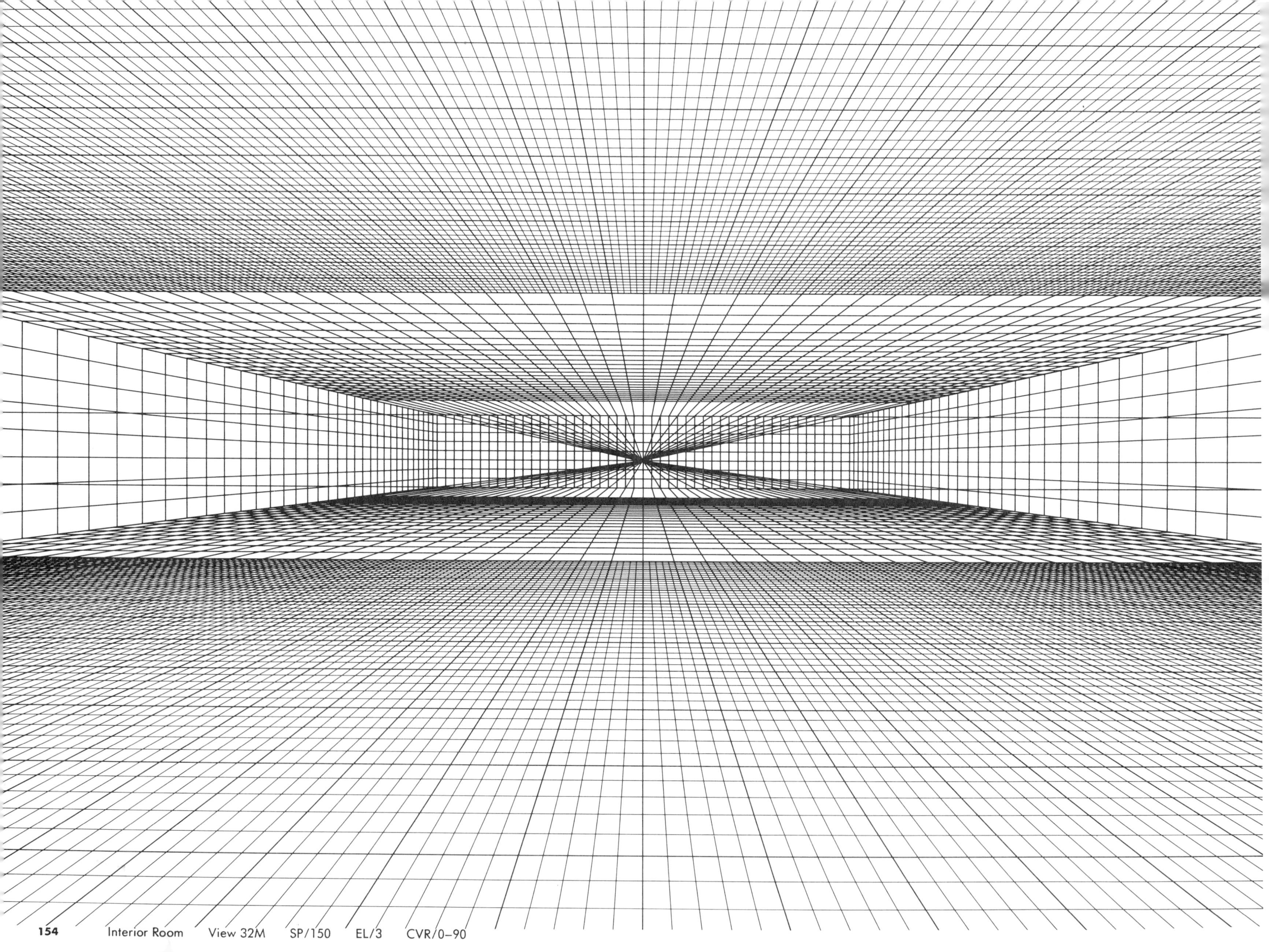

 Interior Room View 32M SP/150 EL/3 CVR/0–90

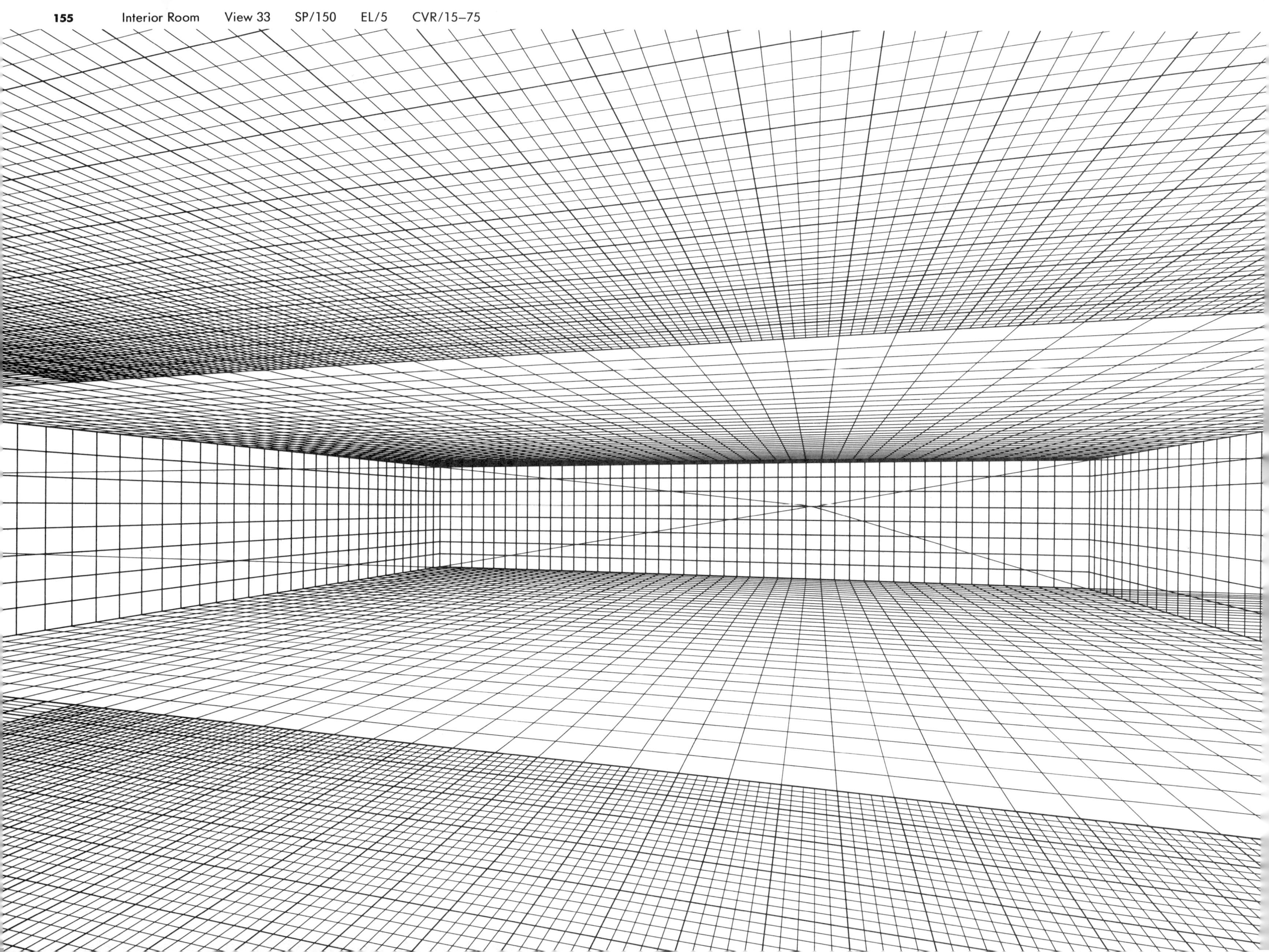

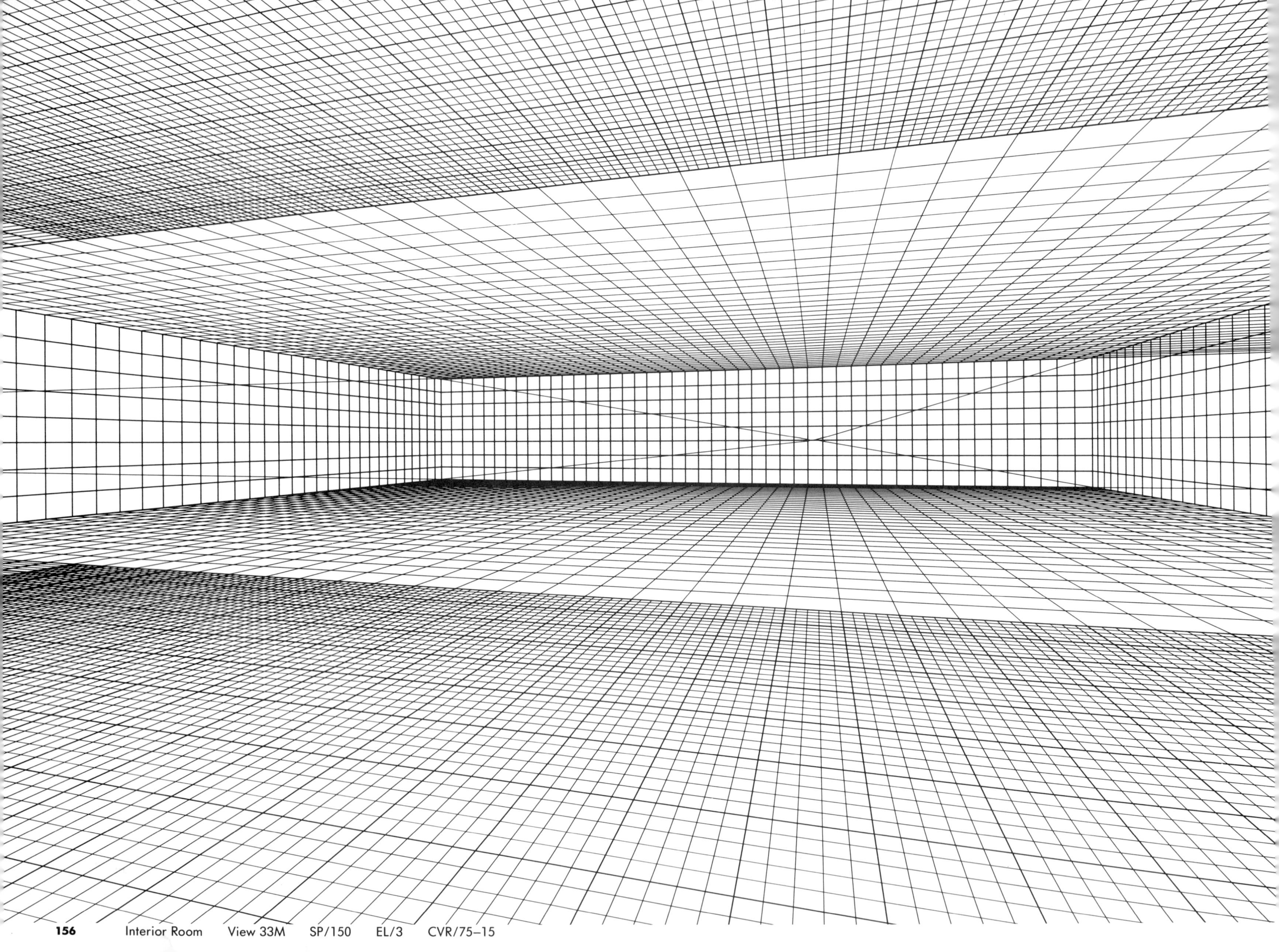

 Interior Room View 33M SP/150 EL/3 CVR/75–15

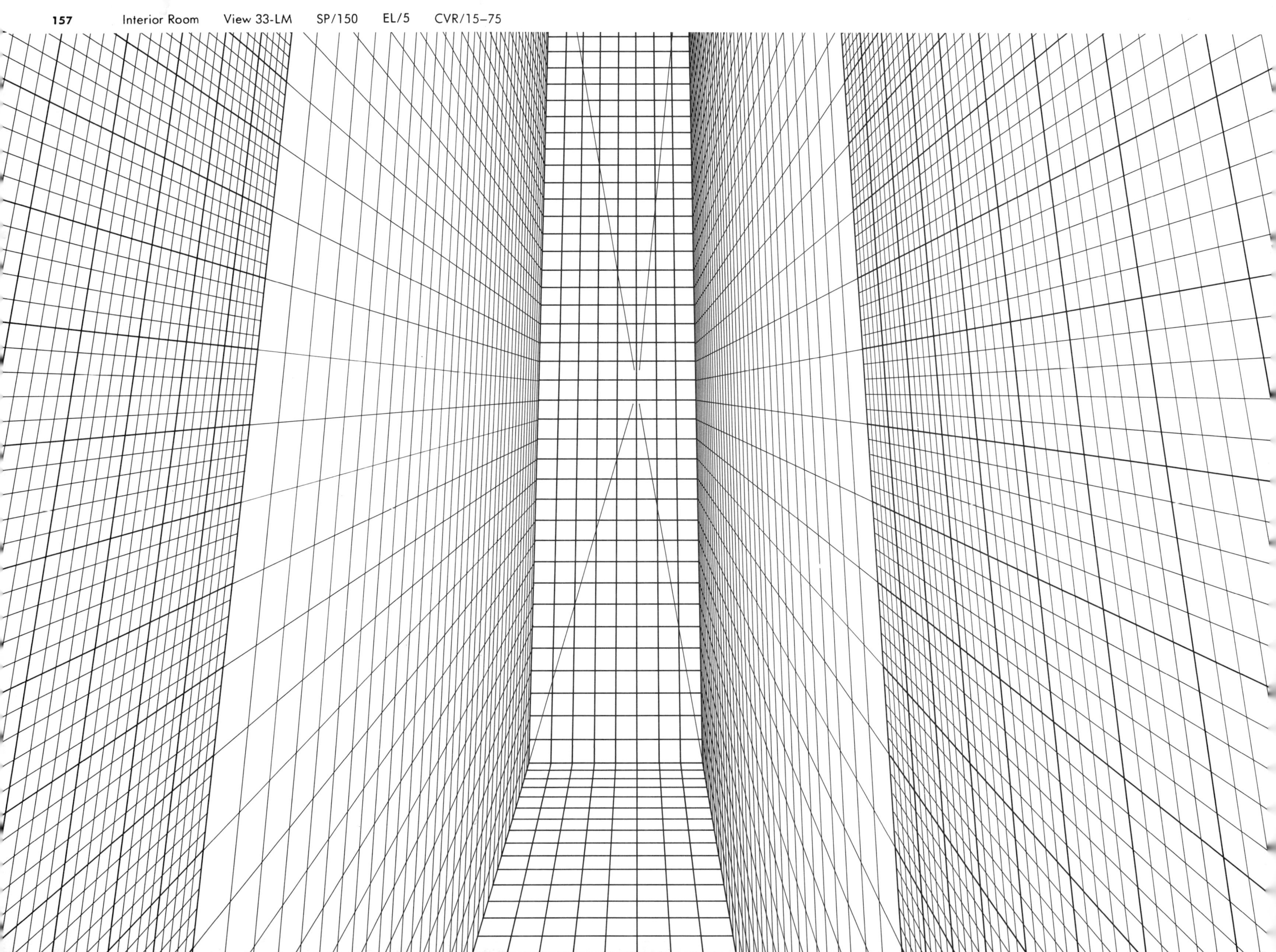

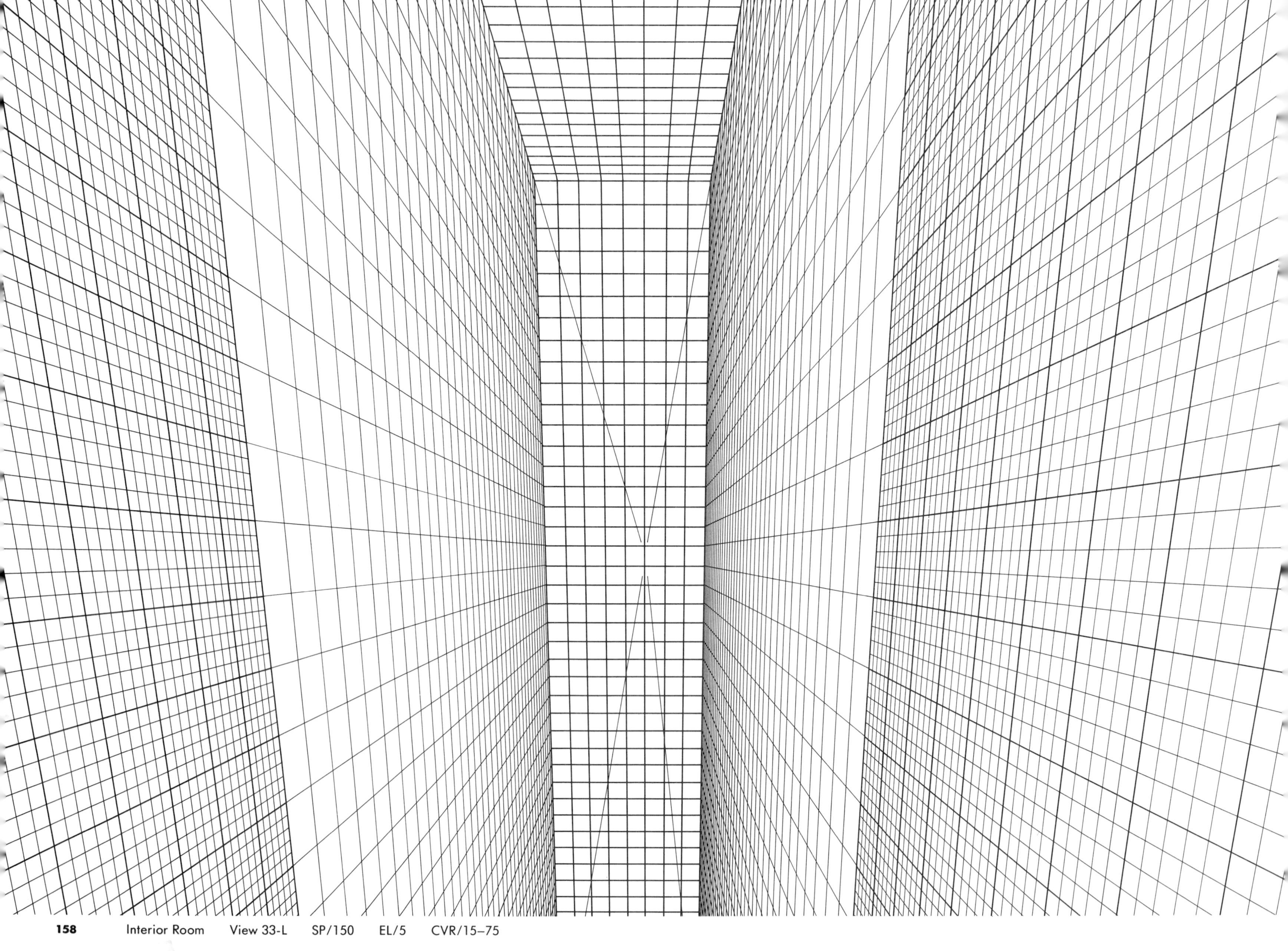



Interior Room View 33-L SP/150 EL/5 CVR/15–75

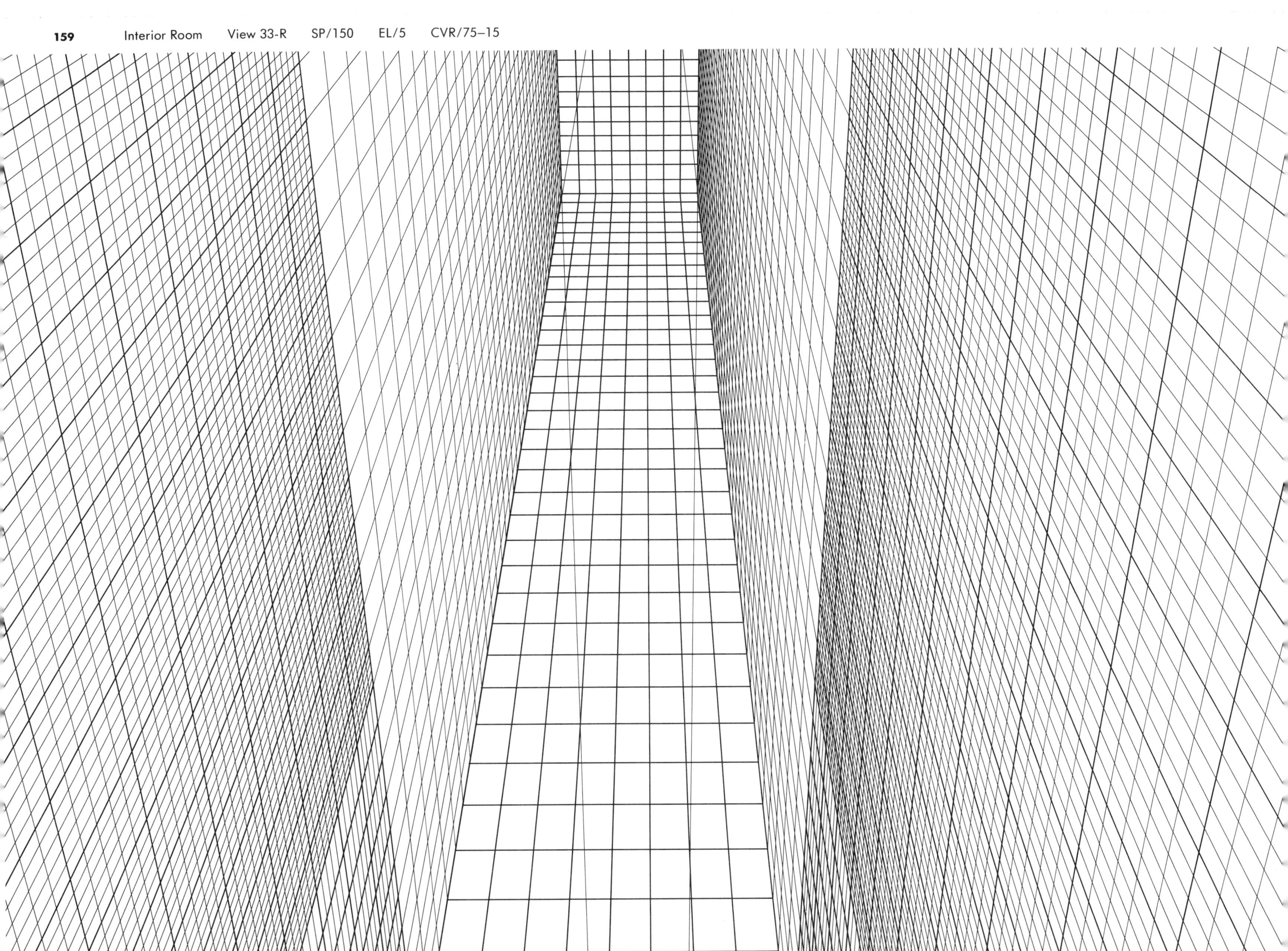

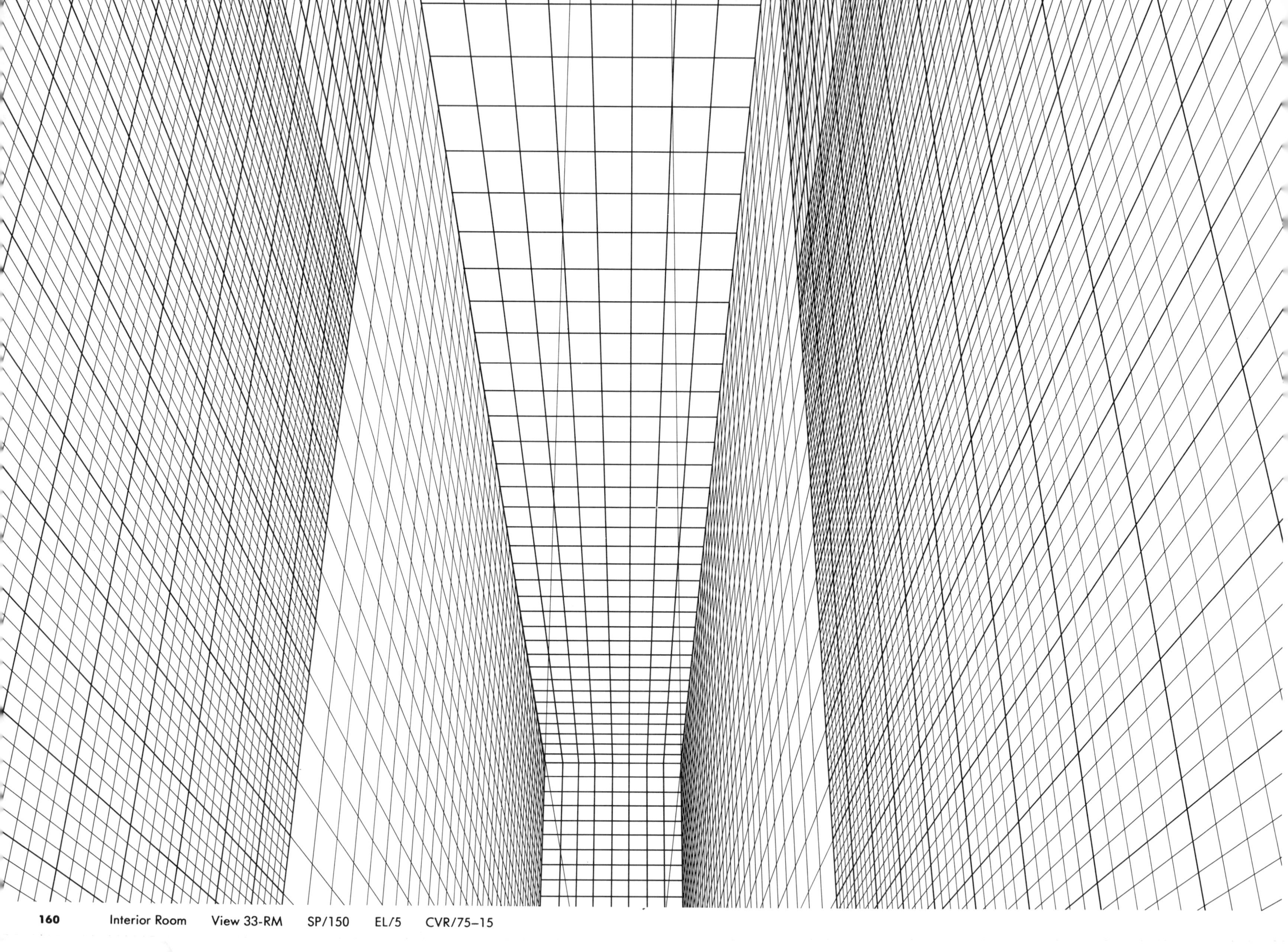



Interior Room View 33-RM SP/150 EL/5 CVR/75–15

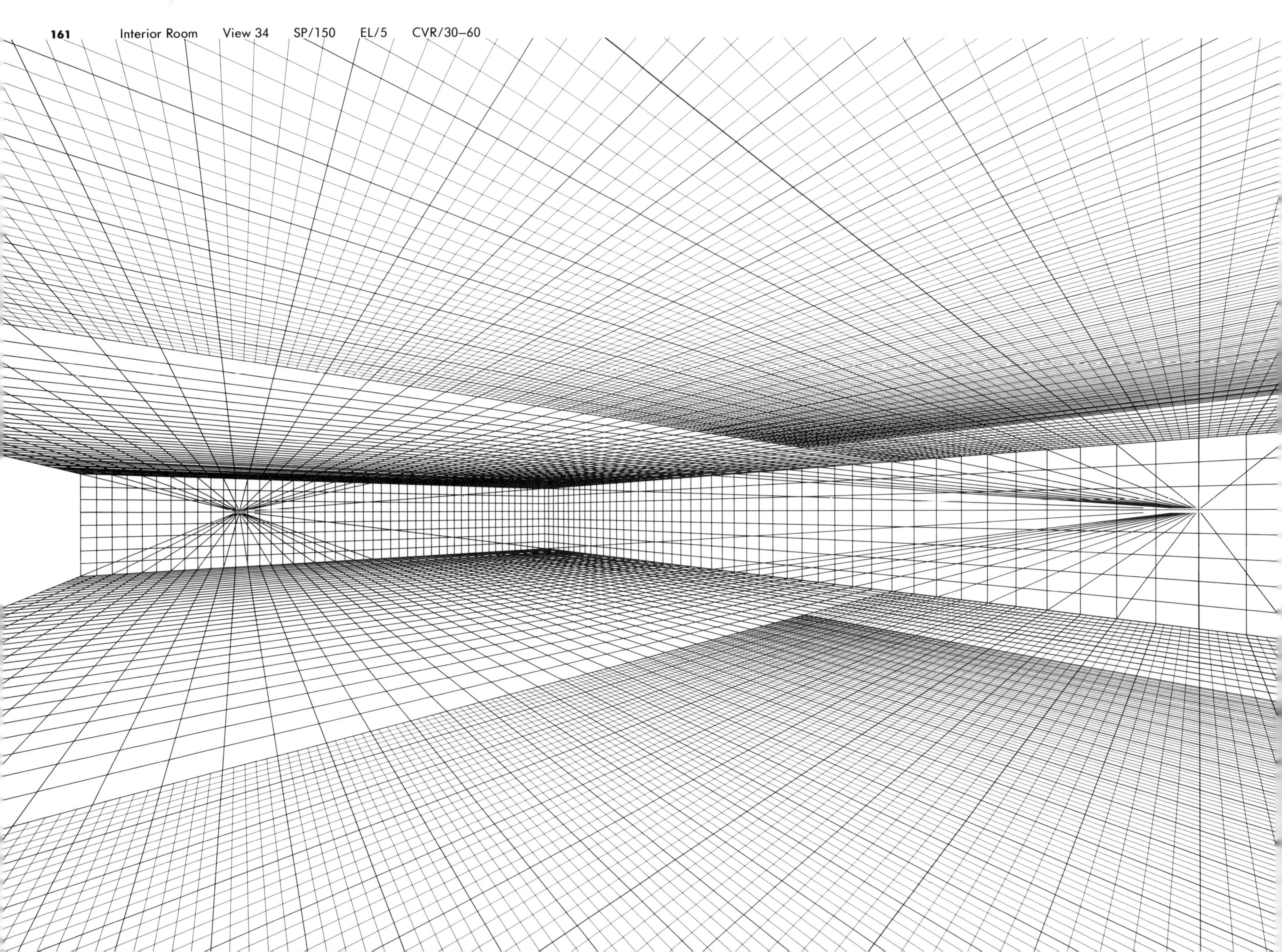

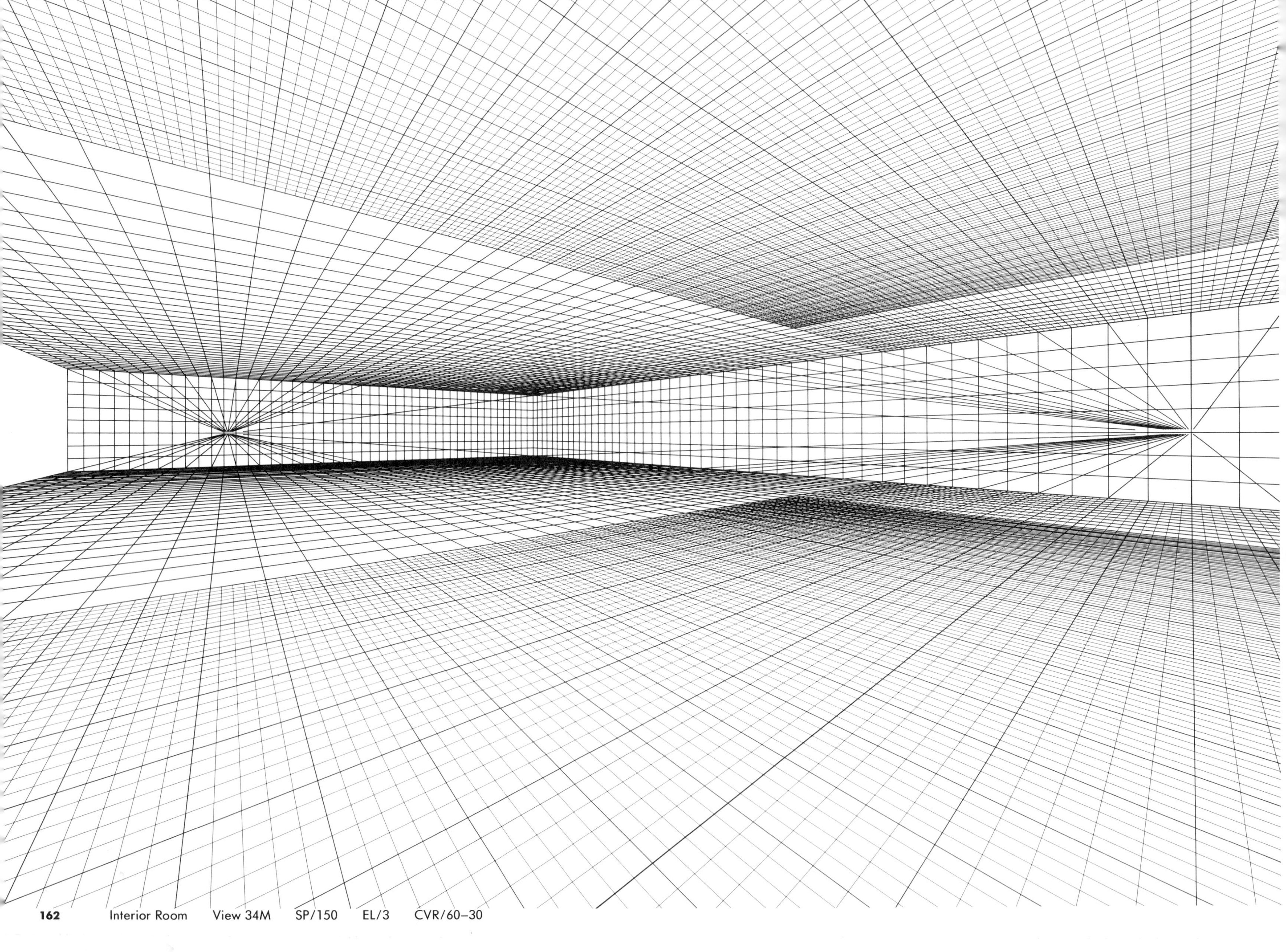

 Interior Room View 34M SP/150 EL/3 CVR/60–30

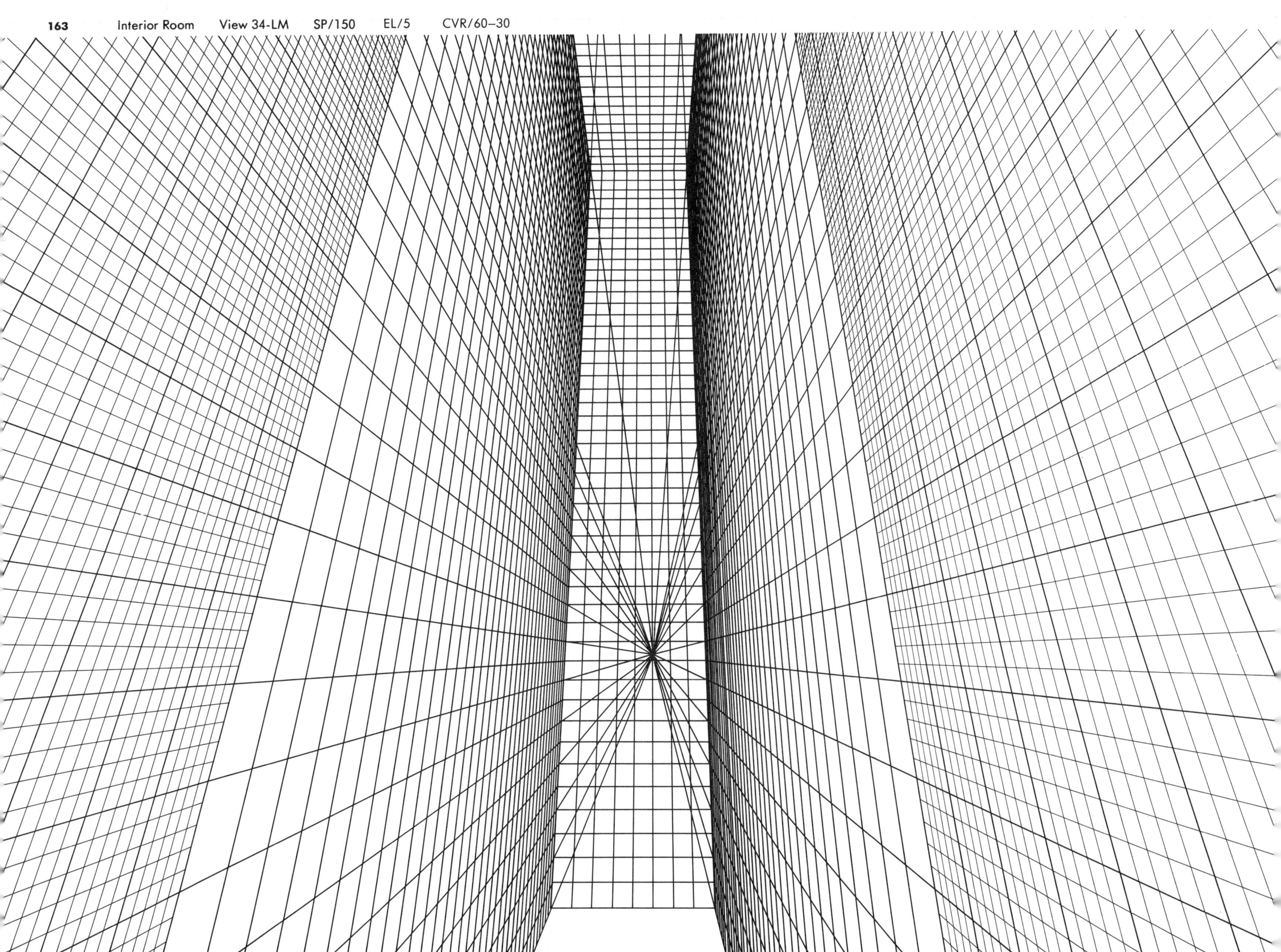

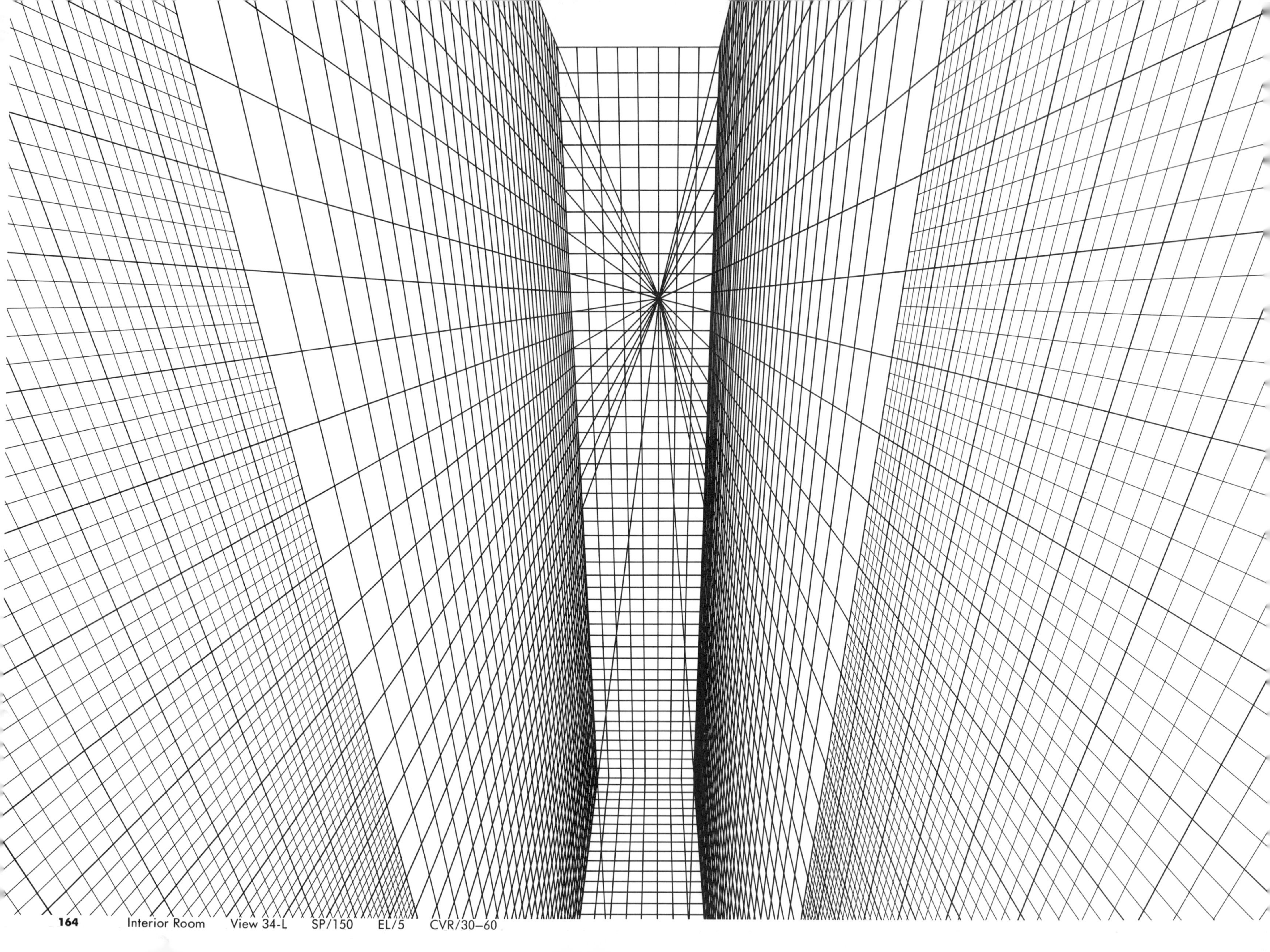

 Interior Room View 34-L SP/150 EL/5 CVR/30–60

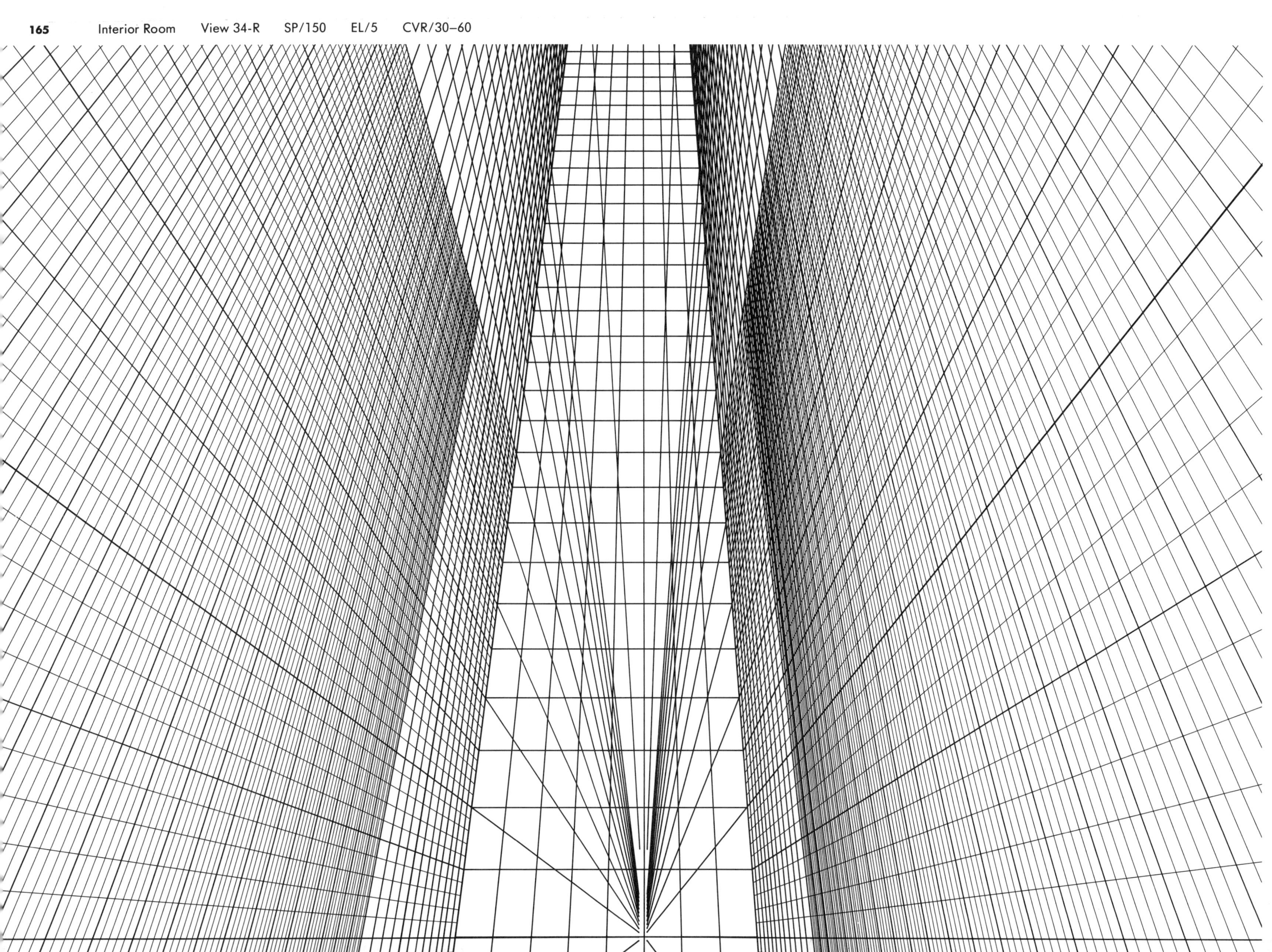

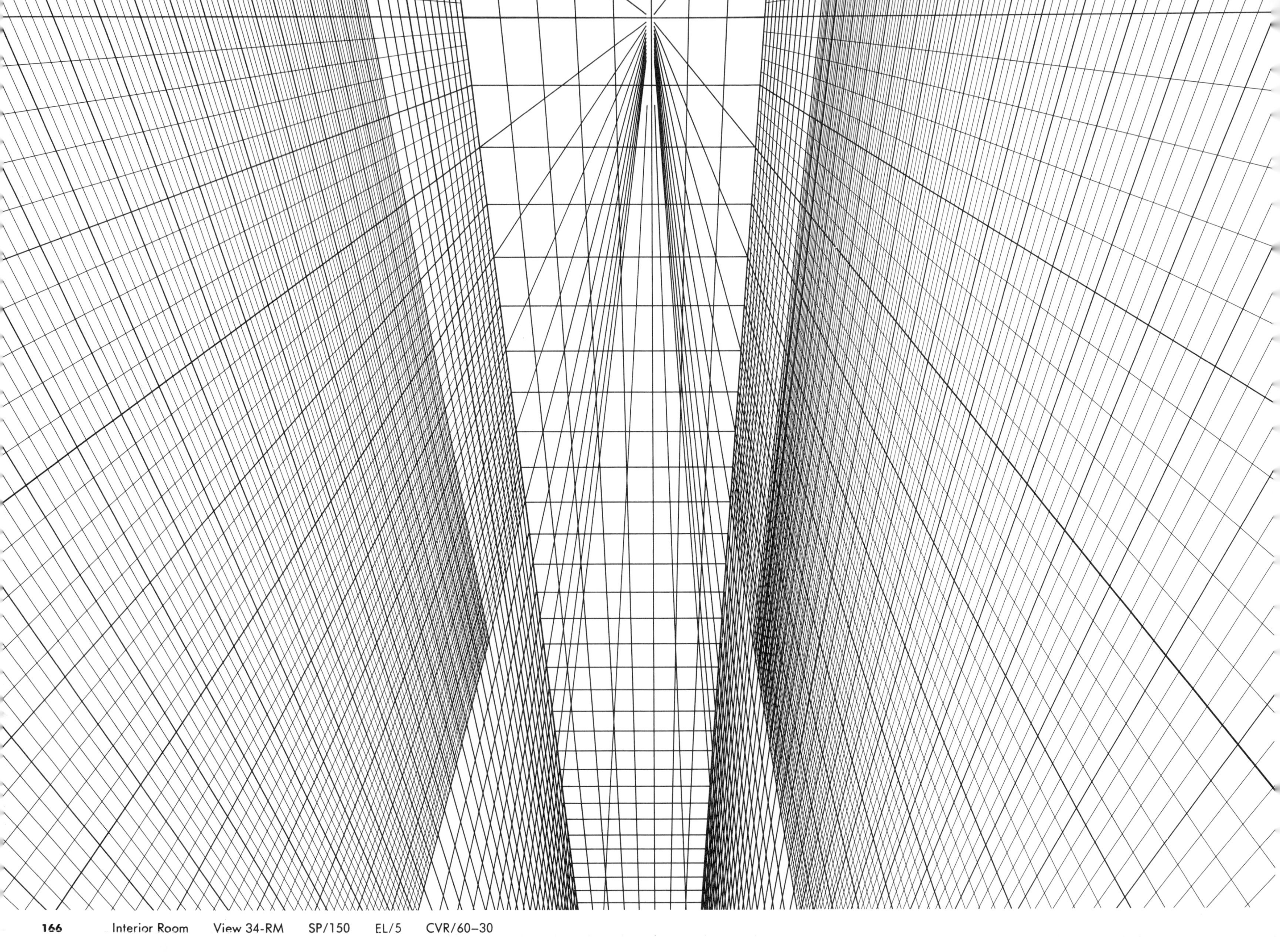



Interior Room View 34-RM SP/150 EL/5 CVR/60–30

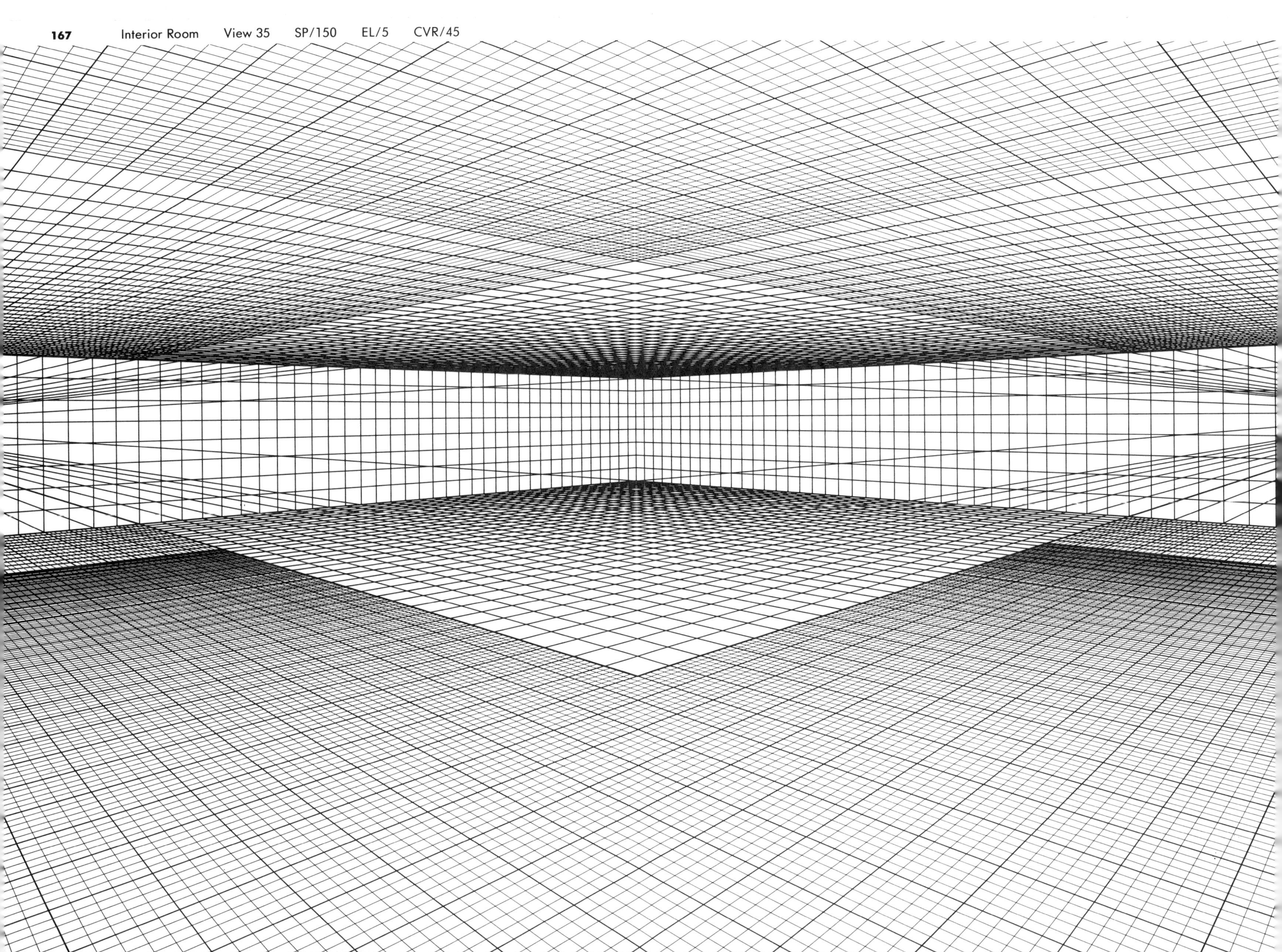

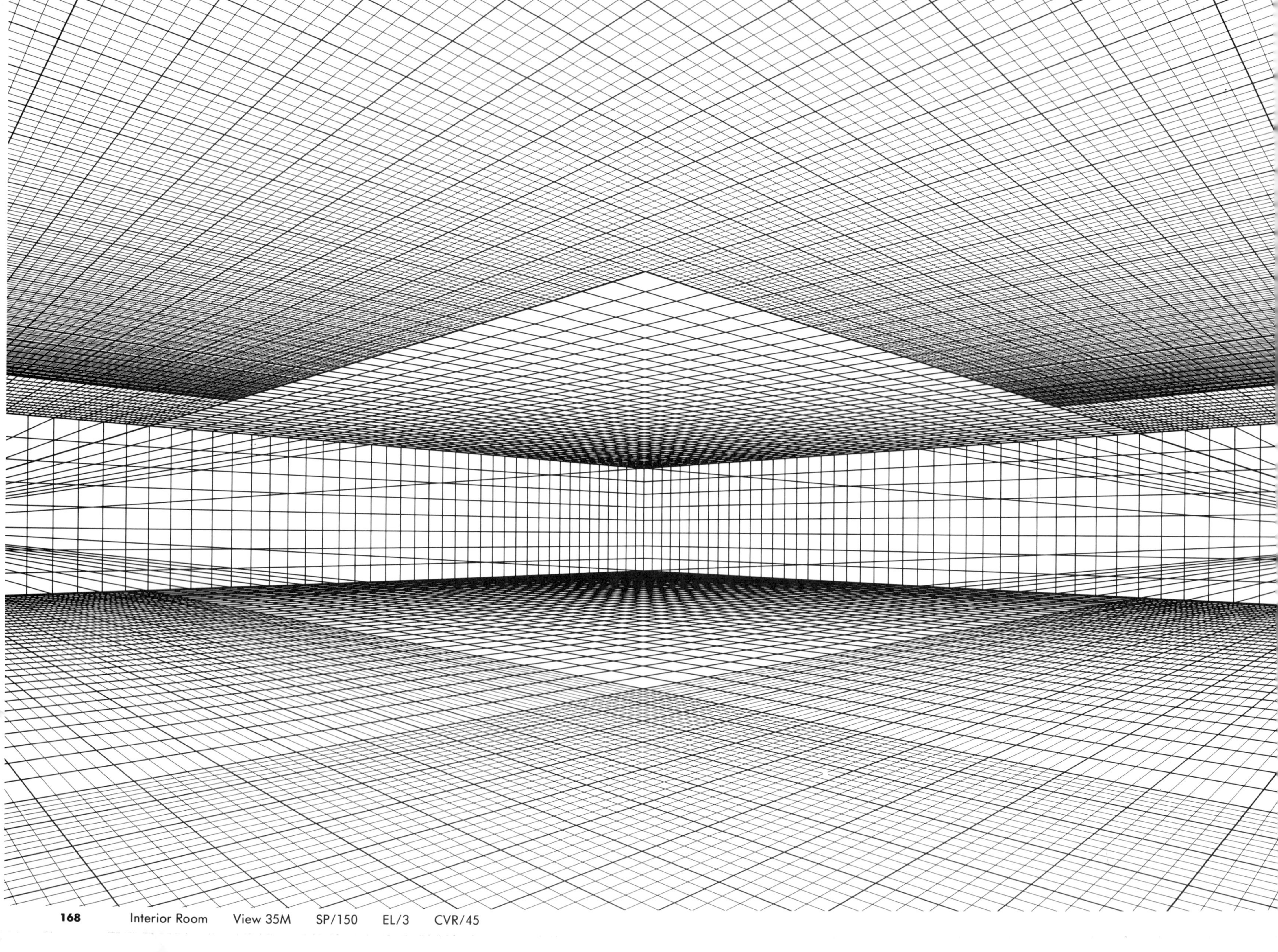

 Interior Room View 35M SP/150 EL/3 CVR/45

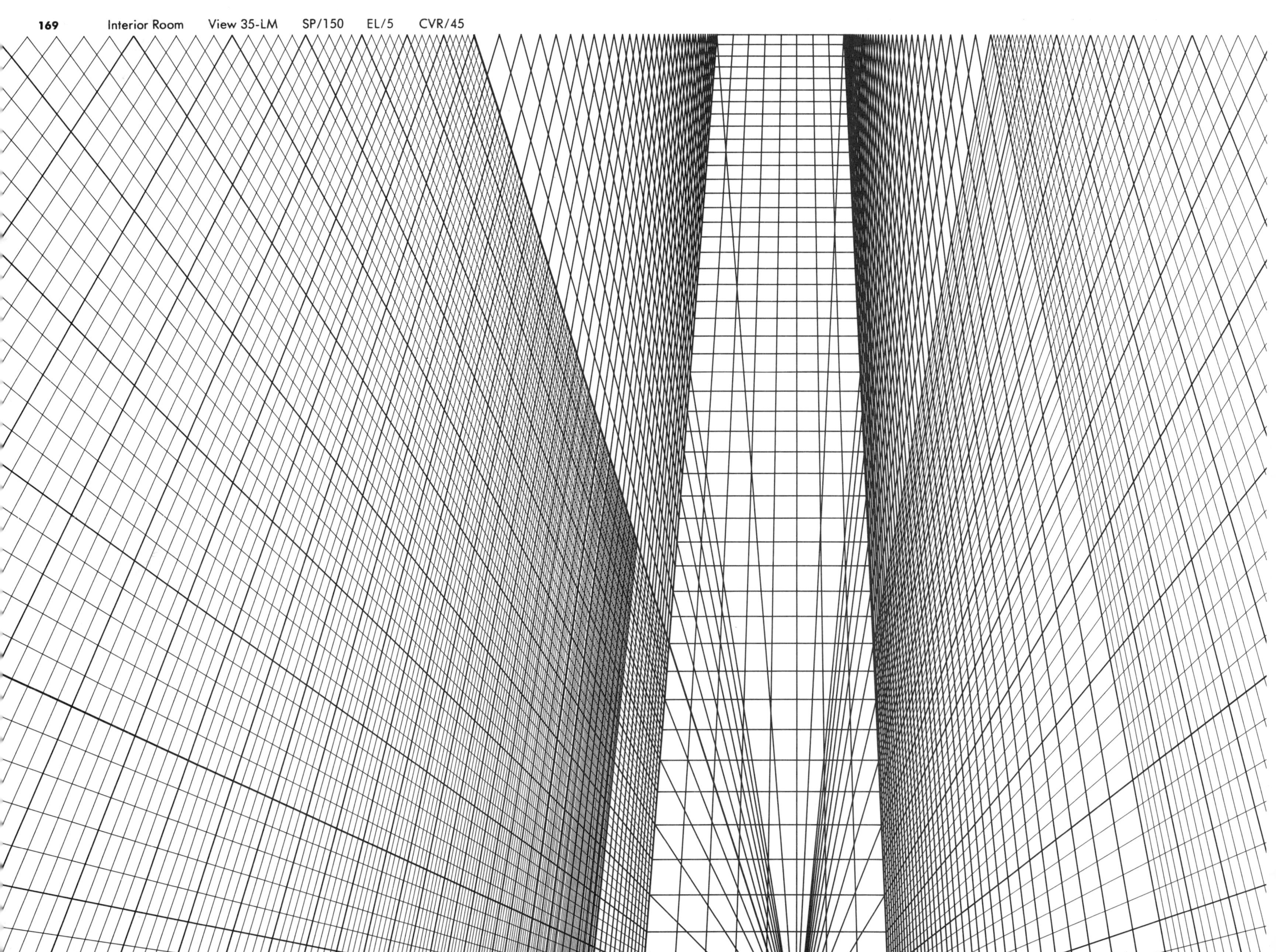

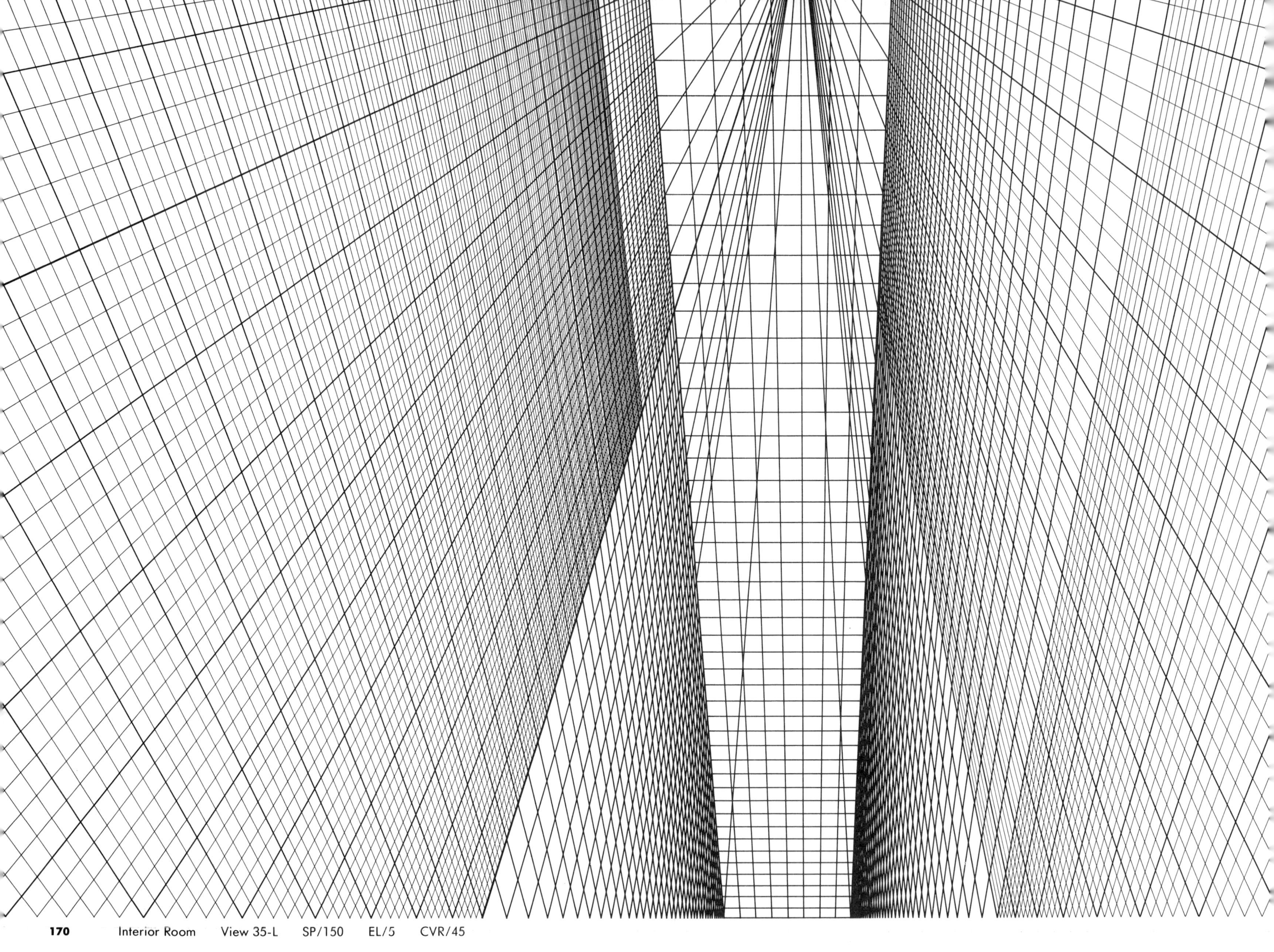

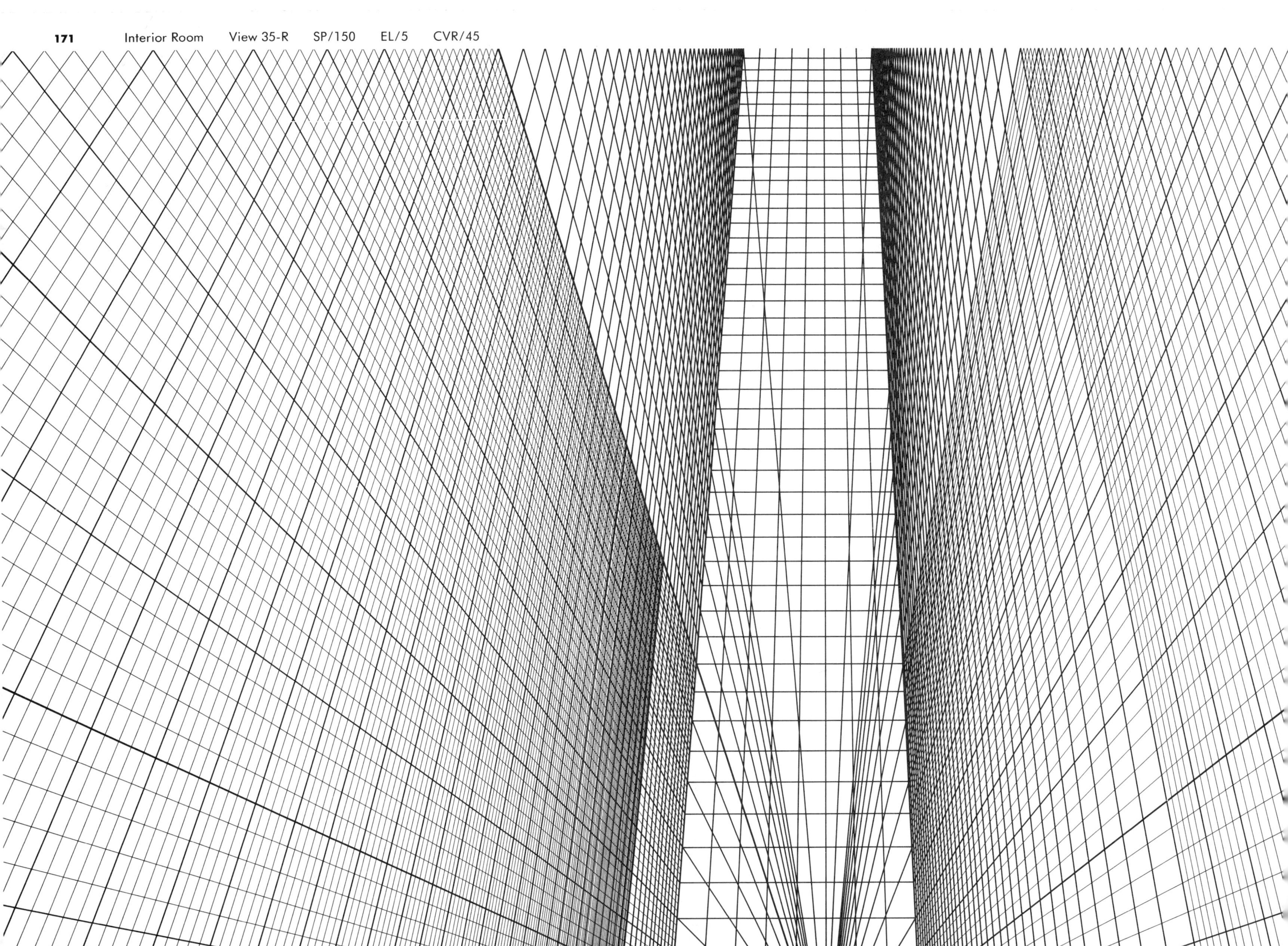

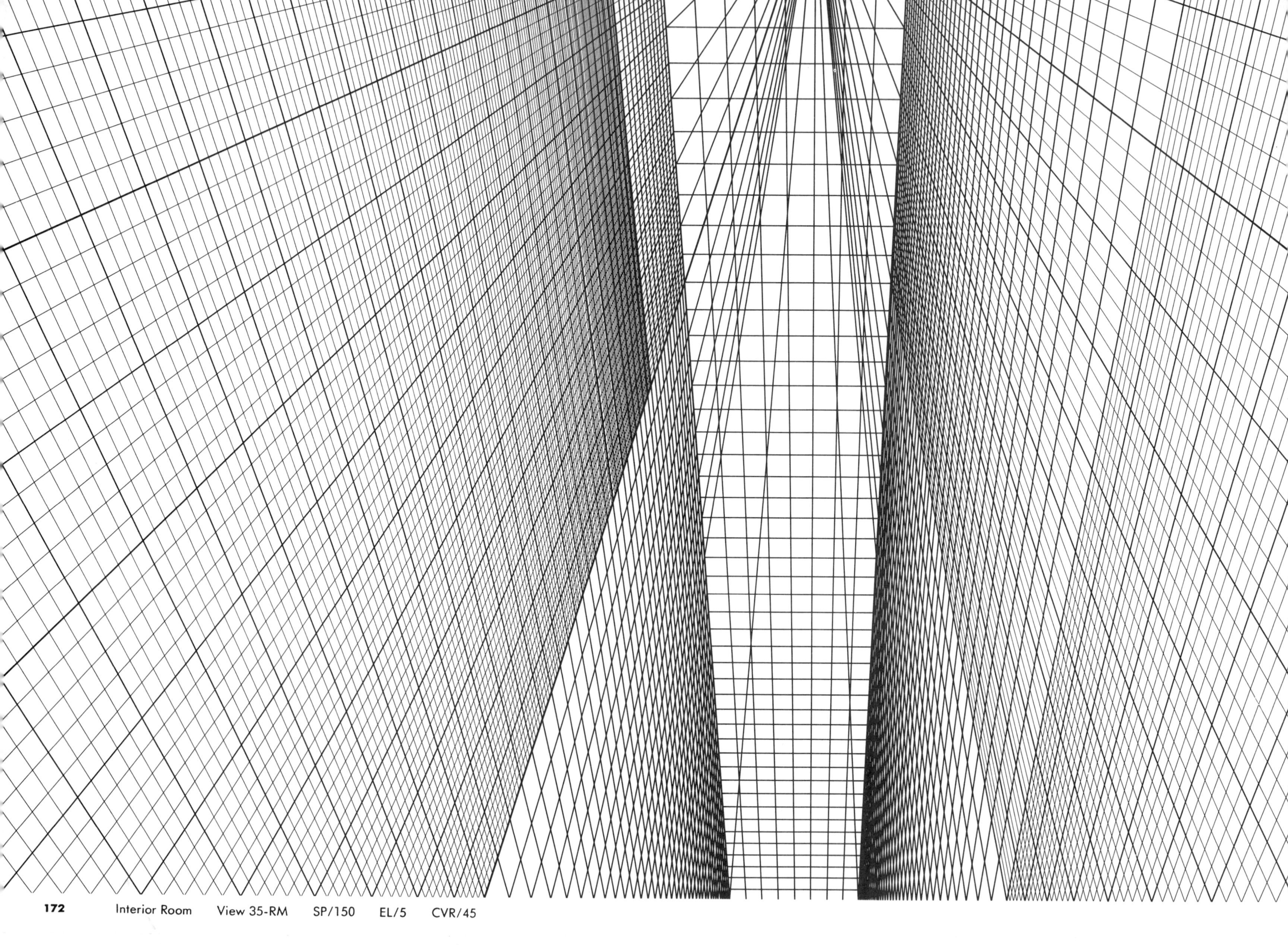



Interior Room View 35-RM SP/150 EL/5 CVR/45

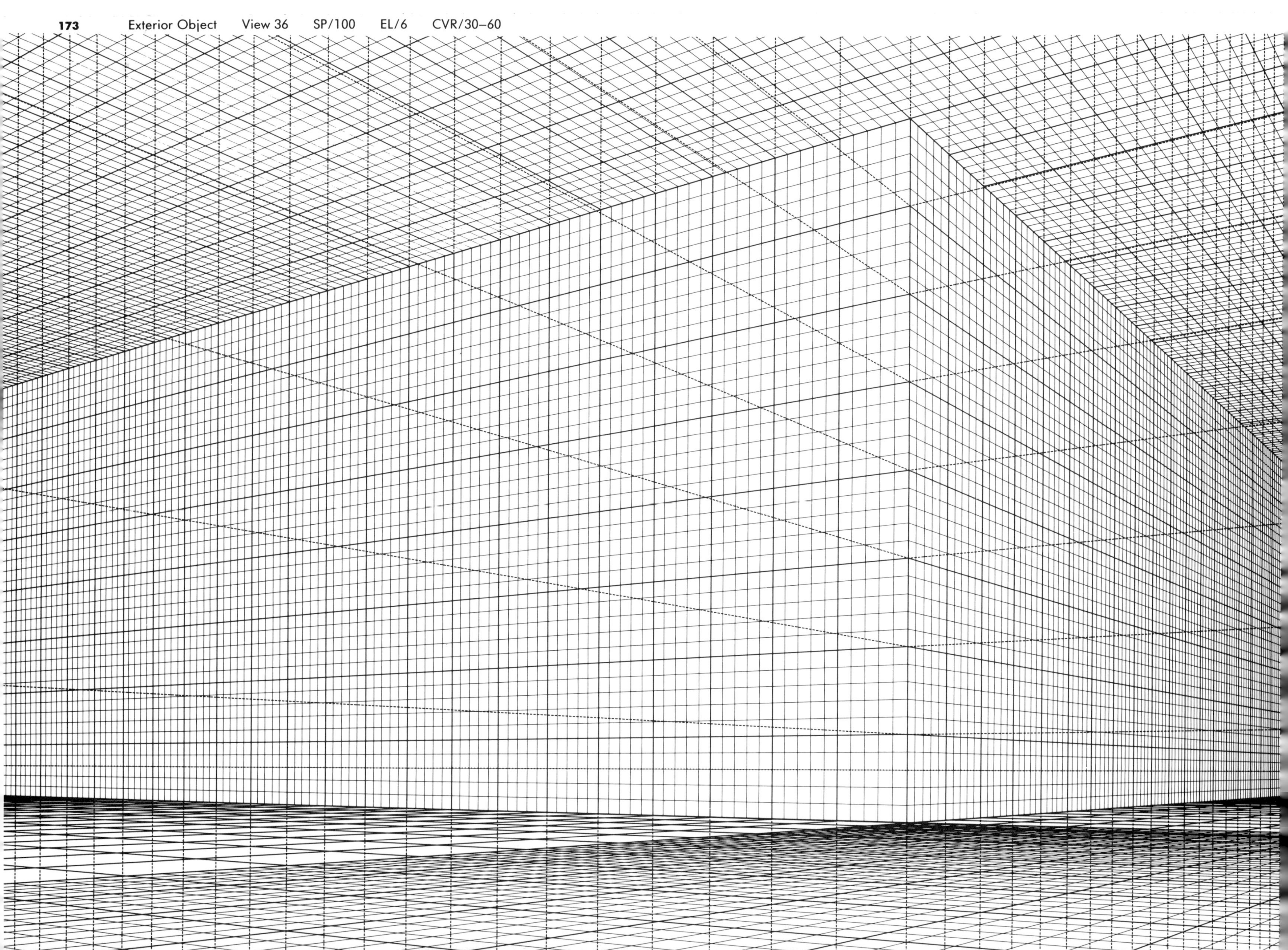

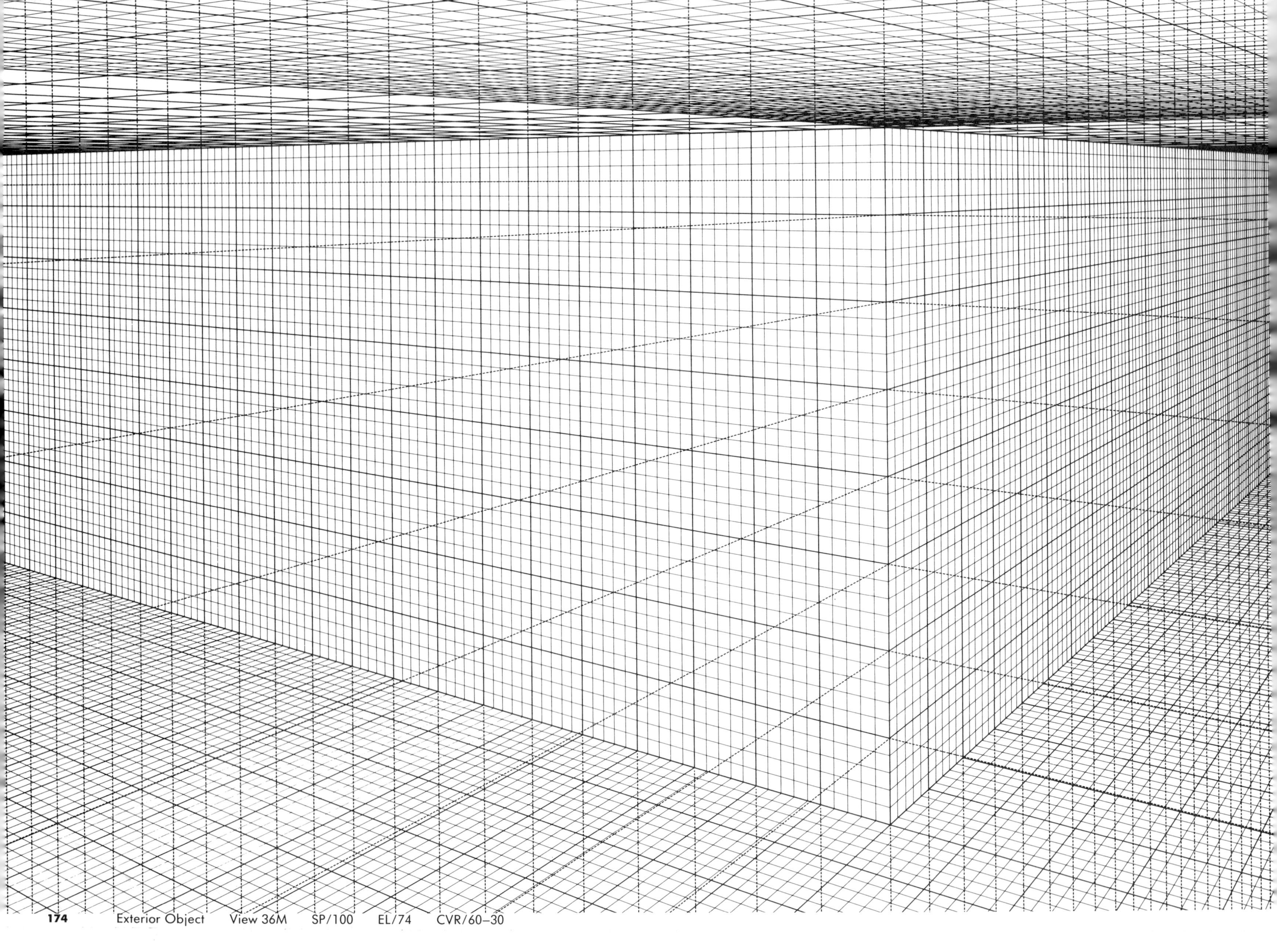

 Exterior Object View 36M SP/100 EL/74 CVR/60–30

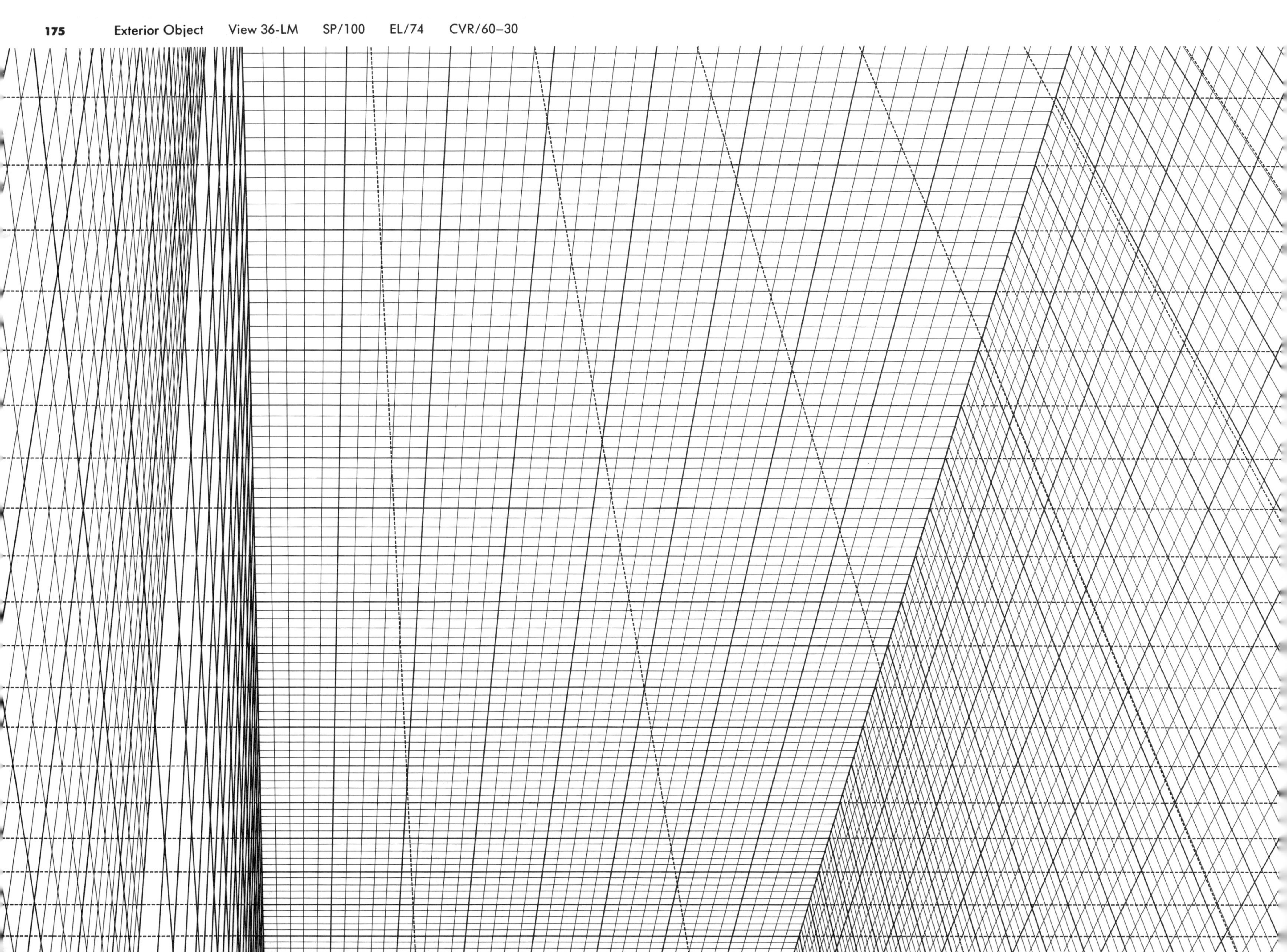

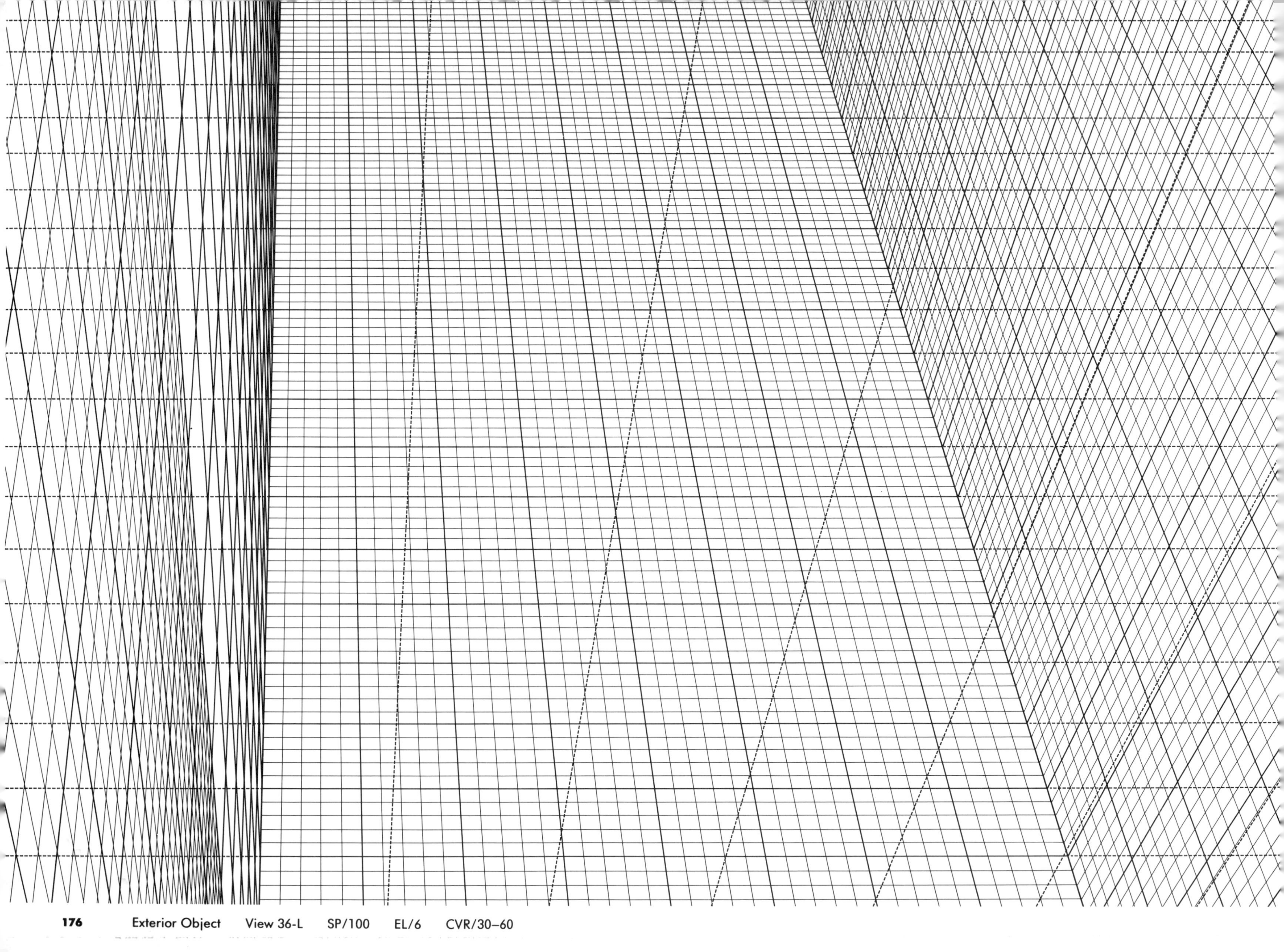

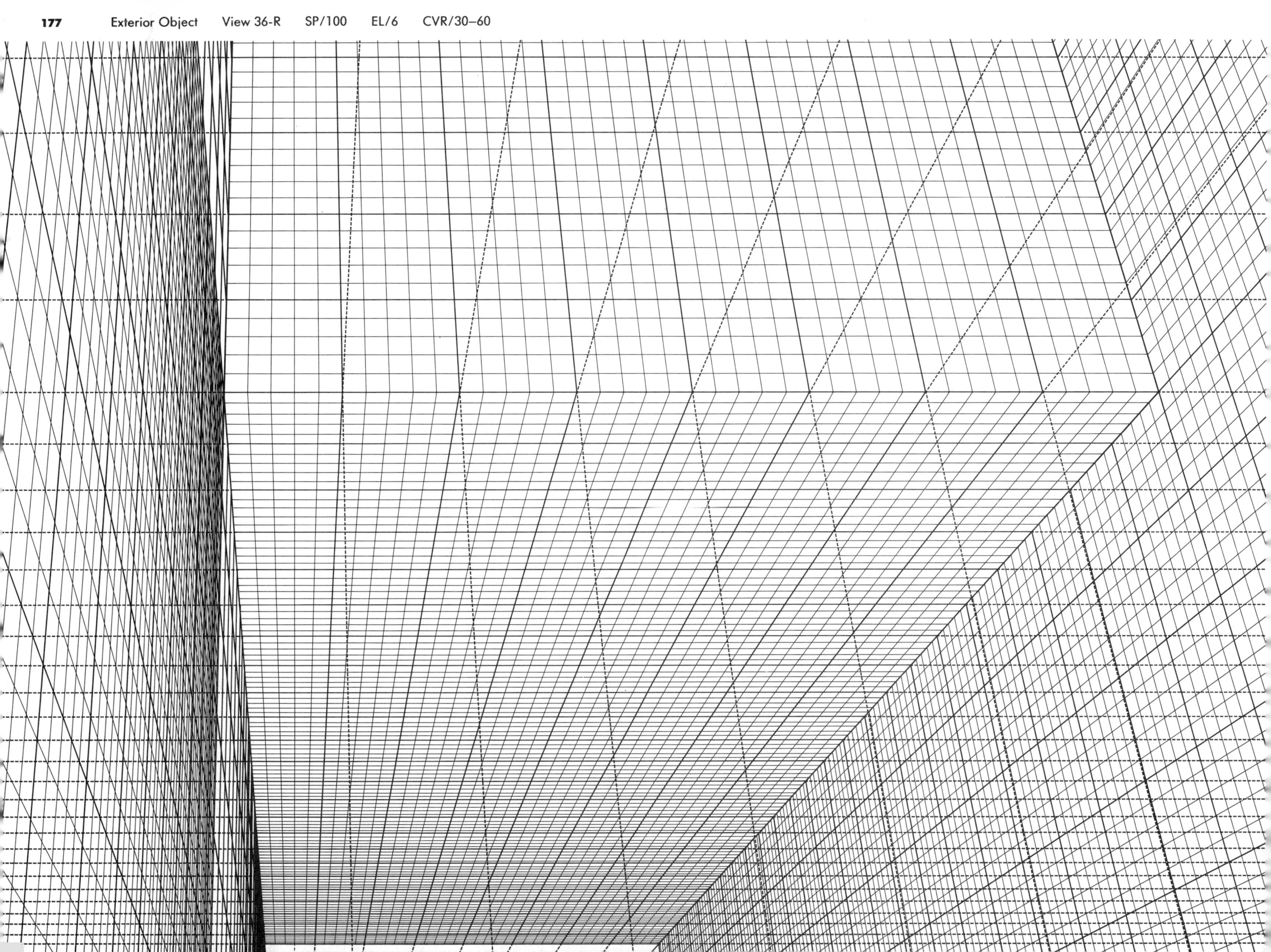

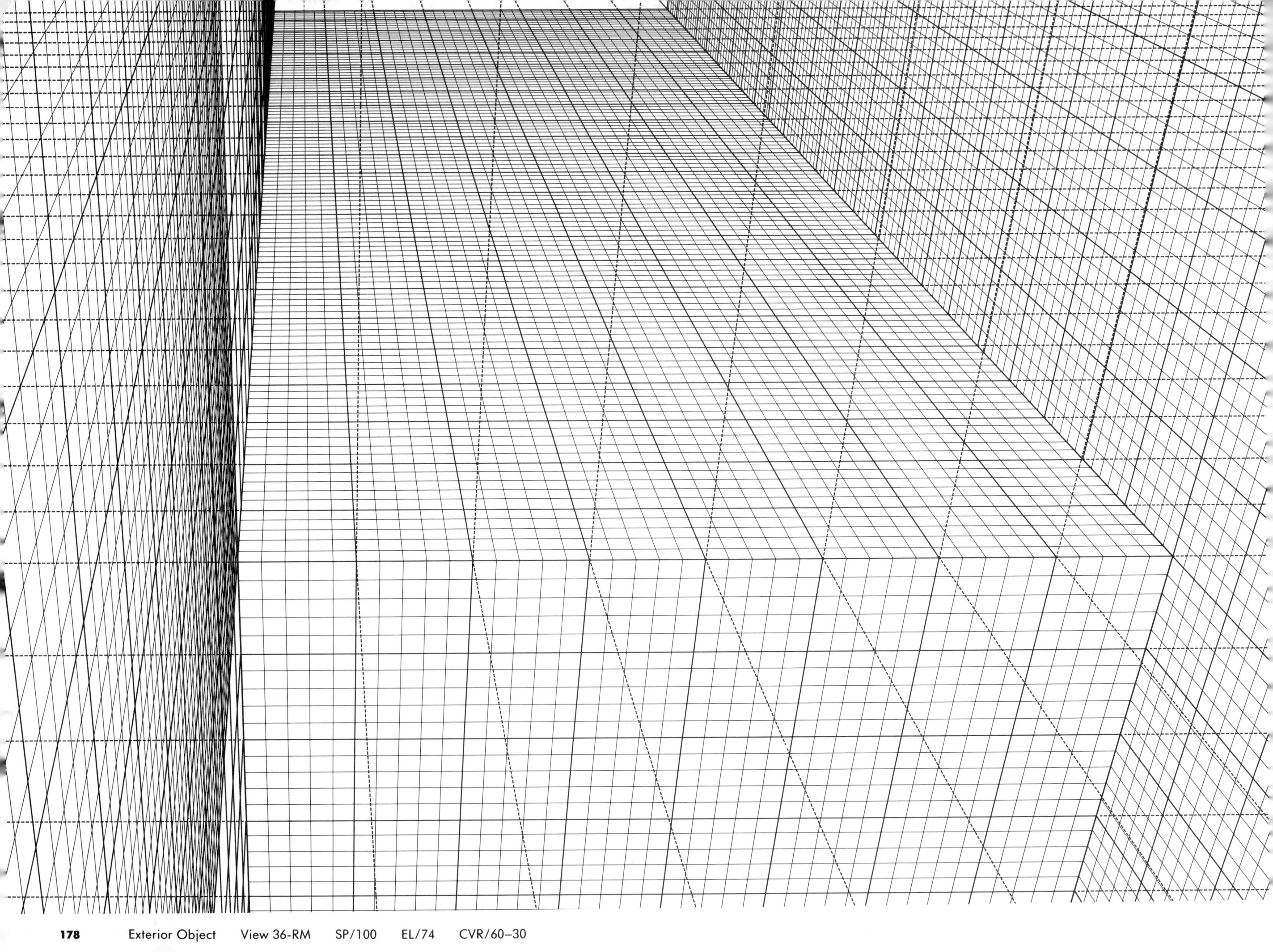

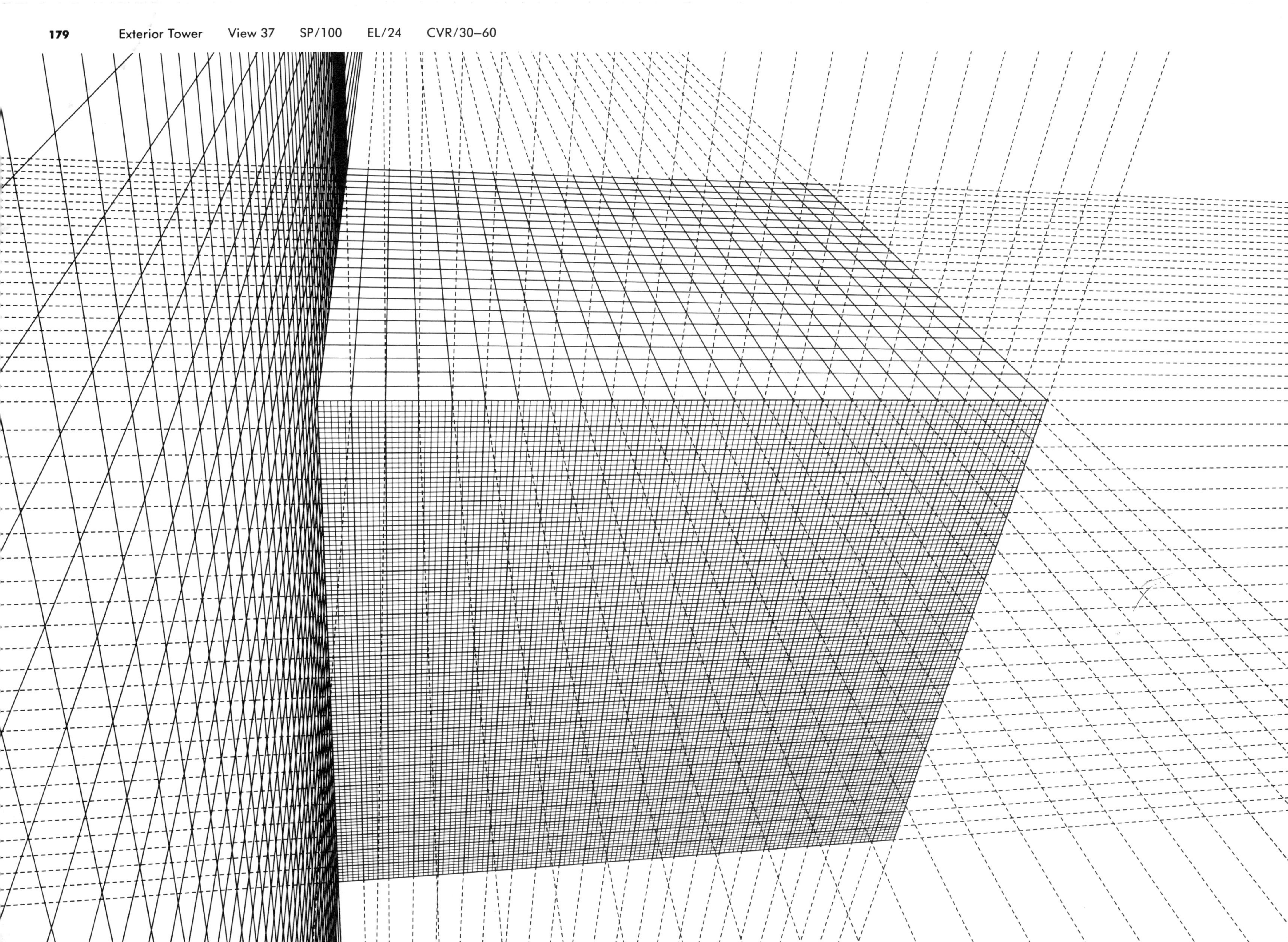

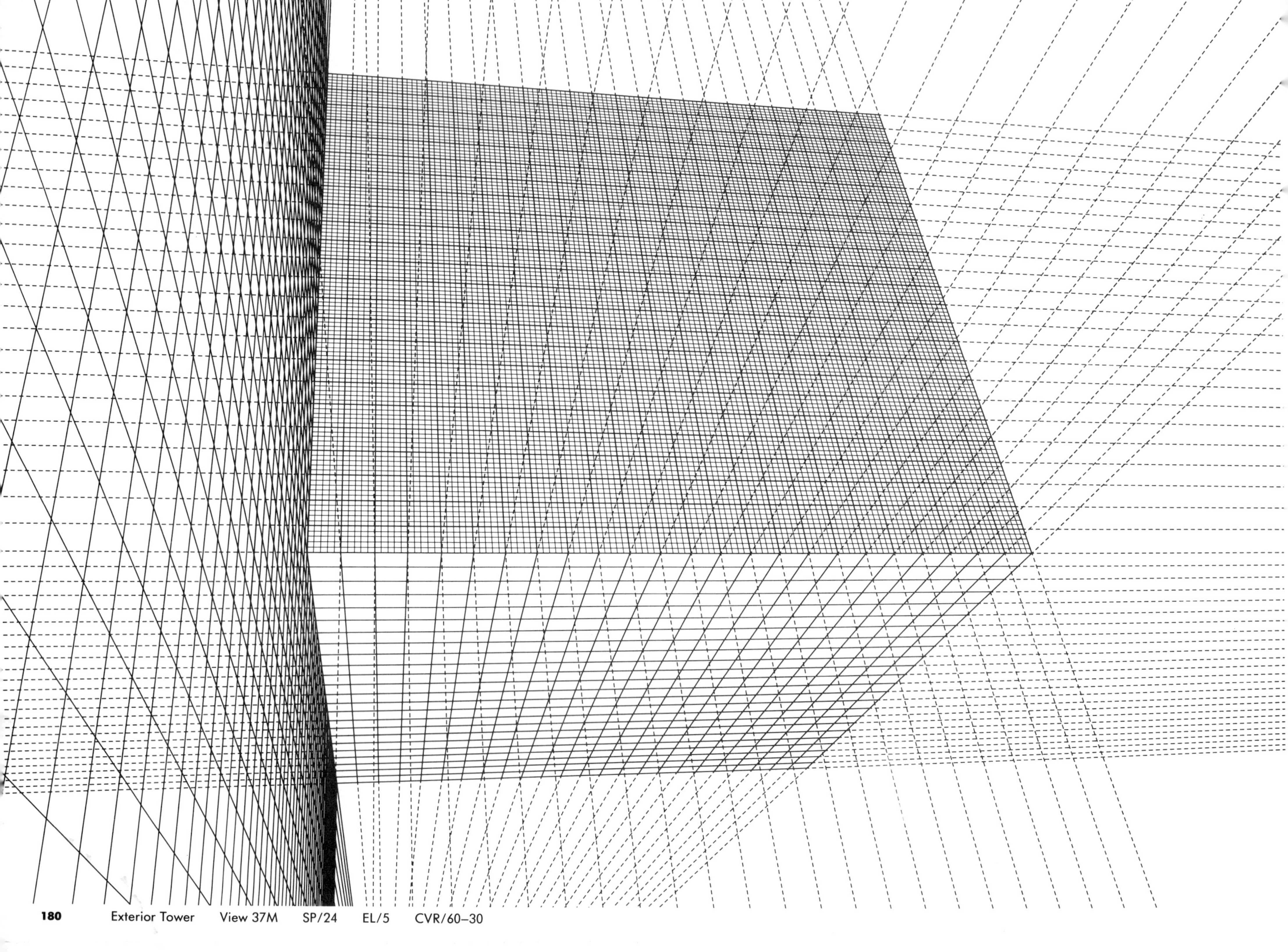

 Exterior Tower View 37M SP/24 EL/5 CVR/60–30

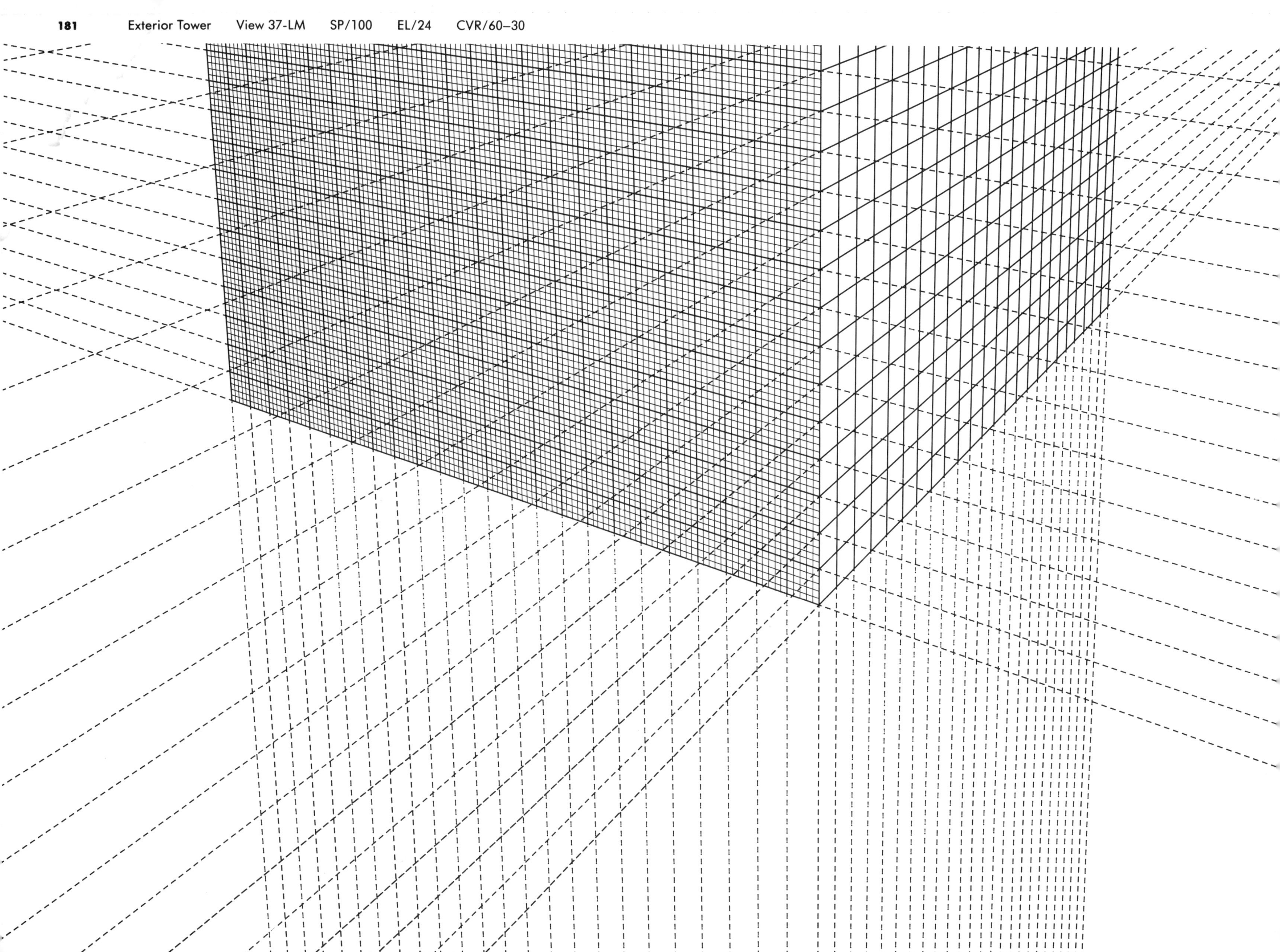

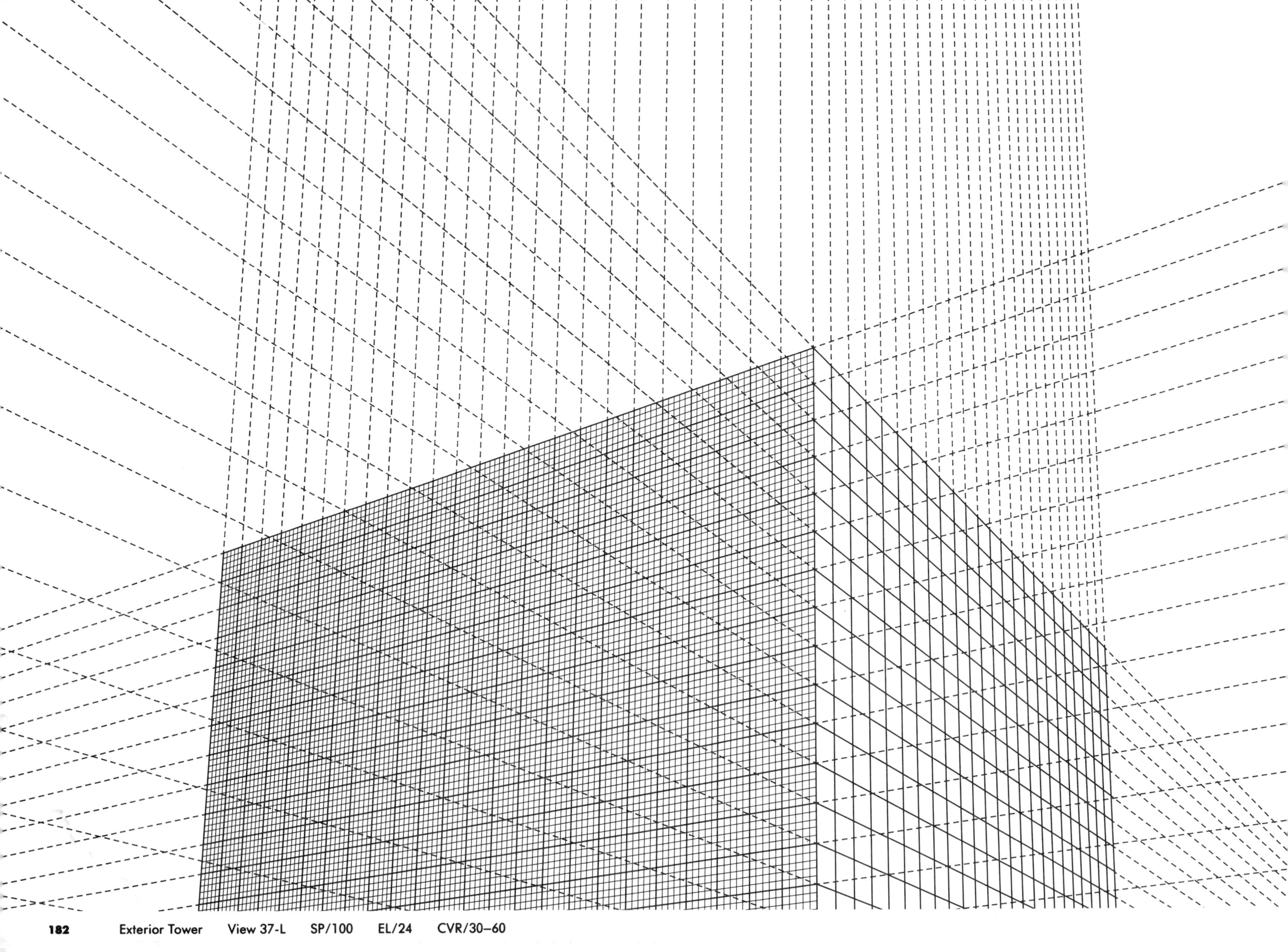

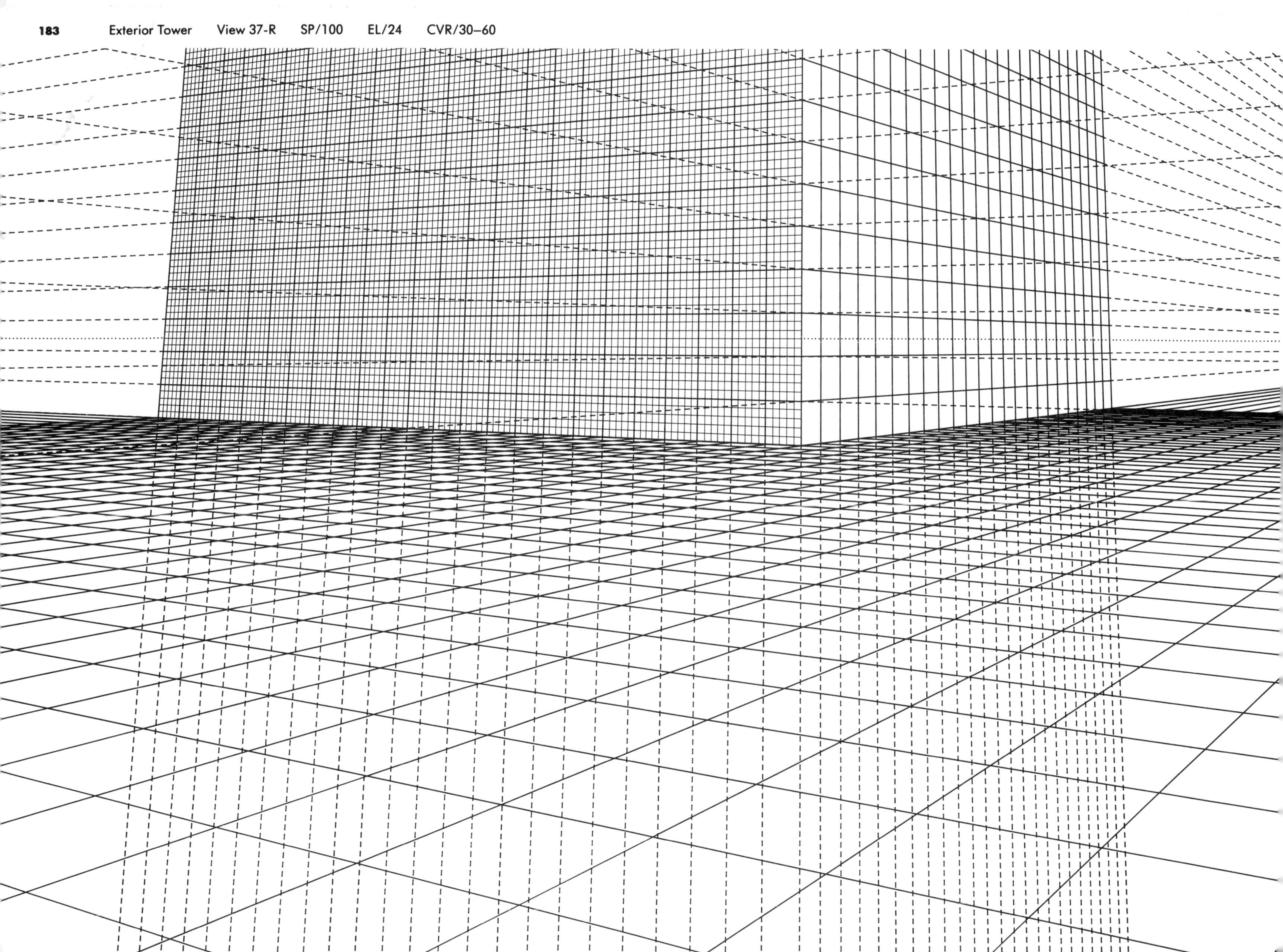

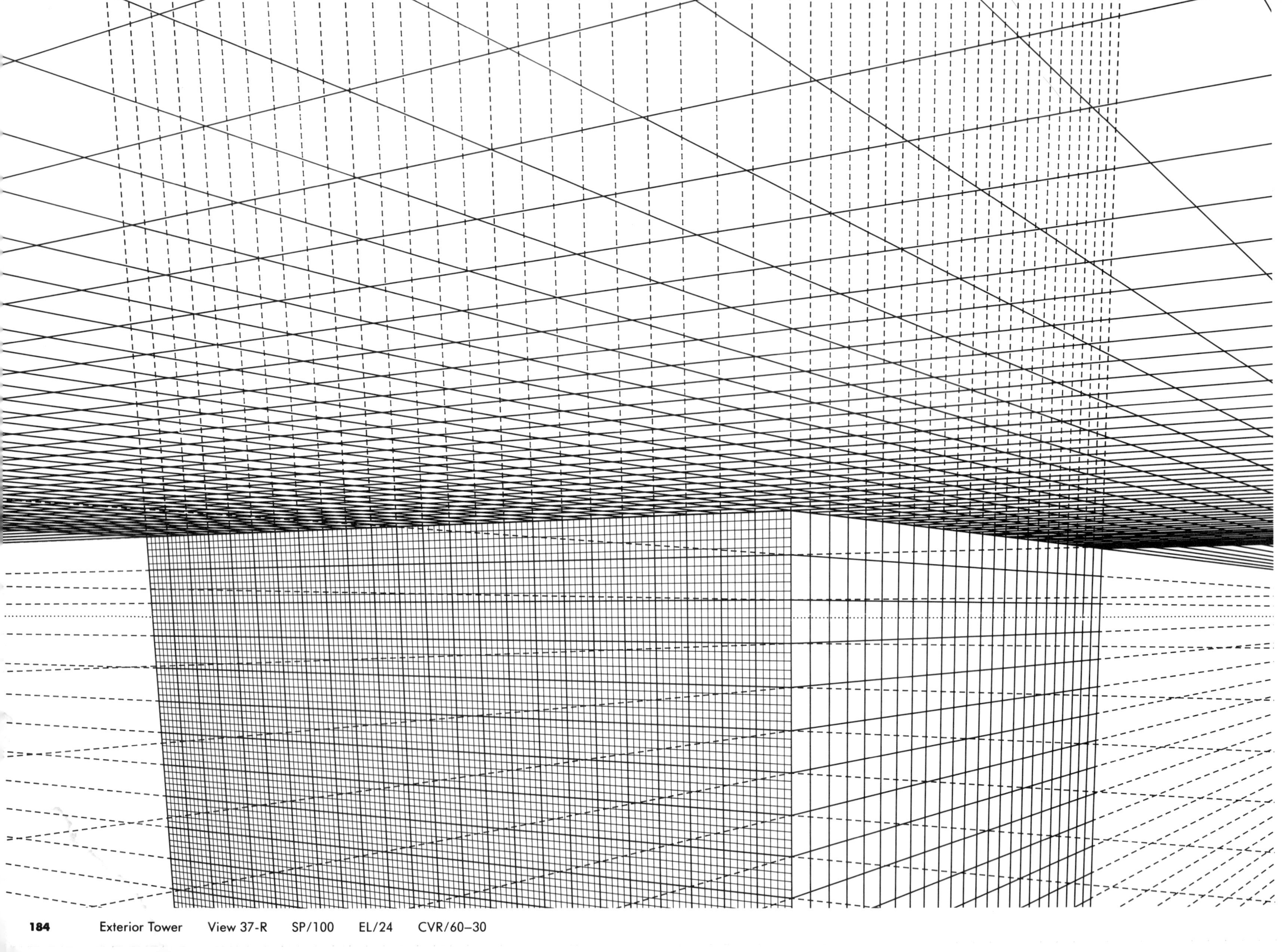

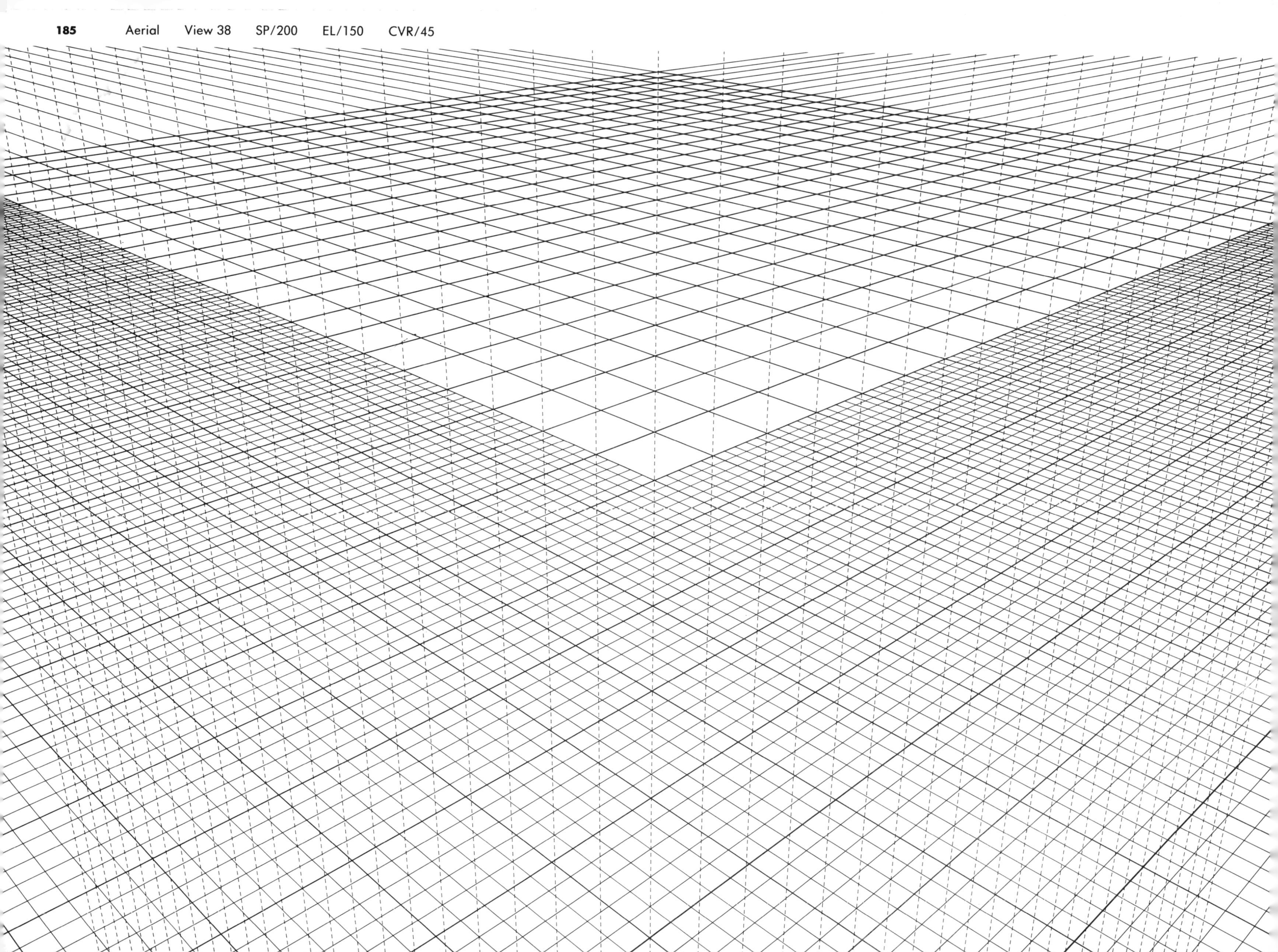

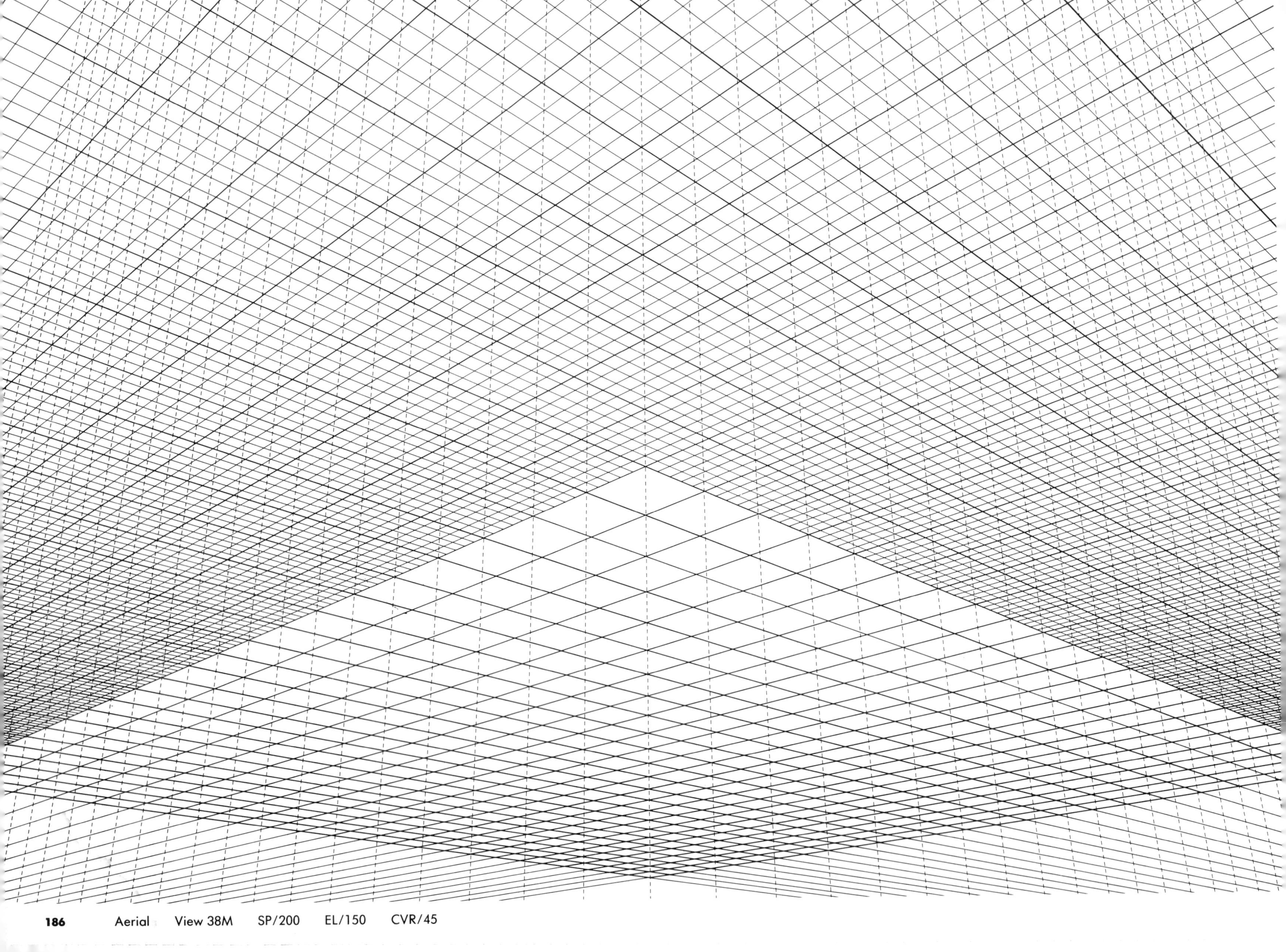

 Aerial View 38M SP/200 EL/150 CVR/45

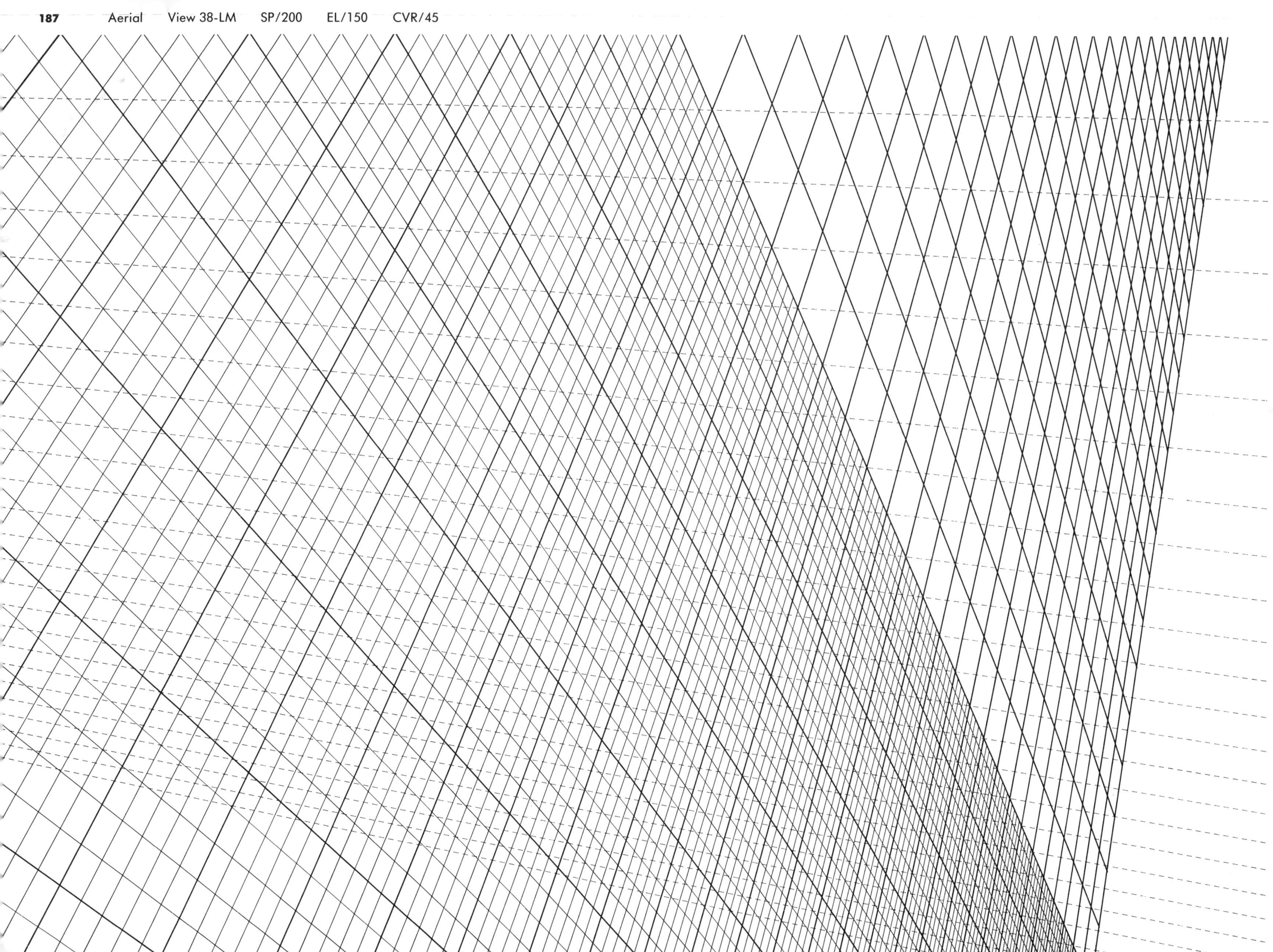

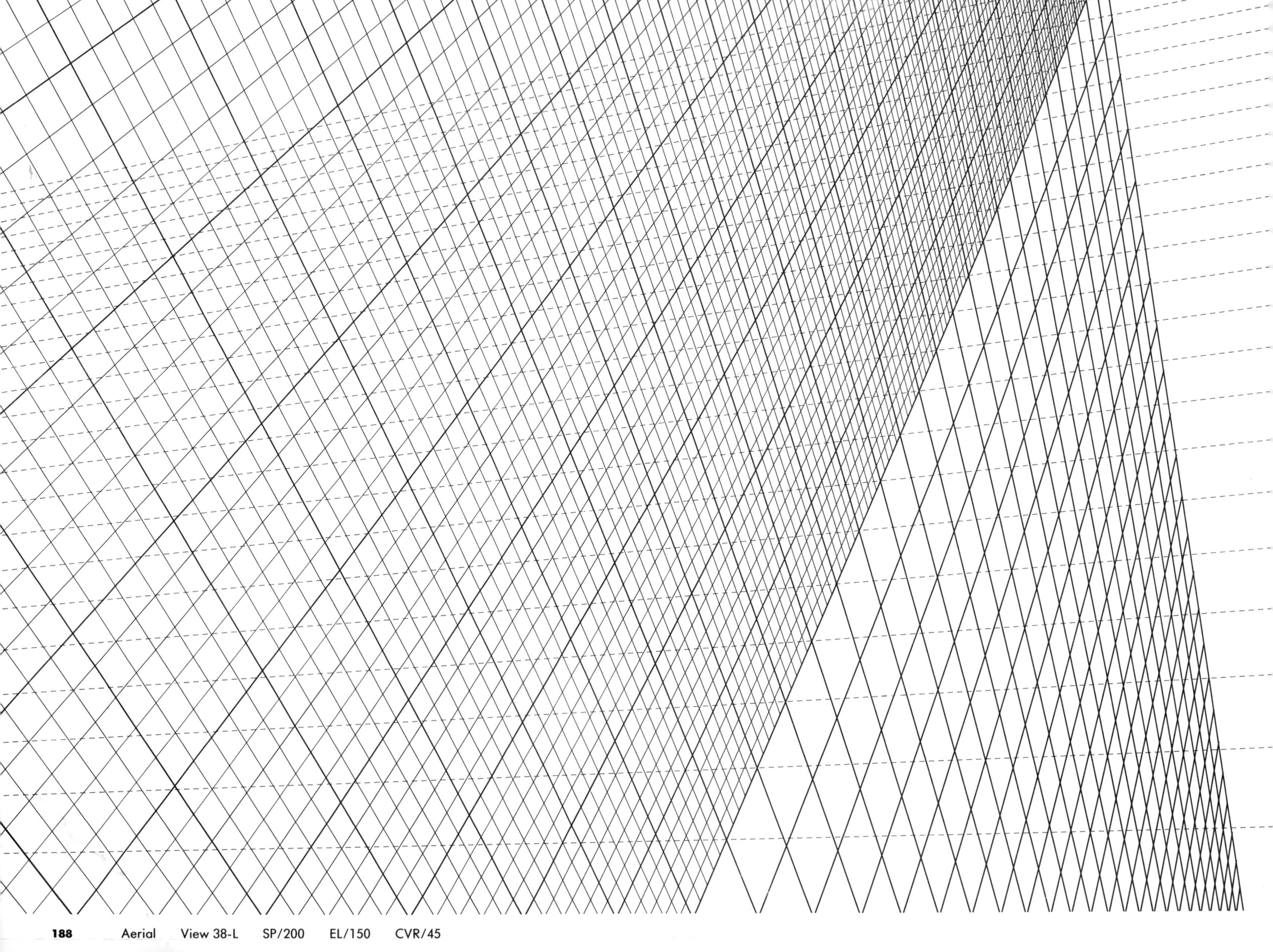

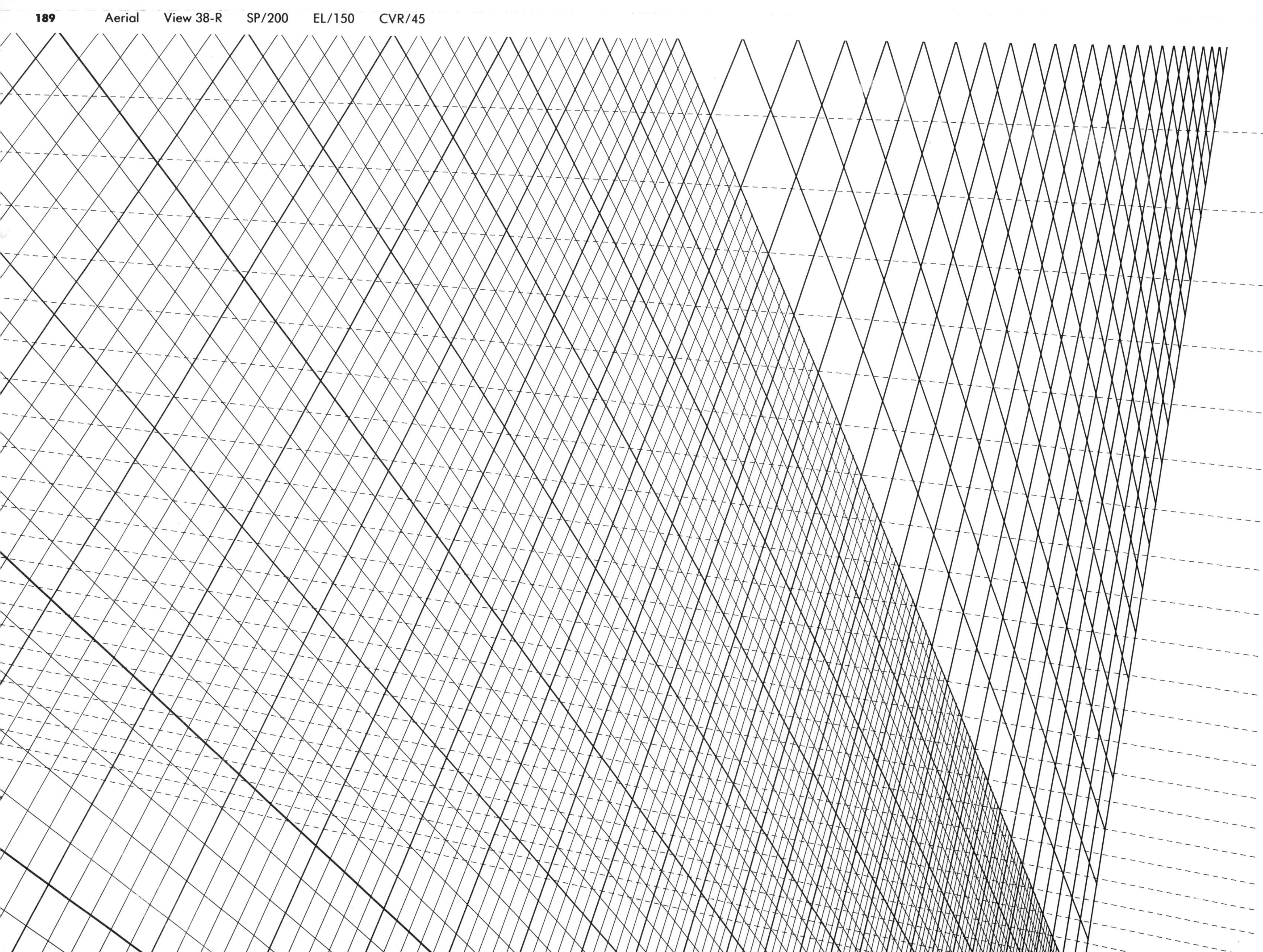

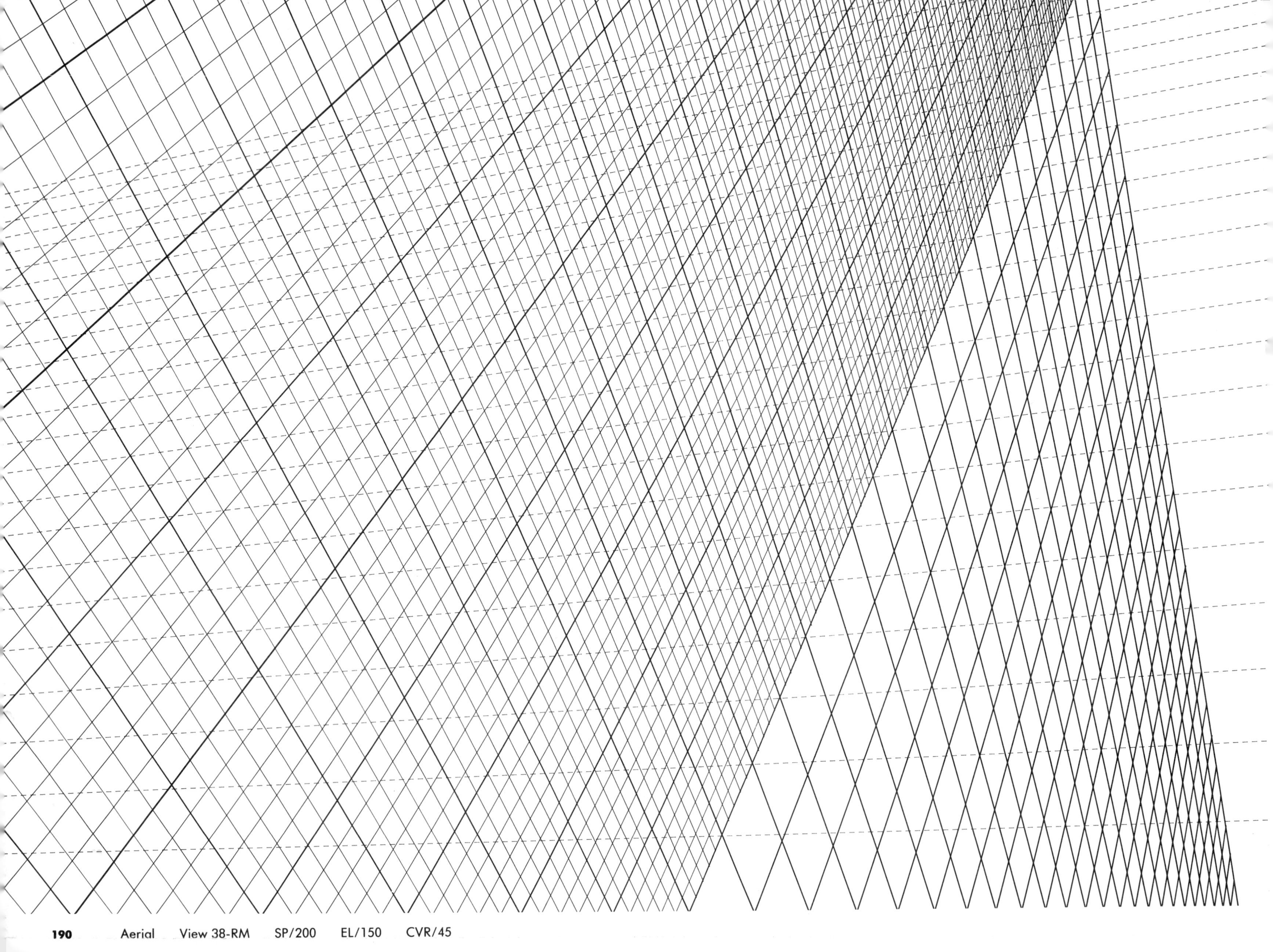

 Aerial View 38-RM SP/200 EL/150 CVR/45